Group Processes

CONTRIBUTORS

ROBERT S. BARON
KENNETH L. DION
FRED E. FIEDLER
J. RICHARD HACKMAN
L. RICHARD HOFFMAN
EDWIN P. HOLLANDER

JAMES W. JULIAN
NORMAN MILLER
ELLIOT G. MISHLER
CHARLES G. MORRIS
MARVIN E. SHAW
NANCY E. WAXLER

Group Processes

Papers from Advances in Experimental Social Psychology

EDITED BY

Leonard Berkowitz

Department of Psychology
University of Wisconsin
Madison, Wisconsin

ACADEMIC PRESS   New York   San Francisco   London   1978
A Subsidiary of Harcourt Brace Jovanovich, Publishers

ACADEMIC PRESS, INC.
111 Fifth Avenue, New York, New York 10003

United Kingdom Edition published by
ACADEMIC PRESS, INC. (LONDON) LTD.
24/28 Oval Road, London NW1 7DX

Library of Congress Cataloging in Publication Data

Main entry under title:

Group processes.

"Papers from Advances in experimental social
psychology."
 Includes bibliographies.
 1. Social groups--Addresses, essays, lectures.
2. Decision-making, Group--Addresses, essays, lectures. 3. Problem solving, Group--Addresses, essays,
lectures. 4. Leadership--Addresses, essays, lectures.
I. Berkowitz, Leonard, Date II. Advances in
experimental social psychology.
HM131.G718   301.18   78-4815
ISBN 0-12-091840-4

CONTENTS

A Contingency Model of Leadership Effectiveness

Recent Developments in Research on the Contingency Model

Why Do Groups Make Riskier Decisions Than Individuals?

A Questionnaire in Search of a Theory

LIST OF CONTRIBUTORS

ROBERT S. BARON (227), Department of Psychology, University of Iowa, Iowa City, Iowa 52240

KENNETH L. DION (227), Department of Psychology, University of Toronto, Toronto, M5S 1A1, Ontario, Canada

FRED E. FIEDLER (167, 209), Department of Psychology, University of Washington, Seattle, Washington 98195

J. RICHARD HACKMAN (1, 57), Department of Administrative Sciences and Psychology, Yale University, New Haven, Connecticut 06520

L. RICHARD HOFFMAN (67, 101), Graduate School of Business, University of Chicago, Chicago, Illinois 60637

EDWIN P. HOLLANDER (115, 153), Department of Psychology, State University of New York at Buffalo, Buffalo, New York 14226

JAMES W. JULIAN (115, 153), Department of Psychology, State University of New York at Buffalo, Buffalo, New York 14226

NORMAN MILLER (227, 301), Social Science Research Institute, University of Southern California, Los Angeles, California 90007

ELLIOT G. MISHLER (363), Psychiatrist, Harvard University, 74 Fenwood Road, Boston, Massachusetts 02115

CHARLES G. MORRIS (1, 57), Department of Psychology, University of
   Michigan, Ann Arbor, Michigan 48109
MARVIN E. SHAW (313, 351), Department of Psychology, University of
   Florida, Gainesville, Florida 32611
NANCY E. WAXLER (363), Harvard Medical School and The Massa-
   chusetts Mental Health Center 02115

PREFACE

All sciences modify their concepts and redirect their attention as research continues, but perhaps none does so more rapidly than social psychology. As someone once said, we have a long past but a short history and all too quickly forget the important ideas and studies of earlier years. However, quite a few of the works published in previous years have retained their significance. This is especially true of the chapters presented in this volume. Although observations have multiplied and theoretical advances have been made since these chapters first appeared, to some degree at least they still pose the questions and issues that confront the latest investigations in their areas and provide the constructs that are often employed in the interpretation of the most recent research findings.

This collection of chapters from *Advances in Experimental Social Psychology* testifies to the continuing importance of these earlier contributions. Social psychology has, of course, had a strong interest in group behavior and performance from its earliest days. Although this interest declined somewhat in the recent past, it is bound to grow again as the field broadens its perspective in response to societal demands. Surely, many of the practical questions addressed to social psychologists deal with the interactions of group members and their productivity, and there is a good chance that every social psychologist will be asked about these matters at one time or another, regardless of his specific professional concerns. Unless he is familiar with many of the major developments in the area of group behavior and performance, he will not be carrying out his professional role as well as he might. Both reflecting and contributing to the developments in social psychology since it began publication in 1964, the *Advances in Experimental Social Psychology* series has reported some of the major theoretical formulations in this area. The present collection gathers these contributions together under one cover and provides a convenient introduction to some of the more significant ideas in the study of group behavior. But even those already well-versed in the literature in this particular area will also find their thinking challenged (and their memories refreshed) by these papers.

One of the central concerns in the analysis of human interaction has to do with the origins and functions of leadership. It is now widely agreed that investigations of leadership cannot be reduced to the study of a few charismatic individuals; who emerges as a leader and how effective he is depends to a considerable extent upon situational conditions. The articles by Fiedler, Hoffman, and Hollander and Julian in this book all deal in various ways with the interaction between the leader, group members, and surrounding conditions as the group confronts its task. The task characteristics and situational demands, as well as the leader's actions, obviously play a major role in determining what the people involved will do. Hackman and Morris discuss how these external conditions can influence group performance. The members' interactions with each other and even their morale and productivity are also affected by the communication channels that are open to them. How freely can they talk to each other? What routes must their messages take in going from one part of the group to another? Shaw provides an important survey of the research on the effects of communication channels on group morale and performance. As they solve their problems or formulate their decisions, the group members must often take some chances. How much assurance of success do they want before adopting a particular course of action? Are the participants' decisions as a group more or less risky than the decisions reached by individuals working alone? Miller and his associates consider these questions and others in their review of group risk-taking. There are all sorts of groups, of course, and important distinctions undoubtedly have to be drawn among them. Nevertheless, some generalizations evidently can be made across a broad spectrum of groups. Waxler and Mishler look at families as groups and apply the concepts and findings of research into group behavior to the analysis of family interactions.

The articles I have just mentioned are clearly important in their own right. We are also fortunate to have many of the writers' latest thoughts on the issues they had faced earlier. In most cases, the original chapter is followed by the author(s)'s comments on recent developments in the particular problem area. These thoughtful essays add to the significance of the original contributions.

Group Processes

GROUP TASKS, GROUP INTERACTION PROCESS, AND GROUP PERFORMANCE EFFECTIVENESS: A REVIEW AND PROPOSED INTEGRATION[1]

J. Richard Hackman

DEPARTMENT OF ADMINISTRATIVE
SCIENCES AND PSYCHOLOGY
YALE UNIVERSITY
NEW HAVEN, CONNECTICUT

Charles G. Morris

DEPARTMENT OF PSYCHOLOGY
UNIVERSITY OF MICHIGAN
ANN ARBOR, MICHIGAN

When decision-makers in public and private institutions in this society are faced with genuinely important tasks, it is likely that they

[1] Preparation of this report was supported in part by the Office of Naval Research (Organizational Effectiveness Research Program, Contract No. N00014-67A-0097-0026); reproduction in whole or in part is permitted for any purpose of the U.S. government. The contributions of Kenneth R. Brousseau, Daniel Feldman, Martin Greller, Janet A. Weiss, and Gerrit Wolf are gratefully acknowledged.

1

will assign those tasks to groups for solution. Sometimes the reason is simply that one individual could not be expected to handle the task by himself (e.g., formulating a new welfare policy, which requires a diversity of knowledge and skills). Other times it is because decision-makers assume that the added human resources available in a group will lead to a higher *quality* product—or will at least lessen the chances that the product will be grossly defective.

Given current knowledge about group effectiveness, the state of affairs described above is not an occasion for optimism. Although literally thousands of studies of group performance have been conducted over the last several decades (Hare, 1972; McGrath & Altman, 1966), we still know very little about why some groups are more effective than others. We know even less about what to do to improve the performance of a given group working on a specific task. Moreover, the few general findings that have emerged from the literature do not encourage the use of groups to perform important tasks. Research has shown, for example, that for many tasks the pooled output of noninteracting individuals is better than that of an interacting group (cf. reviews by Collins & Guetzkow, 1964; Davis, 1969; Lorge, Fox, Davitz, & Brenner, 1958; Mc-Grath & Altman, 1966; Shaw, 1971; Steiner, 1972).

It is tempting to conclude that the "group effectiveness problem" will not be solved in the foreseeable future, and to recommend to de-cision-makers that in the meantime they use groups as infrequently as possible. The present paper explores the possibility that this viewpoint is unduly pessimistic—that the human resources present in groups can, in fact, be harnessed and directed toward more effective performance than would be obtained from individuals alone. We suggest that the key to understanding the "group effectiveness problem" is to be found in the on-going *interaction process* which takes place among group members while they are working on a task. At one extreme, for example, group members may work together so badly that members do not share with one another uniquely held information that is critical to the problem at hand; in this case, the quality of the group outcome surely will suffer. On the other hand, group members may operate in great harmony, with the comments of one member prompting quick and sometimes innovative responses in another, which then leads a third to see a synthesis between the ideas of the first two, and so on; in this case, a genuinely creative outcome may result.

The challenge is to identify, measure, and change those aspects of group interaction process that contribute to such obvious differences in group effectiveness. Toward this end, the chapter is organized into three parts. In Section I we review existing research and thought on the role of group interaction in task-oriented groups, and we suggest that part of the difficulty in understanding the relationship between group interaction and group effectiveness has to do with the nature of existing methodological and conceptual tools. Then, in Section II, we propose an alternative framework for research on group effectiveness. The major functions group interaction serves in enhancing and depressing group effectiveness are explored, and a set of strategies for influencing group interaction and group performance by alteration of "input" factors is proposed within the new framework. The section closes with an argument for a return to action-oriented research as a way to improve simultaneously our understanding of the determinants of group effectiveness and our capability to change and improve it. Finally, in Section III, implications for research and for action are drawn and explored.

I. The Role of Interaction Process in Task-Oriented Groups: Current Thought and Evidence

Although research on group effectiveness rarely includes explicit quantitative assessment of how group interaction affects group performance, it is common for researchers to speculate about the functions of group process when they are developing research hypotheses and when they are interpreting empirical findings. A sampling of such speculations is offered below, both to provide a context for the ensuing discussions, and to reveal the diversity of suggestions that have been made about the functions of group process in task-oriented groups.

Many social psychologists have taken a rather pessimistic. view of tht role of group process—i.e., seeing it as something that for the most part *impairs* group task effectiveness. Steiner (1972), for example, treats group interaction process almost entirely in terms of "process losses" which prevent the group from approaching its optimal or potential productivity. It turns out that in fact the findings of many studies can be predicted by the models Steiner proposes.

Other social psychologists suggest that the interaction among group members helps to catch and remedy errors that might slip by if individuals were doing the task by themselves. Thus, the argument goes, although groups may be slow and inefficient because of process problems, their use is more than justified when solution *quality* (i.e., freedom from errors) is of paramount importance (cf. Taylor & Faust, 1952). Recent work by Janis (1972), however, calls into question the efficacy of group interaction for finding and correcting errors, at least under some circumstances. Janis suggests that "groupthink" may develop as members become excessively close-knit and generate a clubby feeling of "we-ness." Groupthink is evidenced by a marked decrease in the exchange of discrepant or unsettling information, and by a simultaneous unwillingness to deal seriously with such information even when it is forced to the attention of members. Under these circumstances, Janis suggests, the group may develop and implement a course of action that is grossly inappropriate and ineffective. Janis finds that the principles of groupthink help to explain a number of highly significant and unfortunate decisions made by top-level government officials, such as the Bay of Pigs invasion, and Britain's "appeasement" policy toward Hitler prior to World War II. Apparently even for some very important decisions, patterns of group interaction can develop that allow large and significant errors of fact and judgment to "slip through" and seriously impair group effectiveness.

A more optimistic view of the role of group process is offered by Collins and Guetzkow (1964), who propose that in some circumstances interaction can result in "assembly effect bonuses." That is, patterns of interaction may develop in which the individual inputs of group members combine to yield an outcome better than that of any single person—or even than the sum of individual products. The literature reviewed by Collins and Guetzkow, however, offers little help in understanding how to create such bonuses. The "brainstorming" fad of the late 1950's (Osborn, 1957) seemed to offer one clear instance in which the assembly effect bonus led to group outcomes of higher creativity than those obtained by pooling the products of individuals; yet subsequent research failed to reveal any creative bonuses attributable to the group interaction process per se (Dunnette, Campbell, & Jaastad, 1963; Taylor, Berry, & Block, 1958).

Organizational psychologists involved with experiential "training groups" or with "team-building" activities also tend to be optimistic about the possibility of enhancing group task effectiveness by alteration of group process. In general, they assume that members of many task groups are inhibited from exchanging ideas and information and from working together in a concerted fashion to complete the task. Interpersonal training activities are intended at the least to remove some of the emotional and interpersonal obstacles to effective group functioning and thereby to permit group members to devote a greater proportion of their energies toward actual task work. Moreover, when the dysfunctional "process problems" of a group have been dealt with, members may discover new ways of working together which eventually will help them to achieve previously unknown levels of effectiveness (cf. Argyris, 1969; Kaplan, 1973). In effect, the group can *capitalize* on its interpersonal processes in the interest of increased task effectiveness rather than find itself distracted from task work by interpersonal problems.

Research data are not yet available to document the belief that interpersonal training activities lead to positive effects on group task performance. There are substantial data which show that training activities can powerfully affect both the nature of the interaction process in groups and the quality of members' personal experiences. But the few studies that have tested the effects of such changes on actual task performance generally have yielded ambiguous or negative results (for reviews, see Herold, 1974; Kaplan, 1973).

In sum, there is substantial agreement among researchers and observers of small task groups that something important happens in group interaction which can affect performance outcomes. There is little agreement about just what that "something" is—whether it is more likely to enhance or depress group effectiveness, and how it can be monitored, analyzed, and altered. A major purpose of this chapter is to make some headway in developing answers to these questions. As a first step, we propose in the next section an organizing framework which is useful in sorting out the specific relationships among (a) the initial state of a task-oriented group, (b) the group interaction process, and (c) the group's ultimate performance effectiveness.

A. An Organizing Framework

A general paradigm (adapted from McGrath, 1964) for analyzing the role of group interaction process as a mediator of input–performance relationships is depicted in Fig. 1. As used here, "interaction process" refers to all observable interpersonal behavior that occurs between two arbitrary points in time (t_1 and t_2). The state of all system variables potentially may be assessed at any given "slice" in time, and therefore input–output relationships may be examined for periods of time ranging from a few seconds to a year or more. The longer the time between t_1 and t_2, the greater is the amount of interaction intervening between input time and output time, and the more complex the analysis of the role of interaction in mediating input–output relationships becomes. It should be noted that the process depicted in Fig. 1 can and does "recycle" on a continuous basis. That is, many properties of the group and its members (e.g., group communication structures, individual attitudes) both affect the nature of the interaction process and are themselves changed by that process. Such "outcomes" of group interaction then can affect the nature of subsequent interaction, leading to their further modification, and so on. Fortunately, for analysis of task performance

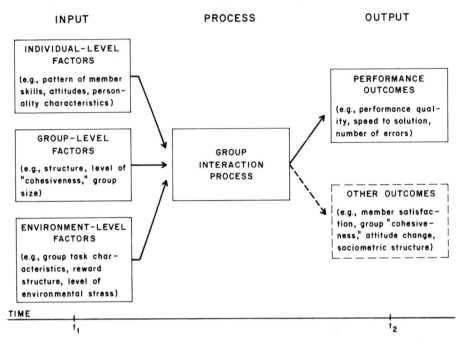

FIG. 1. A traditional paradigm for analysis of group interaction as a mediator of performance outcomes. *Note:* The focus of this chapter is on performance outcomes; as is indicated by the broken-line box in the figure, there are many other kinds of outcomes which also can result from the group interaction process. Adapted from McGrath (1964).

outcomes (the focus of the present discussion) this phenomenon is not problematic; most group tasks have a natural and easily identifiable end point, at which time t_2 measures can be taken with reasonable assurance that "recycling" has not yet occurred.

The fundamental assumption underlying the paradigm in Fig. 1 is that input factors affect performance outcomes *through the interaction process.* Thus, if highly cohesive groups (input at t_1) perform better on some task (outcome at t_2) than less-cohesive groups, it should be possible to explain the performance difference by examining the difference between the interaction processes of the high and the low cohesive groups. That is, the "reason" for obtained input–performance relationships always is available—albeit sometimes well-hidden—in the interaction process itself; by appropriate analysis of interaction process it should be possible to develop a rather complete understanding of input–output relationships in any performance setting.

B. RESEARCH EVIDENCE

In the paragraphs to follow, research on the determinants and consequences of group interaction process is examined in the context of the framework depicted in Fig. 1. Only findings of relevance to group interaction process as a mediator of input–performance relationships in task-oriented groups are included. Findings from three general types of studies are included: (1) those that deal with input–process relationships; (2) those that focus on process–performance relationships; and (3) those that address the full input–process–performance sequence.

1. Input–Process Relationships

It has been well-established that the nature and direction of group interaction process are affected by numerous "input" factors; therefore this research will not be reviewed in detail here. Among the input factors that have been shown to affect group interaction are leader attitudes (Sample & Wilson, 1965), member personality characteristics (Conway, 1967), group size (O'Dell, 1968), group structure (Cohen, Bennis, & Wolkon, 1961), and group history or experience (Hall & Williams, 1966). The nature of the task on which a group is working appears to be a particularly potent influence on group interaction process; almost every study that addressed the question unearthed substantial task–process relationships (Carter *et al.*, 1951; Deutsch, 1951; Hare, 1962; Morris, 1966; Talland, 1955).

2. Process–Performance Relationships

Research that directly relates measured characteristics of group process to performance outcomes is scarce. Relatively systematic attention, however, has been given to one particular aspect of the process–performance relationship—namely, the "weighting" process by which various solution proposals generated by group members are selected and rejected (Kelley & Thibaut, 1954; Steiner, 1972).

Hoffman (1961), for example, proposed that potential solutions to group tasks gain or lose "valence" as supportive or critical comments are made about them in the group discussion. Hoffman and Maier (1964, 1967) developed a behavior category system to tap the process by which the valence of task solutions changes during group interaction. As predicted, specific solutions to the Parasol Assembly Problem (Maier, 1952) tended to be adopted by the group and used in the final product in direct proportion to their valences. Once solutions reached the minimum level of valence necessary for acceptance, they tended to be adopted forthwith, and group members thereafter spent little energy searching for potentially better solutions.

An implication of the valence model is that an obvious or highly salient solution will tend to be adopted more readily than a nonobvious solution. The obvious solution will tend to acquire valence quickly, and perhaps gain adoption before the nonobvious solution is even seriously considered. The Horse Trading Problem (Maier & Solem, 1952) has an obvious (but erroneous) solution, and groups do often solve the problem incorrectly. Presumably if conditions could be arranged so that the initial obvious solution could be kept from achieving substantial valence, and if simultaneously other less obvious solutions could be encouraged, group performance would increase on this task. Maier and Solem attempted to create these conditions in their study by manipulating leader behavior and found that groups with permissive, accepting leaders did tend to avoid the trap of the obvious (but incorrect) solution more often than did groups in a control condition. Although no explicit measurements of group process were made in this particular study, the data do suggest that an input condition (leadership role definition) can affect productivity (selecting the less-obvious but correct solution) through changes in group process (damping tendencies for group members to build up valence for the initially obvious solution). Another study that provides inferential support for the valence model of process–output relationships is reported by Riecken (1958).

One of the few studies that quantifies the relationships between measures of group process and measures of performance effectiveness is reported by Lanzetta and Roby (1960). The task in this study required the group to achieve a particular configuration of lights on an electrical apparatus by appropriate sequencing of group member responses. All communications among members were monitored and recorded. It was found that measures of group interaction predicted task success (time to completion and freedom from errors) better than did measures of members' task-relevant knowledge or various task training procedures. The authors note that "the way the group 'utilizes' its resources and the procedures it employs for communicating essential information are as important, if not more important, than 'knowledge' of the problem for determining its performance" (p. 146).

In summary, these studies suggest that the impact of group interaction on group performance can be analyzed systematically and that the results of such analyses can increase understanding of the reasons why some groups are more effective than others. None of the studies reviewed thus far, however, has addressed explicitly the question of how group interaction *mediates* input–performance relationships. It is to this issue that we now turn.

3. Input–Process–Performance Relationships

The present authors recently attempted to assess explicitly the full input–process–performance sequence. One hundred and eight experimental groups spent 15 minutes on each of four "intellective" tasks. Four hundred and thirty-two separate transcripts of group interaction and 432 group products were obtained. A total of 144 different group tasks were used in the research, 48 each of three task "types": (a) "production" tasks, which require the production and presentation of ideas or images; (b) "discussion" tasks, which require an evaluation of issues; and (c) "problem-solving" tasks, which require specification of a course of action to be followed to resolve some problem. Specific tasks and order of task presentation were appropriately counterbalanced across groups and experimental condition.[2]

Morris (1966) used these data to examine effects of task type and task difficulty on group interaction. Interaction process was measured by a sixteen-category coding system which focuses exclusively on task-oriented interaction among group members. It was found that task type significantly affected interaction in nine of the sixteen behavior categories (which accounted for about 70% of the groups' total interaction). In addition, task difficulty significantly affected five interaction categories (which accounted for approximately 40% of the groups' interaction).

Using the same set of data, Hackman (1968) demonstrated that task type also significantly affected the characteristics of written group products, as measured by six descriptive dimensions (action orientation, length, originality, quality of presentation, optimism, and issue involvement) and two evaluative dimensions (performance adequacy and judged product creativity). The six descriptive dimensions are described in detail by Hackman, Jones, and McGrath (1967).

Thus, Morris established the input–process relationships and Hackman the input–performance relationships on the same set of data. The analyses reported below address the extent to which it can be shown, using the same data base, that group performance varies as a function of group interaction process, thereby completing the input–process–performance chain.

Process–performance relationships were examined for all 432 transcripts together, then separately for each of the three task types (144 transcripts for each). Analysis was by canonical correlation, which ad-

---

[2] Details of methodology are provided by Hackman (1968) and by Morris (1966).

dresses the relationship between group interaction and group performance by considering simultaneously all sixteen interaction categories and all eight product dimensions. In addition to a measure of the overall strength of the relationship between the several process categories and product dimensions, the canonical analysis provides a set of weights for both the predictors (the measures of interaction) and the criteria (the product dimensions). These weights indicate the degree to which each of the variables contributes to the overall process–performance relationship. Thus, the analysis reveals not only "how much" relationship exists between group interaction and group performance measures for the tasks used, but also what specific aspects of the interaction and what specific performance measures contribute most substantially to the obtained relationship.

The canonical correlations are presented in Table I both for the combined sample and separately for each of the three task types. The correlations range from .59 to .68 and are statistically highly reliable. It appears, as had been expected, that substantial variation in group performance on intellective tasks is controlled by the nature of the group interaction process.

Table II shows the canonical weights obtained in each analysis. Unfortunately, these weights provide few clues about the substantive meaning of the process–performance relationships obtained. As anticipated, the weights associated with specific interaction categories and product dimensions do differ for the three task types, but these differences appear not to be interpretable in substantive terms.

In an attempt to clarify the meaning of these results, multiple corre-

TABLE I
CANONICAL CORRELATIONS BETWEEN INTERACTION PROCESS AND
PRODUCT CHARACTERISTICS

	All groups	Groups having production tasks	Groups having discussion tasks	Groups having problem-solving tasks
Canonical correlation	.68	.64	.59	.66
Λ	.271	.179	.242	.192
χ^2	547.8[a]	266.2[a]	186.6[a]	217[a]
N	432	144	144	144

Note: Only the first root (λ_1) for each analysis reached statistical significance and is reported here.

[a] $p < .001$; $df = 128$.

TABLE II
CANONICAL WEIGHTS FOR PREDICTOR (PROCESS) AND CRITERION
(PRODUCT) MEASURES

	All groups	Groups having production tasks	Groups having discussion tasks	Groups having problem-solving tasks
Predictor Weights				
Interaction Process Categories				
1. Structure problem	−.33	−.41	−.29	.05
2. Structure answer	−.14	−.10	−.43	.61
3. Propose solution	.61	.50	.23	.05
4. Clarify	−.17	.03	−.11	.18
5. Defend	−.13	−.28	−.28	−.16
6. Repeat	.24	.28	.53	−.11
7. Agree	−.18	−.02	−.03	.15
8. Disagree	.17	−.27	−.17	.13
9. Seek structuring	.10	.13	−.33	.03
10. Seek solution proposals	−.09	−.10	−.06	.19
11. Seek clarify–defend–repeat	.10	.27	−.20	.40
12. Seek evaluation	.08	−.16	−.01	−.20
13. Procedure	.11	.30	−.12	.11
14. Seek procedure	−.08	−.08	.17	−.12
15. Irrelevant	−.11	−.12	−.04	−.30
16. Fragmentary	−.13	−.06	.04	−.31
Criterion Weights				
Group Product Dimensions				
1. Action orientation	−.08	−.29	.24	−.39
2. Length	.49	.67	.21	.41
3. Originality	.28	.18	−.13	.28
4. Optimism	.03	−.01	−.13	.14
5. Quality of presentation	.01	−.20	.45	−.36
6. Issue involvement	−.56	−.54	−.54	.22
7. Adequacy	−.01	−.03	.10	−.46
8. Creativity	.16	.11	.23	.37

lations were computed predicting each product dimension separately from the sixteen measures of interaction process, and zero-order correlations were obtained between all process measures and all performance measures. Numerous statistically reliable relationships were obtained— but again, the relationships were so scattered and seemingly inconsistent that it was impossible to draw from them any parsimonious explanations of how group interaction affects group performance on intellective tasks. Thus, while the research did provide evidence that group processes are strongly related to group performance outcomes, it failed to shed much light on the *substance* of input–process–performance relationships.

Similar conclusions may be drawn from two other studies which examined the full input–process–performance sequence. Sorenson (1971) varied task type (using intellective tasks very similar to those described above), and measured five aspects of the interaction process (structuring, generating, elaborating, evaluating, requesting). Written group products were judged on originality and overall quality. Significant input–performance, input–process, and process–performance relationships were obtained. Consistent with the results reported above, however, the process–performance relationships were complex and did not vary systematically for different input conditions.

Significant relationships among input, process, and performance measures also were obtained by Katzell, Miller, Rotter, and Venet (1970) using a "twenty questions" task. Manipulated input conditions were leader directiveness, task difficulty, and member compatibility; group process was measured by using Bales' (1950) Interaction Process Analysis. Process–performance relationships in this study were more readily interpretable than those reported previously (i.e., time to solution increased as members sought and exchanged more information), but again these relationships were not moderated by manipulated input conditions.

Some possible reasons why the dynamics of input–process–performance relationships in task-oriented groups have remained enigmatic are proposed in the next section. Then a new approach to research on group process and group effectiveness will be proposed which may lead to more productive attacks on the problem in the future.

C. Problems in Analyzing the Mediating Functions of Group Process

There are at least five reasons why research based on existing methodological and conceptual paradigms has not yet succeeded in determining how group interaction process mediates between input and output states. The first two reasons have to do with the appropriateness of existing methodologies for measuring group process in task-oriented groups; the last three address broader issues of research strategy.

1. Behavior Categories

Most interaction coding systems focus almost exclusively on specific acts of communication which take place among group members. The Bales (1950) system, for example, categorizes such task and socioemotional behaviors as "shows tension" and "asks for orientation"; the Morris

(1966) system categorizes communications directly relevant to carrying out the task such as "proposes solution," "repeats," and "seeks evaluation."

Such systems are appropriate for research programs that aim simply to describe patterns of interaction in task-oriented groups, or that attempt to map the relationships between various input conditions and the resulting group interaction process. They are less likely to be useful in research aimed at understanding how interaction *mediates* the influence of input conditions on group performance, or how different patterns of interaction lead to improved or impaired group effectiveness. The reason is that there is no clear conceptual or operational link between the kinds of behaviors tapped by these systems and the immediate determinants of group effectiveness. It is doubtful, for example, that the "number of questions asked" in a group will directly or indirectly affect the performance of most groups. It appears, rather, that coding systems are needed that derive directly from conceptual propositions about those aspects of group interaction that are crucial in determining group effectiveness for various kinds of group tasks. The content of such theory-based systems, it is argued, would be substantially different from that of most existing systems and would more clearly reveal just what goes on in groups to sometimes facilitate group effectiveness and sometimes impair it.

2. Analytic Models

Most existing interaction coding systems (and analytic models for using system-derived data) yield summary scores that reflect only the frequency or the rate of interaction in various content categories. In some cases (e.g., for monitoring the level of valence associated with various solution proposals in the Hoffman-Maier research) frequency data appear to be fully appropriate for the specific purpose of the research (in this case predicting solution adoption by the group). But in other cases, and perhaps especially for highly complex tasks, frequencies or rates of interaction may be quite inappropriate summary measures to use in attempting to predict group task effectiveness (cf. Morris, 1970). An example provided by H. L. Raush (personal communication) illuminates the point. An understanding of successful chess play surely would not result from a study of the number of times players moved each of the chess pieces during games. Instead, one would almost certainly attempt to discern the *sequences* of moves players made in response to the particular status of the board at any given time. In the context of this illustration, our attempt to predict the characteristics of written group products from simple interaction frequencies does not (in retrospect) seem intuitively reasonable.

Analyses of interaction sequences are difficult to carry out, although they can be quantitatively modeled as stochastic processes (Raush, 1965; Wolf, 1970). To date, however, stochastic modeling of communication processes has been applied primarily to act-by-act exchanges in dyads on a moment-to-moment basis. It appears that if sequential analyses are to be useful in understanding the mediating functions of group interaction process, new analytic techniques will be required. In particular, it will be necessary for interaction sequences to be related directly to the task goals and strategies being pursued by group members, and for procedures to be devised that permit analysis of groups larger than dyads over relatively long periods of time. Although some progress recently has been reported (Swinth & Tuggle, 1971; Wolf, 1974), the development of such methodological and analytic techniques remains a significant research challenge. In the meantime, the researcher interested in assessing how group effectiveness is influenced by interaction process might be well-advised to focus directly on the *molar* strategies that guide the moment-to-moment interactions among members—rather than attempting to chart the component sequences of communicative acts or simply summing the frequencies of interaction in various content categories.

3. Inconsistencies across Tasks

Although few studies have assessed directly the relationship between group process and performance separately for different types of tasks, the McGrath and Altman (1966) review of the field provides implicit support for the proposition that process–performance relationships are likely to be inconsistent across different task types. In almost all studies reviewed in which input–output relationships were tested separately for groups working on different tasks, the form of these relationships was found to vary from task to task. It is not a large leap of inference to conclude that the way in which group processes mediate input–output relationships varies across tasks as well. The data reported in Table II confirm this expectation: Although canonical correlations between measures of group process and measures of group performance were statistically significant for all three types of tasks used, the contribution of specific interaction categories to the prediction of performance outcomes varied substantially (if mysteriously) across task types.

Such findings suggest that it may be unrealistic to work toward achieving a truly general theory of the relationship between group interaction and group performance effectiveness. Instead, it may be necessary to make some a priori distinctions among general classes of tasks and then to delve into process–performance relationships *within* each class.

While this is not as elegant an approach as some might desire, the development of subtheories of process–performance relationships would nonetheless represent a notable improvement over current understanding of group process determinants of group effectiveness.

4. Research Settings

The research settings and methodological strategies used in studies of group effectiveness may themselves constrain the possibility of unearthing significant process–performance relationships. Most research that involves analysis of the interaction process of task-oriented groups takes place in laboratory settings and uses methodological strategies consistent with paradigmatic experimental social psychology. For reasons to be suggested below, such strategies invariably involve substantial (although often implicit) control over both the task of the group and the norms that guide group member behavior.

In a typical social psychological study of groups, for example, the group task is carefully specified a priori, and all groups in the study perform the same single task. Tasks are almost never sampled within an experimental design in a way that allows individual differences among tasks to vary as do individual differences among subjects or groups. Thus, for all practical purposes, the task is held constant within the designs of most social psychological studies of small groups.

Group norms are usually, although unintentionally, held relatively constant as well. Most laboratory studies of group effectiveness use ad hoc groups which are convened for such a short time that the group does not have a chance to develop its own history or its own unique normative structure. And, since most members of experimental groups have a good deal of everyday experience in committees and other small task groups, they are likely to bring to the group rather similar notions about "how one should behave" in a task-oriented group. The net effect is that the norms that guide the behavior of members, like the tasks they are given to perform, are unlikely to vary much from group to group within a given study.

But the research literature on small groups suggests that both group tasks and group norms are very powerful influences on interpersonal behavior (cf. reviews by Hackman, 1975; McGrath & Altman, 1966). Thus, when tasks and norms are held constant (or relatively so) in experimental studies of groups, it is nearly inevitable that the richness and diversity of interpersonal behavior within groups will be reduced substantially.

For some research purposes, it may be quite desirable for tasks and norms to be well-constrained in the experimental setting. Precisely because variation in interpersonal behavior is reduced, groups in the research will be more similar to one another than would be the case if tasks and norms varied widely. This, in turn, will decrease the amount of "error" variance within experimental conditions. As a result, the chances of reliably detecting and describing relationships between input and output variables for these groups will be enhanced. Group processes, which could have substantially muddied the waters that separate input from output states, have been given much less chance to do so.

But if the research purpose is to understand how group processes themselves relate to output states (or how they mediate input–output relationships), exactly the reverse is true: if the research setting and methodological strategy serve to reduce the variation in group interaction process in the groups under study, then the likelihood of obtaining strong empirical correlates of group process measures also will be reduced. In a sentence: The less meaningful variation there is in the interaction among group members, the less one can learn about the role of interaction in affecting group performance. The implication, then, is that studies of process–performance relationships will require research designs that allow interpersonal processes to vary more widely than is the case in traditional small group experiments. And this, we suggest, will require substantially different treatment of group tasks and group norms than has heretofore been the case.

5. Cultural Norms

Cultural norms about appropriate behavior in groups are such that patterns of behavior that might optimize group task effectiveness are unlikely to appear in natural groups. Typically, the norms that guide individual behavior in groups tend to be rather "conservative"—that is, they minimize the chances that members will have to deal with uncomfortable or anxiety-arousing behaviors within the group. Member feelings are expressed circumspectly, if at all; interpersonal risks are taken rarely, and then only when avenues of face-saving retreat are readily available; deviant ideas and behaviors tend to be dealt with swiftly, if gently; and so on (Argyris, 1969; Hackman, 1975). When groups are studied as they develop "naturally," whether in laboratory or field settings, then there is an excellent chance that the behavior of group members will be guided by such norms.

One implication is that research findings from such groups will apply only to other groups in which these norms also are operative. This would seem, however, to be a constraint on generality that is of little consequence because the norms are in fact so widely shared by members of task-oriented groups in this culture.

A second implication seems more serious: What if these norms were in many ways *dysfunctional* for group effectiveness on most tasks? If this were true, then researchers would inevitably discover that group processes serve mainly to *impair* group effectiveness, and in all probability research attention would be turned toward understanding more fully the nature and extent of the apparently widespread group "process losses." Given this possibility, the rather pessimistic conclusions some social psychologists have drawn about the role of group processes in affecting performance outcomes can be viewed in new perspective.

Almost no studies have addressed the more optimistic side of the same coin—i.e., whether groups might perform more effectively if members worked together in ways that are quite *different* from generally accepted norms about appropriate behavior in groups (Argyris, 1969). It is conceivable that increased leverage could be brought to bear on the "group effectiveness problem" by deliberate experimental alteration of the norms that govern interaction in groups, such that new and potentially more functional patterns of behavior emerge for empirical examination. Otherwise, research will necessarily continue to describe and document "the way things are" in groups at present—including the sometimes-unfortunate consequences of traditional norms about behavior in task-oriented groups.

6. Summary

We have suggested that various methodological tools and research strategies typically used in studies of group effectiveness may severely limit the kinds of understandings that can emerge from that research. Indeed, the rather pessimistic conclusions of much psychological research on the role of group interaction process (e.g., that process may operate primarily to keep a group from achieving its potential productivity) may themselves have been predetermined in part by the methodologies used. We turn now to some alternatives for research on small group effectiveness which follow from the above discussion.

II. Toward Some Alternatives for Research and Action

In the pages to follow, we suggest a new, three-pronged approach to research on group effectiveness. The first prong involves a more differentiated view of the functions of group interaction than has typified previous research. The second concerns ways that "input" factors might be altered to improve group effectiveness, in view of the newly proposed functions of group interaction. The third concerns a new kind of research strategy which may be required to learn whether and how such improvement in group performance can be achieved.

A. THE FUNCTIONS OF INTERACTION PROCESS IN INFLUENCING GROUP EFFECTIVENESS

The input–process–output model of group effectiveness discussed earlier suggests that interaction process somehow mediates input–performance relationships. Yet the question of exactly how such mediation takes place remains open and troublesome. In fact, material presented in the previous section suggests that there may be no single or general answer to the question; instead, group process may serve quite different functions, depending on the kind of task facing the group.

In attempting to develop ideas about the ways group interaction can determine group performance, we offer the following general propositions about "what makes a difference" in group performance effectiveness.

1. As Katzell *et al.* (1970) note, the number of factors that can affect group output is so great that managing more than a few factors at a time, either conceptually or experimentally, is nearly impossible. As a strategy for dealing with this mainfold, they suggest using "a single set of mediating variables in order to link conceptually and functionally all kinds of group inputs . . . with various kinds of group outputs" (p. 158). Consistent with this strategy, we propose that a major portion of the variation in measured group performance is proximally controlled by three general "summary variables": (*a*) the *effort* brought to bear on the task by group members; (*b*) the *task performance strategies* used by group members in carrying out the task; and (*c*) the *knowledge and skills* of group members which are effectively brought to bear on the task. It is proposed that, if one could somehow control or influence these three summary variables, one would be able to affect substantially the level of effectiveness of a group working on almost any task.

2. Each of the summary variables can be substantially affected (both positively and negatively) by what happens in the group interaction process. The interaction among group members can, for example, either increase or decrease the level of effort members exert in doing the task, and can affect how well the efforts of individual group members are coordinated. Similarly, group interaction can lead to either effective or ineffective task performance strategies, and to efficient or wasteful use of the knowledge and skills of group members. The specific roles that group interaction plays in a given situation will depend substantially on the task being performed.

3. Different summary variables (or combinations of them) are operative for different types of group tasks. For some tasks, for example, how hard group members work (member effort) will amost entirely determine their measured effectiveness; Ringlemann's group tug-of-war (Ingham, Levinger, Graves, & Peckham, 1974) is an example of such a task. For other tasks, effort will be mostly irrelevant to performance effectiveness, and other summary variables will be operative. For example, on a vocabulary test which is taken collaboratively by group members with no time limit, performance is unlikely to be affected by how hard members "try," but it will be dependent on their collective knowledge of the meanings of words (member knowledge and skill). For some group tasks, of course, measured performance effectiveness may depend on two or on all three of the summary variables. The point is simply that which of the summary variables will "make a difference" in measured group effectiveness is heavily determined by the type of group task on which the group is working.

The remainder of this section extends and elaborates on the above propositions, with particular emphasis on the ways in which group interaction can impair or enhance performance effectiveness via the three summary variables. We shall acknowledge the likelihood of "process losses" in certain kinds of performance situations—but also shall attempt to point toward ways in which interaction process can increase the likelihood of process "gains." Then in the next major section, we address concrete strategies which may be useful in actually improving group performance effectiveness.

1. Member Effort

The first summary variable to be considered is also the most ubiquitous: How hard group members work on a task should be an important determiner of group effectiveness on many different types of group tasks. And, while many personal and situational factors can influence the level of effort the group brings to bear on its task activities, it is proposed that group interaction affects effort primarily in two ways: (a) by affecting the *coordination* of the efforts of individual group members, and (b) by affecting the *level* of effort group members choose to expend working on the group task (their task motivation).

a. Coordination of member efforts. If effectiveness on a given group task is influenced by the amount of effort group members apply to it, then it is important that members coordinate their activities so that efforts of individual members are minimally "wasted." On the Ringlemann tug-of-war, for example, a group will do quite poorly unless the group devises some means of ensuring that members pull at the same time (Ingham *et al.*, 1974). Whatever coordination is achieved among members should be evident in the group interaction process; that is, examination of interaction should at least reveal the nature of the coordination scheme being used by the group, and (especially for tasks unfamiliar to group members) it may show exactly how the group came up with whatever coordination devices it is using.

Steiner (1972) shows that, when the efforts of individual group members must be coordinated to accomplish the task, there will always be some slippage which can only serve to keep a group from achieving its potential productivity (i.e., that which would be obtained if the efforts of each group member were fully usable by the group in the service of task effectiveness). Moreover, the larger the group, the greater will be the process loss (which Steiner calls "coordination decrement"), simply because the job of getting all members functioning together in a coordinated fashion becomes increasingly difficult as the number of members gets larger. Therefore, attempts to increase productivity by helping group members coordinate their activities more effectively can be construed as working toward minimizing inevitable process losses, rather than as creating "process gains."

b. Enhancing or depressing the level of member effort. While individual members usually approach a given group task with some notion about how hard they expect to work on it, what happens in the group can radically alter that expectation in either direction. Presumably an individual will increase his level of effort to the extent that working hard with the other group members leads to the satisfaction of his personal needs or the achievement of his personal goals. If his task-oriented efforts are reinforced, he should work harder on the task; but if his efforts are ignored or punished, his effort should decrease. The point here is that social interaction can importantly affect how much effort an individual chooses to expend in work on the group task, and that the level of effort can easily change over time as the characteristics of the group interaction change.

The depression of member effort has been explored by Steiner (1972) in terms of a "motivation decrement." He suggests, for example, that member effort declines as group size increases. If our paradigm is valid, then the explanation for the relationship between group size and member effort will be found in the patterns of interaction that characterize small vs. large groups. Conceptual and empirical attention recently has been directed toward such group process aspects of motivation decrements (see Steiner, 1972, Chapter 4). As yet, however, systematic attention has not been given to ways in which patterns of interaction might be created in groups which would result in a "motivation increment," encouraging members to work especially hard on the group task. The feasibility of creating such increments in task-oriented groups is explored later in this chapter.

2. Task Performance Strategies

As used here, "strategy" refers to the collective choices made by group members about how they will go about performing the task. Included are choices group members make about desirable performance outcomes and choices about how the group will go about trying to obtain those outcomes. For example, group members might elect to try to make their product funny or elegant or practical; or they might decide to free-associate for a period of time to get ideas about the task before evaluating any of the possibilities that they generated. These all are examples of performance strategies under the voluntary control of group members, and they are related. How we proceed to carry out the task depends in part on what we are trying to achieve, and vice versa.

A number of researchers have demonstrated that the performance effectiveness of a group can be affected markedly by the strategies members use in working on the task (e.g., Davis, 1973; Maier, 1963; Shiflett, 1972; Shure, Rogers, Larsen, & Tussone, 1962; Stone, 1971). What specific strategies will be effective or ineffective in a given performance situation, however, depends on the contingencies built into the task itself. As was the case for effort, there are some tasks for which differences in performance strategies will have relatively little impact on the ultimate effectiveness of the group. The group vocabulary test described earlier, for example, would seem minimally responsive to differences in task performance strategies; almost any strategy that ensures that the group member who knows the answer will communicate it to his peers will suffice. Strategy would be considerably more important for a task requiring the solution of complicated algebra problems. In this latter case, the approach a group takes to the task (e.g., breaking the problem into parts vs. trying to solve the problem in one step) could have a considerable impact on the probability of successful performance. Because of such differences in "what works" for different tasks (or at different stages of work on a single complex task), it has been suggested that perhaps the only universally effective strategy may be an ability and willingness to switch from one specific strategy to another as the need arises (Shiflett, 1972, pp. 454–455).

There are two ways in which group interaction process can affect the performance strategies that a group brings to bear on its task: (a) through implementing preexisting strategies that are shared among group members; and (b) through reformulating existing performance strategies or generating new ones. These two functions of group interaction are examined below.

a. *Implementing existing, shared strategies.* As people gain experience with particular kinds of tasks in the course of their everyday lives, particular strategies for working on these tasks become well-learned. When an individual is given a new task from some familiar, general class of tasks, he need not spend time actually deciding how to work on the task or selecting appropriate outcomes. Instead, he can simply begin to do the new task. The same process occurs when a group of individuals works on a task that is familiar to them. Everyone in the group may know very well the "obviously right" way to go about working on the task, and no discussion of strategy need take place in the group at all.

In such cases, group interaction serves mainly to implement existing strategies already well-learned by group members, and no evidence of the group's "working on its performance strategy" may be visible in the overt interaction among members. This phenomenon is demonstrated clearly in the data collected by the present authors. It will be recalled that the characteristics of the written products prepared by groups in the study were quite strongly affected by the type of task being worked on. Production tasks led to highly original products; discussion tasks prompted tasks high in issue involvement; and problem-solving tasks led to products high in action orientation (Hackman, 1968). In a follow-up study, individuals working by themselves indicated, when asked, that they were confident that a response to a production task "ought" to be original, a response to a discussion "ought" to be heavily issue-involved, and so on.

Yet analysis of interaction transcripts from the original study revealed that these apparently well-learned strategies were rarely discussed in the experimental groups. One hundred of the 432 transcripts (each of a single group working on one task) were randomly selected and analyzed for strategy comments by two judges. A total of only 143 comments about strategy were found—less than 1.5 comments per group. Only 25 of the comments prompted further discussion among group members, and on 36 of the transcripts there was no strategy-relevant interaction at all during the 15-minute work period.

These data support the notion that, at least in some circumstances, group members are both capable and desirous of implementing implicitly-agreed-upon performance strategies without explicit discussion of what they are doing. In these circumstances, the interaction process serves primarily as a *vehicle* for implementing the pre-existing performance strategies. As in the case of coordination of member efforts, group members may encounter interpersonal difficulties which impair the efficiency with which such implementation is actually carried out—in other words, a process loss occurs which results in suboptimal group effectiveness.

b. Developing or reformulating strategic plans. While most tasks do not constrain a group from overtly discussing and reformulating its performance strategies (or from developing new strategies from scratch), there appears to be a pervasive norm in groups *not* to address such matters explicitly (Weick, 1969, pp. 11–12). The low incidence of strategic discussion noted above in the Morris-Hackman data is one possible example of this norm in operation. Another is reported by Shure *et al.* (1962). In that study, it was found that "planning" activities tended to be generally lower in priority than actual task performance activities—even when group members were aware that it was to their advantage to engage in planning before starting actual work on the task, and when it was possible for them to do so without difficulty. A closely related phenomenon is the tendency for group members to begin immediately to generate and evalute solutions when they are presented with a task, rather than to take time to study and analyze the task itself (Maier, 1963; Varela, 1971, Chapter 6).

To the extent that norms against strategy planning exist, the chances are lessened that the preexisting strategies members bring to the group will be altered and improved upon, or that new (and possibly more task-effective) strategies will be generated by group members. This obviously can limit the effectiveness of the group on many types of tasks.

To explore whether group task effectiveness can be *improved* by explicit attention to matters of performance strategy, an additional analysis was made of the interaction transcripts collected by the present authors. The relationship between the frequency of strategy comments and the judged creativity of group products was analyzed for the 100 transcripts described above. Only comments made during the first third of the performance period were included in the analysis, since those made later would be unlikely to have much effect on the final product. Even though relatively little interaction about strategy took place in these groups, the relationship between the number of strategy comments and group creativity was significantly positive ($p < .05$), as shown in Fig. 2.

The data in Fig. 2 are correlational, and they do not permit a conclusion that strategy discussion "caused" increased group creativity. Nevertheless, to search for possible reasons for the positive relationship obtained, all transcripts in which relatively full-fledged exchanges about

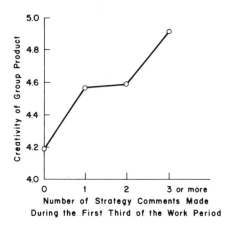

Fig. 2. Relationship between amount of interaction about performance strategy and group product creativity.

strategy took place were reviewed. The transcripts suggest that one function of strategy discussion is to "unfreeze" individuals from traditional, well-learned approaches to the task, and thereby open the possibility of discovering a more task-effective way of proceeding. Strategy discussion often began, for example, after one group member made a suggestion that was deviant from shared ideas about "appropriate" strategy for the task at hand (e.g., suggesting a bizarre solution to a routine problem-solving task). In some cases, in the process of explaining to the deviant why his idea was faulty, members began to explore new ways of proceeding, some of which subsequently were adopted.[3]

In sum, it appears that the functions of interaction guiding the implementation and reformulation of performance strategies may be of considerable importance in understanding and predicting group performance effectiveness. Moreover, the data suggest that overriding existing norms which often discourage explicit strategy planning may be a useful way to help groups improve their performance effectiveness in some circumstances.

3. Member Knowledge and Skill

The knowledge and the skills of group members—and the way these are brought to bear on the group task—are the third general summary

---

[3] Increased group effectiveness as a consequence of explicit planning activities also has been reported by Shure et al. (1962). Davis (1973), on the other hand, suggests that social decision schemes that remain *implicit* may have some adaptive advantages—such as minimizing intermember divisiveness.

variable which may be impacted by group interaction process. Once again, there are some tasks that require only a minimal level of knowledge or skill for effective performance, and there are others for which performance measures will be substantially affected by the level of knowledge and skill group members bring to bear on the task. A task requiring the group to assemble a number of very simple mechanical devices with no time limit should not be very responsive to differences in the knowledge and skill members apply to the task; the group vocabulary test described earlier, on the other hand, should be highly responsive to the way group members assess and apply their knowledge.

It is proposed that group interaction serves two major functions in influencing the effectiveness with which the knowledge and skill of group members are applied to the task: (a) assessing and weighting the possible contributions of different group members—who presumably vary in the level of task-relevant talent they have brought to the group; and (b) creating conditions within the group which will lead to a change (presumably an increase) in the overall *level* of knowledge and/or skill group members have and are able to apply to the task.

a. Assessing and weighting member knowledge and skill. For tasks on which knowledge or skill are important in determining performance, it often is possible to predict how well the group will do solely on the basis of the talents of its members (Davis, 1969; Haythorn, 1968; Kelley & Thibaut, 1969; Steiner, 1972). The specific predictive model required, of course, depends on the task: for some tasks, the group should operate at the level of its most competent member (e.g., as in Steiner's "disjunctive" model); for others, the group would be expected to perform at the level of the "average" member; for still others, group performance should be determined by the least competent member (e.g., Steiner's "conjunctive" model).

In general, empirical tests of such predictive models have been reasonably successful. Of special interest for present purposes, however, is the recurrent finding that, when actual group productivity is at variance with predictions, it is usually because the model has *over*-predicted group performance. That is, given the level of member talent in the group, the group "should" have performed better than it actually did. The implication is that the interaction process of the group, through which the talents of members are assessed, weighted, and brought to bear on the task, must have been in some way inadequate.

For some tasks, such process losses should not be substantial. For example, when the specific knowledge or skill required is obvious, and when obtaining the solution does not involve complex teamwork among members, sophisticated or subtle social processes are not required to identify the necessary talents and to apply them to the task. Instead, group interaction may serve merely as a vehicle for exchanging data, and for informing other members that one "knows the answer." There is little opportunity here for process foul-ups.

Such apparently was the case in a study reported by Laughlin, Branch, and Johnson (1969) which used the Terman Concept Mastery Test as the group task. A large number of students were given the Terman test, and were trichotomized on the basis of their test scores. Triads were formed with different combinations of member talent (i.e., high-high-high, high-high-medium, high-high-low, etc.). All ten possible combinations of talent were used. The Terman test was then readministered, but this time it was taken collaboratively by triad members. The relative level of performance of triads could be predicted quite accurately from the overlap of member talent within each type of triad. All predictions were made solely from the preinteraction test scores of triad members; group interaction processes within triads did not enter into the data or theorizing at all, and apparently would have contributed little to predictions of group effectiveness on this particular task. Similar findings predicting group performance from member knowledge or skills with minimal attention to matters of interpersonal process have been reported by Egerman (1966), Goldman (1965), Johnson and Torcivia (1967), Laughlin and Branch (1972), and Laughlin and Johnson (1966).

On other tasks, however, the mediating role of group process may be more substantial and the risk of process losses substantially greater. Consider, for example, tasks on which the knowledge or skills required for successful performance are complex and subtle, and on which considerable teamwork is required to coordinate and apply member talents. In such circumstances, our ability to predict group effectiveness simply from measures of individual talent without knowledge of group process should be diminished.

A novel case in point is the prediction of the performance of professional athletic teams from data about the skills of individual team members (Jones, 1974). As would be expected, Jones found substantial relationships between measures of individual skill and team performance; teams with better athletes did better. However, the *level* of prediction attained was higher for some sports than for others. For example, nearly 90% of the variation in baseball team effectiveness was predictable from measures of team member skill, as compared to only about 35% for basketball teams. As the author notes, success in basketball is especially dependent upon personal relations and teamwork among players. Thus, process losses might be more likely to impair basketball team effectiveness than would be the case for other team sports.

We have suggested above that, when the primary functions of group interaction are to assess, weight, and apply member talent, process losses are inevitable. For some tasks (i.e., those involving complex skills and high levels of teamwork) the potential losses are greater than for others. In every case, however, group process considerations will determine to some degree how near a group comes to its potential performance, given the capabilities of its members.

b. Affecting the level of talent available to the group. Group interaction process can, at least potentially, serve as a means for actually increasing the total amount of member talent available to the group for work on the task. The issue here is *not* the simple exchange or coordination of existing knowledge and skill, as discussed above; that function of group interaction (while relatively easy to observe and document) does not result in a net increase in the total supply of talent available to the group. Instead, the present focus is on how group members can do more than merely share among themselves what they already know— and instead work as a group to gain knowledge or generate skills that previously did not exist within the group.

Virtually no controlled research has been carried out on this latter function of group interaction. The "training group" approach to the development of interpersonal skills (Argyris, 1962; Bradford, Gibb, & Benne, 1964; Schein & Bennis, 1965) postulates that group members can effectively use one another as resources to increase member interpersonal competence and thereby increase the level of competence in the group as a whole. But the social processes through which such learning takes place are only beginning to be illuminated (cf. Argyris & Schon, 1974), and additional research on the talent-enhancing functions of group interaction is much needed.

4. Summary

In this section, we have examined the impact of group interaction process on each of three summary variables: (*a*) the level of effort brought to bear on the task; (*b*) the task performance strategies implemented by group members in carrying out the task; and (*c*) the level of knowledge and skill at the disposal of the group for task work. The impact of group interaction on each of the three summary variables is summarized in Table III.

The table emphasizes that the functions interaction process serves are quite different for each of the three summary variables. The implication is that a researcher who is attempting to understand the process determinants of group performance will have to examine different aspects of the group process, depending on which of the summary variables

TABLE III

SUMMARY OF THE PROPOSED FUNCTIONS OF GROUP INTERACTION

Summary variables postulated as important in affecting performance outcomes	Impact of interaction process on the summary variables	
	(A) Inevitable process losses	(B) Potential for process gains
Member effort brought to bear on the task	Interaction serves as the less-than-perfect means by which member efforts are coordinated and applied to the task	Interaction can serve to enhance the level of effort members choose to expend on task work
Performance strategies used in carrying out the task	Interaction serves as a less-than-perfect "vehicle" for implementing preexisting strategies brought to the group by members and (often) shared by them	Interaction can serve as the site for developing or reformulating strategic plans to increase their task-appropriateness
Member knowledge and skills used by the group for task work	Interaction serves as a less-than-perfect means for assessing, weighting, and applying member talents to the task	Interaction can serve as a means for increasing the total pool of knowledge and/or skill available to the group (i.e., when the group is the site for generation of new knowledge or skill by members)

is operative in his particular task situation. By the same token, the approach an interventionist would take in attempting to help group members create more task-appropriate patterns of interaction would vary, depending on the summary variables operative for the task being performed.

As noted in column A of Table III, there are inevitable process losses associated with each of the three summary variables. A group can never handle the process issues in column A perfectly; the group's performance therefore will depend in part on how successful members are in finding ways to minimize these process losses. At the same time (column B) there are potentially important (but often unrecognized) process *gains* associated with each of the summary variables. That is, at least the possibility exists for group members to find and implement new, task-effective ways of interacting which will make it possible for them to achieve a level of effectiveness which could not have been anticipated from knowledge about the talents and intentions of group members prior to the start of work on the task. In the section to follow, we explore possibilities for achieving such performance-enhancing process gains.

B. Recasting the Role of "Input" Factors as Determiners of Group Effectiveness

The second thrust of our proposed three-pronged approach to research on group effectiveness is explicitly change-oriented. We shall attempt to show how group effectiveness can be *improved* above the level expected from column A of Table III by alteration of various "input" factors (cf. Fig. 1). The specific input factors considered are (a) the structure of the norms which guide group member behavior; (b) the design of the group task; and (c) the composition of the group— i.e., the characteristics and histories of group members. All these factors affect the interaction process of the group, and all can be adjusted or "set" prior to the start of actual task performance activities.

We propose that each of the three summary variables discussed in the previous section is especially responsive to changes in one of the input factors identified above. In particular, we explore below the possibility that (a) performance strategies can be made more task-appropriate by modification of group *norms;* (b) member effort can be increased by redesign of the group *task;* and (c) the level and utilization of member knowledge and skill can be improved by altering the *composition* of the group.

1. Task Performance Strategies

The task performance strategies used by members of a group, while always potentially open to member-initiated change, often are well-codified as behavioral *norms* of the group. Group members typically share a set of expectations about proper approaches to the task, and to some degree the group enforces member adherence to those expectations. Such norms often short-cut the need explicitly to manage and coordinate group member behavior: Everyone knows how things should be done, and everyone does them that way. If a person deviates from a norm about strategy, perhaps by suggesting an alternative way of proceeding or by behaving in a way inconsistent with the norm, he often is brought quickly back into line so that he does not further "disrupt the group" or "waste time" (Hackman, 1975).

Ideally, the presence of such norms should contribute to the task effectiveness of the group—simply because little time would have to be spent on moment-by-moment behavior-management activities, leaving more time for actual task work. However, this advantage will accrue only if the norms that guide the selection and use of performance strategies are fully task-appropriate. If existing norms about strategy are dysfunctional for effectiveness, then performance is likely to suffer unless they are changed—despite their time-saving advantages.

As was discussed earlier, the problem is that reconsideration of strategic norms in task-oriented groups rarely occurs spontaneously, even when there is clear evidence available to members that the group may be failing at the task (e.g., Janis, 1972).[4] Member views about what is and is not an "appropriate" way to approach a given task often are firmly held, widely shared, and very resistant to change. Even attempts by a group leader or outside interventionist to change strategic norms may be resisted, because such changes often involve explicit consideration of group process issues or because they involve alteration of familiar ways of doing things. Neither of these is easy or comfortable for most people to do.

The challenge, then, is to create conditions that encourage group members to reconsider (and possibly change) their norms about performance strategy when existing norms appear to be suboptimal for the task at hand. Three approaches to meeting this challenge are presented and evaluated below.

a. Diagnosis-feedback. Because group members often are unaware of the nature and impact of the norms that govern their choice and use of task performance strategies, it may be necessary to help them gain increased understanding of these norms before they will be able to change them. Of considerable potential use in this regard is the Return Potential Model proposed by Jackson (1966). This model addresses the distribution of potential approval (and disapproval) group members feel for various specific behaviors which might be exhibited in a given situation. It is represented in a two-dimensional space: The ordinate is the amount of approval and disapproval felt; the abscissa is the amount of the given behavior exhibited. A "return potential curve" can be drawn in this space, indicating the pattern and the intensity of approval and disapproval associated with various possible behaviors. Data to plot such a curve can be obtained either directly from reports of group members or indirectly by observing patterns of approval/disapproval which occur in the group.

An example of a return potential curve is shown in Fig. 3. This

---

[4] The problem may not be as great when the group is dealing with a strange or unfamiliar task. In such cases, few cues will be present in the task materials to engage members' learned views about how they "should" proceed on the task—and they may be forced to discuss strategic options in order to arrive at a shared and coordinated plan for dealing with the task.

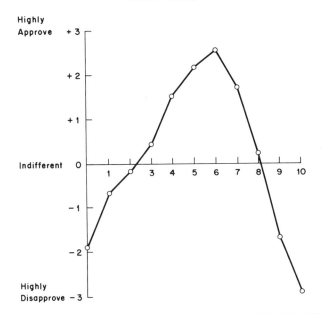

FIG. 3.   Schematic representation of the Return Potential Model (RPM) of normative structure. *Note:* The ordinate is an axis of evaluation; the abscissa is an axis of behavior. Adapted from Jackson (1965, p. 303).

particular curve might reflect the norms of a group regarding the amount of talking an individual does during a group meeting. Both too little and too much talking, in this case, would be disapproved, but the intensity of the disapproval is somewhat stronger for someone who talks too much than for someone who talks too little. (The units of behavior in the example in Fig. 3 are arbitrary; in practice, the abscissa would be scaled using units appropriate to the behavior in question.) A return potential curve can, theoretically, assume any shape. For behavior such as "raising questions about task performance strategies," the curve might begin near the "indifferent" point on the approval/disapproval axis, and move sharply downward for increasing amounts of that behavior.

   Much of the elegance of the Return Potential Model lies in its usefulness as a vehicle for generating quantitative measures of the characteristics of norms (Jackson, 1965, 1966). These measures can be provided to group members to increase their understanding of the existing norms of the group, and to help them decide whether (and how) they wish to change their norms. In addition to its uses in diagnosis and change activities, the Return Potential Model should be useful for research on norms, since the quantitative measures can be used to assess the conditions under which groups actually do alter their norms, and what the effects of these changes are on group process and group effectiveness. At present, unfortunately, the potential of the model either for research or as a basis for normative interventions has not received systematic test.

b. Process consultation. A second general approach to changing group norms about strategy involves the use of an outside consultant to help group members discover and implement new, more task-effective ways of working together. In its most flexible and general form, process consultation involves joint work by the consultant and the group to diagnose the state of the group, and to plan what to do on the basis of that diagnosis (Schein, 1969). Thus, process consultation involves a flexible and often ad hoc set of interventions. And, since there is no set of standard procedures to guide the behavior of the consultant, it demands a great deal of clinical sensitivity and skill on his part.

Few studies have explicitly examined the effectiveness of process consultation as a strategy for helping group members revise their internal processes, although a number of case reports suggest that the approach may be effective. One study (Kaplan, 1973) has addressed explicitly the usefulness of process consultation as a technique for altering group norms about how interpersonal relationships in the group are managed. In the experimental condition, Kaplan helped group members develop norms which encouraged direct and public processing of social and emotional phenomena as they emerged in the group. His findings showed that process consultation did result in alteration of the "target" norms of the group, resulting in numerous changes in interpersonal relations in the group and in member satisfaction with the group experience.

In the Kaplan study (and in most applications of process consultation as a change technique) the consultant spent considerable time with group members, helping them examine existing norms and experiment with new ways of behaving. In this respect, broad-gauge process consultation is not a very efficient technique, at least in the short term. The hope, of course, is that over the long term group members will develop their own diagnostic and action skills which will reduce their dependency on the consultant and, in effect, allow group members to consult with themselves about the most effective task and interpersonal processes to use in various situations.

An alternative consultative approach which would appear more useful in the short term is to educate group members in specific strategic alternatives or techniques for carrying out the task. A number of such techniques have been proposed, such as setting time limits for discussion, temporally separating the generation of ideas from their evaluation, focusing on analysis of the task before beginning to perform it, setting time aside to locate facts and potential obstacles to implementation of the solution, devising specific structures for exchanging ideas and information among group members, and so on (cf. Kepner & Tregoe, 1965; Maier, 1963; Osborne, 1957; Varela, 1971). Some of the techniques proposed are based on research findings; others derive more from intuitive considerations. All are intended to provide strategies for proceeding which will be immediate aids to group effectiveness.

There are at least two problems with technique-based approaches to changing preexisting group norms about strategy. First, the value of any given technique depends on the task; yet there is very little information available which would help group members to select those strategic approaches likely to be particularly useful for a given task. Second, some of the techniques involve the use of group relations skills which may not be valued or well-practiced by group members. Maier and Thurber (1969), for example, assessed the effectiveness of three different strategies for working on a problem-solving task. None of the strategies was found to be of much help in improving performance effectiveness, but the authors attribute the results more to the unwillingness or inability of group members to follow the instructions provided, than to the effectiveness of the strategies themselves. Thus, regardless of the potential usefulness of a given strategy for working on a task, the strategy cannot help the group unless members are both motivated and sufficiently skilled to use it appropriately. At present little is known about how to introduce such techniques and to train members in their effective utilization.

Probably the most straightforward process consultative technique for helping members consider and possibly revise their norms about performance strategy is simply to provide the group with a "preliminary group task" prior to their beginning work on the primary task. This preliminary task would require members to discuss the task performance strategies they plan to use on the main task—and to consider revising or replacing them if warranted by the discussion. Thus, the intervention capitalizes on the tendency of group members to follow rather slavishly the demands of tasks they see as legitimate (March & Simon, 1958, p. 185; Shure et al., 1962). Since in most normal circumstances the immediate demands of the primary task apparently drive out tendencies toward strategy-planning activities, introduction of a preliminary group task could serve to "hold off" such immediate task demands until strategy planning has been completed.

The "preliminary task" approach was used with some success in a recent study which will be described in more detail later (Hackman, Weiss, & Brousseau, 1974). Group members were given a preliminary task which explicitly required them to discuss the performance strategy they would use on the main task. The experimenter discussed with the group the importance of formulating a task-appropriate performance strategy, answered any questions members had about the preliminary task, and then left the room. Groups in the experimental condition did in fact follow the requirement of the preliminary task and discussed their strategic options extensively. As predicted, control groups engaged in virtually no spontaneous discussions of strategy; they proceeded directly to work on the primary task in ways that were consistent with their private, a priori notions about how such a task should be done.

While a preliminary group task gets strategy discussions underway with a minimum of personal involvement by the outside consultant, it does not guarantee that any new performance strategies which are developed in such discussions will be more task-effective than members' private, prior hypotheses about how the task should be done. In particular, it appears that relatively unstructured discussions of strategy should facilitate task performance only when (a) members' a priori preferences about strategy are suboptimal for the task at hand and/or (b) the task is sufficiently complex or subtle that overt discussion of strategy will reveal ways of proceeding which otherwise would not have been considered.

c. Task redesign. Both the diagnosis-feedback and the process consultation approaches require group members to address directly the norms of the group about performance strategy, and both require some intervention by an outside consultant or researcher. A third general approach to the change of group norms about strategy—task redesign—deals with norms more indirectly, and minimizes the role of "outsiders" in the group itself.

In particular, the group task can be designed so that it requires, suggests, or provides cues which prompt specific ways of going about performing the task (e.g., Solem, 1974). For example, if a task which requires the assembly of small mechanical devices is physically laid out in a linear fashion, with chairs and equipment for group members placed along the side of a long narrow table, members will almost certainly assume that they should form an assembly line and will proceed to operate as if that is the optimal way of doing the task. But if materials are arranged around a circular table, with a full complement of equipment provided for each member, a strategy of individual assembly probably will be adopted instead, with just as little overt discussion.

At the extreme, of course, a task can be designed so that *no* discretion about strategy is left to the group; members are informed not only what their goal is, but exactly how to proceed to achieve that goal.

If the task-provided instructions are indeed optimal, the group will be well-protected against ineffective strategies. This is the approach often taken by consultants who attempt to increase the creativity of groups. In the synectics approach, for example, group members are provided with tasks and exercises which specify exactly what strategies group members are to use in working on the task—strategies that are designed explicitly to facilitate the production of original solutions (Prince, 1970). The ultimate success of the group, of course, depends partly on the adequacy of those task-specified strategies.

While, from an "engineering" perspective, task redesign offers considerable appeal as a device for helping group members utilize more task-effective performance strategies in their work, responsibility for the strategies remains outside the group itself. The task serves simply to get potentially task-effective patterns of behavior underway; whether such behaviors will "stick" and become incorporated into the normative structure of the group depends in large part on whether members find the behaviors instrumental for achieving their goals. Diagnosis-feedback and process consultation, on the other hand, probably are less efficient in any given instance, but have the advantage that the group itself "owns" the new procedures it has devised. Moreover, as they use these techniques, members may learn some group relations skills or develop some norms that can be usefully applied to other tasks, or in other groups.

2. Member Effort

There is considerable evidence that the effort members expend on a group task, like the performance strategies they use, is powerfully affected by the norms of the group, especially when members value their membership in the group (e.g., Schachter, Ellertson, McBride, & Gregory, 1951; Seashore, 1954). But less is known about what determines the *direction* of group norms—that is, whether the norms will encourage high or low effort on the task (Vroom, 1969).

We suggest that whether a group develops a norm of high or low effort depends substantially on the quality of the experiences members have as they work on the task—and that these experiences in turn are largely determined by the task itself. For example, if members find the task activities frustrating and unpleasant, they are likely, after some time, to notice the adverse attitudes of others in the group—and perhaps to share these reactions verbally. Gradually, through such interaction, group members may come to an implicit or explicit agreement that the best way to minimize the unpleasantness they experience is to minimize the energy invested in doing the task. If, on the other hand, mem-

bers find their work on the task exciting, fulfilling, or otherwise rewarding, these experiences also are likely to be shared with one another,
and a norm of high effort may be the result.

To the extent that member experiences do depend partly on the
task itself, then it may be useful to consider task redesign as a strategy
for increasing member effort, rather than attempting to address directly
norms about effort. To do the latter, in many cases, would be attacking
the outcropping of the problem rather than the problem itself.

The question, then, is what task characteristics are likely to lead
to high commitment by group members to work hard on the task. Research data are available that identify the characteristics of jobs in organizations that prompt individuals to become personally motivated to
work hard and effectively (cf. Hackman & Lawler, 1971; Hackman &
Oldham, 1974). In particular, it has been shown that individual task
motivation is enhanced for many people when their jobs are high on
five job dimensions: (a) skill variety—the degree to which the individual
does a number of different things on the job involving use of his valued
skills; (b) task identity—the degree to which he does a whole and
visible piece of work; (c) task significance—how much the results of
work on the job will affect the psychological or physical well-being
of other people; (d) autonomy—the personal initiative and discretion
the individual has on the job; and (e) feedback—the degree to which
the person learns as he is working how well he is doing (Hackman
& Oldham, 1974).

If a group task were designed with these or similar characteristics
in mind, one might expect to observe an increase in the motivation
of individual group members to exert high personal energy in carrying
out the task. And, over time, group norms about effort should change
to become consistent with the level of commitment individual members
feel for their work. The final result should be a considerable increase
in the overall level of effort the group expends on the task (a "process
gain"), and this increase should be reinforced by the emergent normative
structure in the group. Again, as with most of the possibilities for enhancing group effectiveness introduced in this section, research tests
remain to be done.

3. Member Knowledge and Skill

Consider now tasks for which the utilization of member knowledge
and skill strongly determines group effectiveness. As was suggested earlier, the single most powerful point of leverage on group effectiveness
for such tasks is simply group composition; a group made up of competent people will do better than a group composed of less-competent

members (cf. Varela, 1971, pp. 153–157). But if it is assumed that the group originally has been composed to maximize the level of member task-relevant talent present, what can be done to maximize the utilization and development of that talent in the service of the group task? How, for example, can the group operate to minimize the inevitable process losses that occur when information is combined and member contributions are evaluated? Or to increase the level of knowledge and skill of individual members so that the total pool of talent available to the group also increases? Or to develop patterns of interaction that increase the group's capability to deal effectively with similar tasks in the future?

Achieving such states of affairs in a group is neither a straightforward nor a short-term proposition. Groups usually have difficulty dealing effectively with individual differences in competence. When weighting of individual contributions is done in the group, difficult issues of inter-personal competitiveness, evaluation, and differential status come very quickly to the fore. Dealing with such issues openly is, for most members of most groups, highly threatening and anxiety-arousing. Group members are likely to erect protective shells around themselves in such circumstances, and as a result the group as a whole loses access to much of the talent already present within its boundaries. And the chances of members using one another to learn genuinely innovative patterns of behavior—or to seek out and internalize knowledge that initially is foreign to them—are very slim indeed.

How might a group break out of such a self-defeating pattern of behavior? Possibly the group task could be structured to require explicit and overt treatment of individual differences in knowledge and skill; or perhaps interventions could be made to help members become aware of (and possibly change) existing group norms specifying how such matters are handled in the group. Such interventions can ensure that issues of individual differences in knowledge and skill are brought to the attention of group members—and can prompt explicit discussion of them. But successful resolution of such matters once they have surfaced may be quite another matter (cf. Bion, 1959). Consider, for example, a group in which (for whatever reasons) members resolved to "deal openly with individual differences in competence on the task, and evaluations of the worth of the contributions of all members." In all likelihood, this resolve would serve more to drive group members into their protective shells than to open the hoped-for new avenues of personal and interpersonal learning.

What are needed, it seems, are interventions that will help group members learn *how* to deal effectively with issues of individual differences within the group, and to create a climate that supports and facilitates learning and sharing of learning. And this suggests the need for a rather long-term program of process consultation (or "team building") in which members gradually build a climate of interpersonal trust within the group, leading to a reduction in the level of personal threat they experience in the group setting. As such a climate develops, members may become better able to experiment with new forms of behavior and become increasingly ready to engage in the usually risky and always-anxiety-arousing activities required to extend one's knowledge and skills in a public setting. Even in the long term, however, there is no guarantee that the group will develop into a site for individual learning and heightened sharing among group members. The process is a fragile one, and fragile things break.

What is relatively certain is that, if such a long-term team-building program is successful, the members themselves will almost invariably be changed as a consequence—that is, they will perhaps become more risk-taking and experimenting in their behavior, and more willing to tolerate stress and anxiety in the interest of increasing and sharing their personal knowledge and skill. When such a point is reached, if it is reached, the group will have become "re-composed," not by the removal of incompetent members and substitution of more competent ones, but instead by changes in the attitudes, skills, and behavioral styles of the existing members.

So we come full circle for tasks on which group effectiveness is strongly determined by the level and utilization of member knowledge and skill. At the first level, we have noted that the most efficient and straightforward means of improving group effectiveness on such tasks is through group composition: Put good people in the group. To move beyond that level, we believe, also requires attention to the composition of the group, but now through changes within the group itself—changes in the attitudes and interpersonal styles of individual members. The characteristics of individual group members are important to the success of groups faced with this type of task. And we believe that the configuration of those characteristics often can be improved—both by the way members are selected for the group, and by the kinds of learning experiences they are provided after they are in it.

4. Summary

Table IV shows graphically the relationship between (a) the three summary variables which have been postulated as important proximal causes of measured group performance effectiveness, and (b) the three "input" factors which have been proposed as points of leverage for influencing the summary variables. As is suggested by the figure, any one of the input factors potentially can effect a change in any of the

TABLE IV
THE THREE SUMMARY VARIABLES AND "INPUT" FACTORS WHICH
MAY BE ALTERED TO INFLUENCE THEM

	"Input" factors which are manipulable to change group process and performance		
	Be-havioral norms	Task design	Group com-position
Task performance strategies	▓		
Summary variable(s) operative for the focal group task — Effort		▓	
Knowledge and skill			▓

Note: Shaded cells represent especially promising sites for change aimed at improving group performance effectiveness.

summary variables. Yet when the intent is to generate improvements in group effectiveness by influencing what happens in the interaction process of the group, some input factors appear to be more useful than others. In particular, we have focused on how task performance *strategies* are especially open to improvement by alteration of group *norms;* how the *effort* members bring to bear on the task is especially influencable by the design of the group *task;* and how the effective utilization and development of task-relevant *knowledge and skill* within the group may be especially dependent upon group *composition.*

The first thrust of our proposed three-prong approach to research on group effectiveness focused on the functions of group interaction; the second examined the role of input factors in changing group interaction and group effectiveness. We now turn to the final theme, which addresses the kind of research that we believe is needed to extend knowledge about ways to improve the effectiveness of task-oriented groups.

C. EXPERIMENTALLY CREATING NONTRADITIONAL STRUCTURES AND PROCESSES IN GROUPS

1. Rationale

Hoffman (1965), in a chapter on group problem solving in an earlier volume of this series, noted the emphasis in small group research on identifying and studying "barriers" to group creativity. While acknowledging that such barriers must be overcome if creative problem solving is to be promoted, he also argued for efforts directed to "inventing and testing new ways of encouraging creative group problem solving" (p. 127). Such research, he suggested, should not only advance our knowledge of group process but, when successful, should be of considerable practical value to society as well.

In the decade since Hoffman made his plea, almost no systematic research has addressed the question of "inventing and testing" ways of improving group effectiveness. The suggestions and speculations we have offered in the previous section, then, can be seen as a reprise of Hoffman's theme.

Many of our proposals, if implemented, would require group members to engage in activities of a nontraditional nature. Suggestions were made, for example, for the design of nonstandard types of tasks, for the creation of norms that differ from widely shared expectations about "appropriate" behavior in groups, and for the development within the group of new personal and interpersonal styles for dealing with individual differences in member competence.

The impact of such structures and styles on interaction process and on group effectiveness must remain, for the moment, a matter for speculation, because systematic research on their functions and dysfunctions has not been carried out. Moreover, their effects usually will not even be observable in existing, naturally occurring groups, simply because such structures and styles evolve rarely, if at all, without outside stimulation or assistance.

The implication, then, is that to learn about the positive and negative effects of these (and other nontraditional) structures and styles, it will first be necessary to create these conditions within the groups under study. And this, of course, requires a research strategy that does not accept as given either the kinds of structures within which groups typically operate, or the ways group members naturally behave within these structures. We are, in effect, suggesting a resurrection of the "action research" tradition which proved so vital to small group research earlier in this century.

2. An Example

An example of change-oriented research, in this case involving alteration of performance strategies by manipulation of the norms of the group, has been reported by Hackman et al. (1974) and is summarized below.

a. Design. The study was designed to determine (*a*) whether or not task-appropriate discussions of performance strategy could be induced in groups through a relatively straightforward intervention into the norms of the group, and (*b*) the effects of such discussions (if successfully created) on group productivity and member relationships for two different tasks.

Three intervention conditions were created in four-person groups. In the "strategy" condition group members were asked to spend the first 5 minutes of the (35-minute) performance period explicitly discussing their goals and how they might optimally work together to maximize their productivity. As a guide to this discussion, members were given a "preliminary group task" which provided general guidelines for how a group might go about discussing performance strategy. In the "anti-strategy" condition, group members were asked explicitly *not* to "waste any time" in discussions of procedure or strategy, but instead immediately to begin productive work on the task. In the control condition, no special instructions were given, other than the exhortation (given in all conditions) to try to maximize group performance.

The experimental task required group members to assemble various kinds of electrical components described on "task order lists" given each group member. Each type of component had a dollar value, and productivity was measured by the total worth in dollars of the components produced by the group. Group members were informed that, since they could not complete everything on all lists in the time period allowed, they would have to make some decisions about which components to produce.

The basic task was identical in all conditions; i.e., the dollar value of each type of component was the same on all task order lists, and the total number of components of each type on the lists given each group was the same. The way the task information was *distributed* among group members, however, was experimentally varied. In the "unequal information" condition, the task order lists of group members contained different quantities of the various types of components that could be produced. In the "equal information" condition, the quantities specified on the lists of group members were identical. It was expected that any beneficial effects of discussions of performance strategy should appear much more powerfully in the "unequal" than in the "equal" task condition, since in the "equal" condition each member had all relevant information personally at hand. Therefore, members presumably could decide on their own the most profitable components to produce. In the "unequal" condition, on the other hand, group productivity potentially could be impaired if members did not share information and coordinate their decisions about what components to produce.

In sum, the design involved three intervention conditions (strategy, antistrategy, and control) crossed by two task conditions (unequal information and equal information). Dependent measures (in addition to productivity) included observations of the group process and member reports of their experiences in the group.

b. Findings. The strategy intervention did successfully alter the interaction process of the experimental groups. Measures of strategy-relevant interaction revealed that nearly all the groups in the strategy condition spent the first portion of their work time in a discussion of task strategy, and virtually no strategy discussion occurred in the anti-strategy groups. Moreover, the groups in the control condition also did not discuss strategy spontaneously, thus confirming the prediction that groups rarely engage in discussions of strategy on their own initiative.

The effects of the interventions on group performance effectiveness are depicted in Fig. 4. As predicted, groups that received the strategy induction performed especially well in the unequal task condition. Groups that received the antistrategy induction performed especially well in the equal task condition. Of special importance is the finding that performance was substantially *lowered* for the strategy groups in the equal task condition, and for the antistrategy groups in the unequal task condition. Control groups were low for both task conditions.

It appears, therefore, that the strategy intervention was indeed helpful to group performance effectiveness, but only when the task actually required coordination and sharing among group members; when the task could be done equally well without such coordination, strategy discussions led to a deterioration of performance—perhaps because such discussions served little useful purpose and simply wasted group members' time. Similarly, exhortation to a hard-working, task-oriented set (as induced in the antistrategy condition) increased effectiveness when there

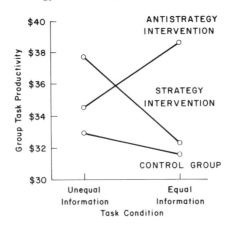

FIG. 4.   Dollar productivity of groups in the strategy study.

was no objective need to coordinate and share among group members, but impaired performance when such a need was present (i.e., as in the unequal information condition).[5]

Both positive and negative "spin-off" effects of the strategy induction were noted. While groups in the strategy condition reported more conflict and more interpersonal problems than did other groups, they showed higher flexibility in their approach to the task and a considerable ability to change procedures in midstream if the group found itself doing poorly. In addition, in the strategy condition all group members tended to see themselves as high on personal leadership and influence, whereas in both the control and the antistrategy groups, members viewed themselves as not having much influence on the group. Thus, the strategy intervention seems to have created a condition of participative leadership in which all members had considerably more "say" in what the group did than was the case for the other conditions.

In sum, the research suggests that it is in fact possible to create new, nontraditional norms in groups (at least having to do with matters of task performance strategy) and to trace the effects of those norms on group effectiveness. Moreover, observations of the interaction of the groups in this study showed both that group process was powerfully affected by the interventions made, and that changes in interaction process led directly to measurable differences in group productivity.

III. Conclusions and Implications

In this concluding section, we suggest some implications of the proposals made and findings reported in this paper. Out comments are arranged in two groups: those having to do with matters of research and theory on group effectiveness, and those relevant to action steps or interventions aimed at improving group effectiveness.

A. IMPLICATIONS FOR RESEARCH AND THEORY

1. The Elusiveness of General Theory

While there have been numerous attempts to integrate findings about group effectiveness and to draw general-level conclusions about behavior in groups, so far no general theory of small group effectiveness has appeared.

---

[5] It is important to note that the "strategy" groups had the methodological cards stacked against them; they spent a least 5 minutes (one-seventh of the total work time) in planning activities that were not directly productive, while other groups were using this time to actually produce components. Thus, the performance of these groups reflects a *rate* of productivity that greatly exceeded that of other groups once the assembling of components actually began.

We suggest here the possibility that no single theory can encompass and deal simultaneously with the complexity of factors that can affect group task effectiveness. Instead, it may be necessary to settle for a number of smaller theories, each of which is relevant to a specific aspect or phase of the performance process, or to performance effectiveness under certain specified circumstances. One intent of the present paper has been to help structure the domain within which such smaller theories might be developed. In particular, we have attempted to examine in some depth (*a*) the role of group interaction process as a major determinant of group productivity; (*b*) some selected "input" variables which we see as powerful influences on group performance and thus as useful points of leverage for changing performance—whether directly, or through the group process; and (*c*) three "summary variables" (effort, performance strategies, knowledge and skill) which are proposed as devices for summarizing the most powerful proximal causes of group task effectiveness.

A general framework suggesting how these three classes of variables interact in the task performance sequence is shown in Fig. 5. By further researching this input–process–output sequence for different types of tasks (here classified in terms of the summary variables), we believe that additional understanding can be achieved which will aid both in predicting and in changing group effectiveness in a large number of performance settings. But in any event, a general and unified theory of group

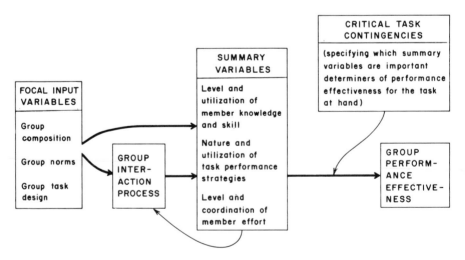

Fig. 5. Framework showing the relations among the focal input variables, group interaction process, and the three summary variables in affecting group performance effectiveness.

effectiveness, we believe, is currently out of reach—and is likely to remain so.

2. Research on Group Effectiveness

In the introductory section of this chapter, we pointed out that many social psychologists have reached rather pessimistic conclusions about the efficiency and performance effectiveness of interacting groups. A working assumption of this chapter, however, has been that it is in fact possible to conduct research that will demonstrate how groups can be designed and managed so that they perform much more effectively than they would "naturally." To design and carry out such research, we believe the following will be required:

1. It will be necessary to attempt to *create* effective groups in order to understand their dynamics. Merely describing what happens in existing, natural groups is unlikely to generate knowledge useful for improving group effectiveness because some of the most critical ingredients of truly effective groups may never appear spontaneously in groups allowed to develop naturally. As a start toward the design of such research, we have proposed several ways in which "input" factors might be experimentally modified to see if they generate more task-effective group processes and higher quality outputs. In particular, we have suggested that by appropriate alteration of group task design, of group norms, and of the way groups are composed, it may be possible to increase the coordination and level of member effort, the task-appropriateness of group performance strategies, and the utilization and development of the task-relevant knowledge and skill of group members.

2. It will be necessary to design experiments in which group processes are allowed to vary more widely than typically is the case either in laboratory experimental groups or in naturally occurring groups in field settings. We have suggested that, while group interaction process is a crucial mediator of input–output relationships in groups, its functions often are not visible because group tasks and group norms tend powerfully to constrain the richness of the patterns of interaction which can emerge. Fuller understanding of the determinants and effects of group interaction will require both "opening up" or revising the norms people bring to the group about what is and is not appropriate behavior in task-oriented groups, and more attention to the sampling and experimental variation of group tasks than heretofore has been the case in experimental research.

3. Finally, it will be necessary to adopt a more differentiated view of the functions of group interaction as a mediator of input–performance relationships than previously has been the case. Given that group interac-

tion may serve quite different functions in affecting performance effectiveness for different kinds of tasks, methods of coding and analyzing interaction process will have to be developed which are specifically appropriate to the kind of group task on which the group is working.

3. Methodological Issues

If research is to be carried out according to the suggestions outlined above, a number of new methodological tools and techniques will have to be developed, especially in the areas of task description and group process measurement.

a. Interaction coding systems. We have proposed that interaction coding systems, to be useful for tracing how group processes mediate between input conditions and group effectiveness, should have two attributes. First, they should be capable of dealing with interaction process as it develops and changes over time, rather than generating simple frequency tallies or summaries averaged over an entire performance period. And second, they sould address at a relatively molar level those aspects of interaction which are uniquely important in affecting performance outcome for the class of task being performed. In terms of the summary variables outlined in the present chapter, for example, one should use systems that assess the activities of group members (*a*) to coordinate and increase member efforts, or (*b*) to select and implement appropriate performance strategies, or (*c*) to assess, weight, and develop the knowledge and skills of members. Which of these aspects of the interaction process would actually be measured, of course, would depend on which (or which combination of) summary variables is operative for the task being performed. While systems having the attributes described above do not presently exist, we believe that their development would be a most useful addition to the set of tools available to researchers interested in studying small group effectiveness.[6]

b. Task classification systems. The problem of task description is heavily intertwined with the problem of measuring group interaction, as Roby pointed out in 1963:

> The thesis will be argued here, first, that any major advance in theory and research on small group problems will depend very heavily on progress in measuring task properties and group processes. Second, it will be argued that these measurement problems are closely interrelated—that

---

[6] While not focusing specifically on the functions of interaction specified above, the systems developed by Alderfer, Kaplan, and Smith (1974) and by Argyris (1965) offer at least a point of departure for analyses to assess the degree to which members have developed nontraditional patterns of interpersonal behavior in the group (for example, by creating conditions for heightened behavioral experimentation in the group, or by generating high levels of intermember sharing).

is, a clarification of the essential attributes of group tasks will contribute significantly to a better understanding of the important aspects of group process, and conversely [pp. 1–2].

Despite repeated calls for increased attention to task description and classification—and despite a number of attempts to develop schemes for differentiating among group tasks—no satisfactory methodology for describing group tasks has yet emerged (cf. Davis, 1969, Chapter 3; Hackman, 1969; Roby & Lanzetta, 1958; Shaw, 1971, Chapter 9; Zajonc, 1965). One of the difficulties in describing tasks is that they serve, simultaneously, at least two functions which must be described in different terms.

(1) *Task as stimulus.* Through direct instructions about what is to be done, and through cues present in the task materials, tasks affect member behavior in the group setting. Examples earlier in this paper have shown that task design can prompt high or low effort on the part of group members, attention to matters of performance strategy or ignoring these issues, and so on. The stimulus properties of group tasks can be described on an almost endless number of dimensions, in terms ranging from the molecular (e.g., the way lights in a stimulus display are configured) to the molar (e.g., the overall judged challenge of a task). The job of task description is difficult, and will inevitably involve a good deal of "bootstrapping," because a strictly empirical/inductive approach is too large and cumbersone, and because theories that can specify a priori the most critical stimulus characteristics of tasks for given research questions are not yet available.

(2) *Task as moderator.* As is noted throughout this paper, tasks moderate process–performance relationships in groups; that is, what kinds of behaviors serve to increase or decrease task effectiveness depend, often to a substantial extent, on the nature of the task itself. Thus, any attempt to understand process effects on performance must take account not only of group interaction, but also of the *contingencies in the task* that determine the kinds of behaviors that contribute to effectiveness for that task (cf. Hackman, 1969).

Such contingencies can be referred to as "critical task contingencies"—i.e., they specify what types of behavior are critical to successful performance on the task in question. Which of the three summary variables discussed in this paper are relevant to performance effectiveness on a given task depends on these task-based contingencies, as is shown in Fig. 5. Thus, if a researcher were attempting to understand or change group effectiveness in a given instance, he would deal only or mainly with those summary variables that were objectively important in determining effectiveness for the task at hand.

In practice, it often is possible to determine which of the three summary variables is likely to be most important for a given task simply by inspection—e.g., by asking, "If greater effort (or different performance strategies, or different levels of knowledge and skill) were brought to bear on this task, would performance effectiveness be likely to change markedly?" But ultimately it will be necessary to know what it is in the *task itself* that determines whether such questions are answered affirmatively or negatively. For only when critical task contingencies can be described in terms of the task itself will it become possible to generate unambiguous and objectively operational propositions about the interactions among task characteristics, group processes, and group effectiveness.

B. IMPLICATIONS FOR ACTION

1. The Elusiveness of General Interventions

There is no dearth of small group intervention techniques available to the practitioner interested in trying to change group behavior and task effectiveness. Yet, just as we have argued that there is not likely to be a general theory of group effectiveness, we also eschew the notion that there can be any single intervention package that will be universally helpful in improving group effectiveness.

Consider, for example, small group "team building," a popular intervention technique which focuses on the interpersonal relationships and social climate present in the group. Team building may be of great use in helping group members develop the capability to utilize member knowledge and skill effectively on a task. For this reason, the technique may aid performance effectiveness on tasks for which this summary variable is operative. But team building may be actively dysfunctional, at least in the short term, for tasks where the operative summary variable is effort, because the energies of group members are siphoned away from the task itself and applied instead to the exciting and involving interpersonal processes which take place during the team-building process. Similarly, redesign of the group task may aid effectiveness on tasks where effort is an important determinant of productivity, but be much less relevant and useful as an intervention if the performance-impeding problem is one of faulty performance strategies.

In general, intervention techniques which have been offered as devices for improving group effectiveness fall into two classes: (a) interpersonal techniques, often utilizing experiental learning devices intended to improve the quality of the relationships among group members; and

(b) procedure-oriented techniques, which provide group members (often via the group leader) with specific strategies for going about their work on the task in a more effective manner.

Relatively little research has been done to assess the value of such techniques for improving group task effectiveness. In general, however, research suggests that interpersonal interventions are powerful in changing patterns of behavior in the group—but that task effectiveness is rarely enhanced (and often suffers) as a consequence (Deep, Bass, & Vaughan, 1967; Hall & Williams, 1970; Hellebrandt & Stinson, 1971; Kaplan, 1973; Wagner, 1964). Procedure-oriented interventions, on the other hand, often may be helpful in improving effectiveness for the task immediately at hand, but rarely can they be incorporated readily into the on-going process of the group (Hackman et al., 1974; Maier, 1963; Maier & Maier, 1957; Maier & Solem, 1952; Shure et al., 1962; Varella, 1971, Chapter 6).

What seems needed, then, are the following:

1. Development of interventions that are effectiveness-enhancing in the short term, *and* that simultaneously lead to alterations of the normal processes of the group such that the overall competence of the group as a performing unit is enhanced.

2. Development of a taxonomy of groups, tasks, and situations that specifies the potential utility of various interventions for different types of performance situations. Recent work by Herold (1974) offers some promising leads toward such a taxonomy. In the meantime, however, we believe that interventionists will have to rely on especially careful personal *diagnoses* of the task, the group, and the situation—and tailor their change-oriented activities to what those diagnoses reveal.

2. Toward Increased Self-Management by Task Groups

As techniques for modifying situations and intervening in group processes become known and tested, it is tempting for a consultant to a group (whether outsider or group leader) to use this knowledge, perhaps covertly, to move the group toward greater effectiveness.[7] And,

---

[7] For reasons of space and energy, we deliberately have not addressed the question of "leadership" in this chapter. Suffice it to say here that our view of effective leadership is quite similar to what we mean when we describe an effective consultant—i.e., someone who is a sensitive and skilled diagnostician of the group and its environs, and who helps create conditions in the group and the situation which facilitate effective group behavior. Thus, when we speak of the "consultant" in this section, we do *not* mean to imply that such a person is not a group member; indeed, we see consultative activities as among the most important aspects of an effective leader's role.

indeed, this "engineering" approach has been advocated and used with apparent success in some situations (Varela, 1971, Chapter 6).

We believe, however, that in the long term it is better if the group members themselves develop the skills and the understanding to manage their *own* development as a productive unit. This will lessen the reliance of the group on the continued expertise of the consultant, and often may increase the commitment of group members to the group and its goals—because they come increasingly to "own," and therefore care about, its processes and its products. Moreover, the problems of a group often are highly idiosyncratic, hinging on rather unique coincidences of people, tasks, and situations. Relying on specific advice and assistance from an outside professional on an on-going basis would, it appears, be a grossly inefficient way to improve the long-term effectiveness of the group.

There are two major—and quite different—hurdles to be overcome if a group is to gain increased self-control over its own task performance processes and increased competence in managing its performance activities. The first is development of heightened awareness of the determinants of group processes and group performance. The second is developing the competence (both technical and interpersonal) to respond adaptively to the newly understood problems and opportunities.

Outside assistance would seem to be critical in helping a group overcome both of these hurdles. Members must break out of the reactive stance most people assume in task performance situations and take a more active, seeking, and structuring orientation toward their task and interpersonal environment. This is unlikely to occur spontaneously, for reasons discussed at several points throughout this paper. It is, however, likely to become self-perpetuating, once the consciousness of group members about the determinants of their behavior is raised, assuming that they decide that they do in fact wish to take a more active stance toward the task and social environment. With the aid of a competent leader or consultant, the traditional implicit norm of reactivity can be replaced by a new norm of proaction—on the part of both individual group members, and the group as an interacting unit.

Group members themselves, for example, probably should be as involved as possible in diagnostic activities aimed at determining the demands of the task and the resources the group has at hand to work on the task. By participating in such diagnoses, members should achieve the fullest possible awareness and understanding of the factors that affect their own performance activities and their effectiveness as a group. They should, therefore, become increasingly well-prepared to engage in new activities intended to reduce their process losses—and to chart avenues for realizing previously unrecognized possibilities for process gains.

An especially critical point is reached when group members have become aware of the need for change and have developed the motivation to initiate and carry through specific changes. At this point, they may be in rather desperate need of assistance to learn how, competently, to do what they already want to do—and they may not be aware of many of the problems that they will face in implementing their plans. This is especially important for changes involving the internal process of the group: Merely wanting to be "less punitive in dealing with ideas," for example, while often an admirable goal and one that may be task-effective on many tasks, is extraordinarily difficult for most members of most groups to carry off successfully. Similarly, changing the task of the group—something most members of most groups take as an unalterable given—requires the use of personal and interpersonal skills which are not well-practiced. Again, members will require assistance in finding ways to carry out their intentions.

The challenge to the consultant is to help the group members raise their collective consciousness about what "might be" and to learn how to achieve their newly found aspirations; the challenge to the small group researcher is to provide, for the consultant and the group members alike, the knowledge and the tools that will help them get there from here.

REFERENCES

Alderfer, C. P., Kaplan, R. E., & Smith, K. K. The effect of variations in relatedness need satisfaction on relatedness desires. *Administrative Science Quarterly*, 1974, 19, 507–532.

Argyris, C. *Interpersonal competence and organizational effectiveness*. Homewood, Ill.: Irwin-Dorsey, 1962.

Argyris, C. Explorations in interpersonal competence. *Journal of Applied Behavioral Science*, 1965, 1, 58–83.

Argyris, C. The incompleteness of social psychological theory: Examples from small group, cognitive consistency, and attribution research. *American Psychologist*, 1969, 24, 893–908.

Argyris, C., & Schon, D. *Theory in practice*. San Francisco, Calif.: Jossey-Bass, 1974.

Bales, R. F. *Interaction process analysis*. Cambridge, Mass.: Addison-Wesley, 1950.

Bion, W. R. *Experiences in groups*. New York: Basic Book, 1959.

Bradford, L. P., Gibb, J., & Benne, K. (Eds.) *T-group theory and laboratory method*. New York: Wiley, 1964.

Carter, L. F., Haythorn, W., Meirowitz, B., & Lanzetta, J. The relation of categorizations and ratings in the observation of group behavior. *Human Relations*, 1951, 4, 239–254.

Cohen, A. M., Bennis, W. G., & Wolkon, G. H. The effects of continued practice on the behaviors of problem-solving groups. *Sociometry*, 1961, 24, 416–431.

Collins, B. E., & Guetzkow, H. *A social psychology of group processes for decision-making.* New York: Wiley, 1964.

Conway, J. A. Problem solving in small groups as a function of "open" and "closed" individual belief systems. *Organizational Behavior and Human Performance,* 1967, 2, 394–405.

Davis, J. H. *Group performance.* Reading, Mass.: Addison-Wesley, 1969.

Davis, J. H. Group decision and social interaction: A theory of social decision schemes. *Psychological Review,* 1973, 80, 97–125.

Deep, S. D., Bass, B. M., & Vaughan, J. A. Some effects on business gaming of previous quasi-T group affiliations. *Journal of Applied Psychology,* 1967, 51, 426–431.

Deutsch, M. Task structure and group process. *American Psychologist,* 1951, 6, 324–325. (Abstract)

Dunnette, M. D., Campbell, J., & Jaastad, K. The effect of group participation on brainstorming effectiveness for two industrial samples. *Journal of Applied Psychology,* 1963, 47, 30–37.

Egerman, K. Effects of team arrangement on team performance: A learning-theoretic analysis. *Journal of Personality and Social Psychology,* 1966, 3, 541–550.

Goldman, M. A comparison of individual and group performance for varying combinations of initial ability. *Journal of Personality and Social Psychology,* 1965, 1, 210–216.

Hackman, J. R. Effects of task characteristics on group products. *Journal of Experimental Social Psychology,* 1968, 4, 162–187.

Hackman, J. R. Toward understanding the role of tasks in behavioral research. *Acta Psychologica,* 1969, 31, 97–128.

Hackman, J. R. Group influences on individuals. In M. D. Dunnette (Ed.), *Handbook of industrial and organizational psychology.* Chicago, Ill.: Rand-McNally, 1975.

Hackman, J. R., Jones, L. E., & McGrath, J. E. A set of dimensions for describing the general properties of group-generated written passages. *Psychological Bulletin,* 1967, 67, 379–390.

Hackman, J. R., & Lawler, E. E., III. Employee reactions to job characteristics. *Journal of Applied Psychology Monograph,* 1971, 55, 259–286.

Hackman, J. R., & Oldham, G. R. *Motivation through the design of work: Test of a theory.* Technical Report No. 6. New Haven, Conn.: Dept. of Administrative Sciences, Yale University, 1974.

Hackman, J. R., Weiss, J. A., & Brousseau, K. *Effects of task performance strategies on group performance effectiveness.* Technical Report No. 5. New Haven, Conn.: Dept. of Administrative Sciences, Yale University, 1974.

Hall, J., & Williams, M. S. A comparison of decision-making performances in established and ad hoc groups. *Journal of Personality and Social Psychology,* 1966, 3, 214–222.

Hall, J., & Williams, M. S. Group dynamics training and improvised decision making. *Journal of Applied Behavioral Science,* 1970, 6, 39–68.

Hare, A. P. *Handbook of small group research.* New York: Free Press of Glencoe, 1962.

Hare, A. P. Bibliography of small group research: 1959–1969. *Sociometry,* 1972, 35, 1–150.

Haythorn, W. W. The composition of groups: A review of the literature. *Acta Psychologica,* 1968, 28, 97–128.

Hellebrandt, E. T., & Stinson, J. E. The effects of T-group training on business game results. *Journal of Psychology*, 1971, **77**, 271–272.

Herold, D. M. (Georgia Institute of Technology) Group effectiveness as a function of task-appropriate interaction processes. Paper presented at the American Institute for Decision Sciences meetings, Atlanta, 1974.

Hoffman, L. R. Conditions for creative problem solving. *Journal of Psychology*, 1961, **52**, 429–444.

Hoffman, L. R. Group problem solving. In L. Berkowitz (Ed.), *Advances in experimental social psychology.* Vol. 2. New York: Academic Press, 1965.

Hoffman, L. R., & Maier, N. R. F. Valence in the adoption of solutions by problem-solving groups: Concept, method and results. *Journal of Abnormal and Social Psychology*, 1964, **69**, 264–271.

Hoffman, L. R., & Maier, N. R. F. Valence in the adoption of solutions by problem-solving groups. II. Quality and acceptance as goals of leaders and members. *Journal of Personality and Social Psychology*, 1967, **6**, 175–182.

Ingham, A. G., Levinger, G., Graves, J., & Peckham, V. The Ringlemann Effect: Studies of group size and group performance. *Journal of Experimental Social Psychology*, 1974, **10**, 371–384.

Jackson, J. Structural characteristics of norms. In I. D. Steiner & M. Fishbein (Eds.), *Current studies in social psychology.* New York: Holt, 1965.

Jackson, J. A conceptual and measurement model for norms and roles. *Pacific Sociological Review*, 1966, **9**, 35–47.

Janis, I. L. *Victims of groupthink: A psychological study of foreign-policy decisions and fiascos.* Boston, Mass.: Houghton, 1972.

Johnson, H. H., & Torcivia, J. M. Group and individual performance on a single-stage task as a function of distribution of individual performance. *Journal of Personality and Social Psychology*, 1967, **3**, 266–273.

Jones, M. B. Regressing group on individual effectiveness. *Organizational Behavior and Human Performance*, 1974, **11**, 426–451.

Kaplan, R. E. *Managing interpersonal relations in task groups: A study of two contrasting strategies.* Technical Report No. 2 New Haven, Conn.: Administrative Sciences, Yale University, 1973.

Katzell, R. A., Miller, C. E., Rotter, N. G., & Venet, T. G. Effects of leadership and other inputs on group processes and outputs. *Journal of Social Psychology*, 1970, **80**, 157–169.

Kelley, H. H., & Thibaut, J. W. Experimental studies of group problem solving and process. In G. Lindzey (Ed.), *Handbook of social psychology*, Reading, Mass.: Addison-Wesley, 1954.

Kelley, H. H., & Thibaut, J. W. Group problem solving. In G. Lindzey & E. Aronson (Eds.), *The handbook of social psychology.* (2nd ed.) Reading, Mass.: Addison-Wesley, 1969.

Kepner, C. H., & Tregoe, B. B. *The rational manager: A systematic approach to problem solving and decision making.* New York: McGraw-Hill, 1965.

Lanzetta, J. T., & Roby, T. B. The relationship between certain group process variables and group problem-solving efficiency. *Journal of Social Psychology*, 1960, **52**, 135–148.

Laughlin, P. R., & Branch, L. G. Individual versus tetradic performance on a complementary task as a function of initial ability level. *Organizational Behavior and Human Performance*, 1972, **8**, 201–216.

Laughlin, P. R., Branch, L. G., & Johnson, H. H. Individual versus triadic performance on a unidimensional complementary task as a function of initial ability level. *Journal of Personality and Social Psychology*, 1969, 12, 144–150.

Laughlin, P. R., & Johnson, H. H. Group and individual performance on a complementary task as a function of initial ability level. *Journal of Experimental Social Psychology*, 1966, 2, 407–414.

Lorge, I., Fox, D., Davitz, J., & Brenner, M. A survey of studies contrasting the quality of group performance and individual performance, 1920–1957. *Psychological Bulletin*, 1958, 55, 337–372.

Maier, N. R. F. *Principles of human relations.* New York: Wiley, 1952.

Maier, N. R. F. *Problem solving discussions and conferences: Leadership methods and skills.* New York: McGraw-Hill, 1963.

Maier, N. R. F., & Maier, R. A. An experimental test of "developmental" versus "free" discussion on the quality of group decision. *Journal of Applied Psychology*, 1957, 41, 320–323.

Maier, N. R. F., & Solem, A. R. The contribution of a discussion leader to the quality of group thinking. *Human Relations*, 1952, 5, 277–288.

Maier, N. R. F., & Thurber, J. A. Limitations of procedures for improving group problem solving. *Psychological Reports*, 1969, 25, 639–656 (Monogr. Suppl. 1-V25).

March, J. G., & Simon, H. A. *Organizations.* New York: Wiley, 1958.

McGrath, J. E. *Social psychology: A brief introduction.* New York: Holt, 1964.

McGrath, J. E., & Altman, I. *Small group research: A synthesis and critique of the field.* New York: Holt, 1966.

Morris, C. G. Task effects on group interaction. *Journal of Personality and Social Psychology*, 1966, 5, 545–554.

Morris, C. G. Changes in group interaction during problem-solving. *Journal of Social Psychology*, 1970, 81, 157–165.

O'Dell, J. W. Group size and emotional interaction. *Journal of Personality and Social Psychology*, 1968, 8, 75–78.

Osborn, A. F. *Applied imagination.* (Rev. ed.) New York: Scribner's, 1957.

Prince, G. M. *The practice of creativity.* New York: Harper, 1970.

Raush, H. L. Interaction sequences. *Journal of Personality and Social Psychology*, 1965, 2, 487–499.

Riecken, H. W. The effect of talkativeness on ability to influence group solutions to problems. *Sociometry*, 1958, 21, 309–321.

Roby, T. B. Process criteria of group performance. Paper presented as part of the Task and Criterion work group of the Small Groups in Isolation project of the Neuropsychiatric Division, NMRI, Bethesda, Maryland, 1963.

Roby, T. B., & Lanzetta, J. T. Considerations in the analysis of group tasks. *Psychological Bulletin*, 1958, 55, 88–101.

Sample, J. A., & Wilson, T. R. Leader behavior, group productivity, and rating of least preferred co-worker. *Journal of Personality and Social Psychology*, 1965, 3, 266–269.

Schachter, S., Ellertson, N., McBride, D., & Gregory, D. An experimental study of cohesiveness and productivity. *Human Relations*, 1951, 4, 229–238.

Schein, E. H. *Process consultation.* Reading, Mass.: Addison-Wesley, 1969.

Schein, E. H., & Bennis, W. *Personal and organizational change through group methods.* New York: Wiley, 1965.

Seashore, S. *Group cohesiveness in the industrial work group.* Ann Arbor: Institute for Social Research, University of Michigan, 1954.

Shaw, M. E. *Group dynamics*. New York: McGraw-Hill, 1971.

Shiflett, S. C. Group performance as a function of task difficulty and organizational interdependence. *Organizational Behavior and Human Performance*, 1972, 7, 442–456.

Shure, G. H., Rogers, M. S., Larsen, I. M., & Tassone, J. Group planning and task effectiveness. *Sociometry*, 1962, 25, 263–282.

Solem, A. R. The effect of situational vs. behavioral problem statements on solution quality. Unpublished manuscript, University of Minnesota, 1974.

Sorenson, J. R. Task demands, group interaction, and group performance. *Sociometry*, 1971, 34, 483–495.

Steiner, I. D. *Group process and productivity*. New York: Academic Press, 1972.

Stone, T. H. Effects of mode of organization and feedback level on creative task groups. *Journal of Applied Psychology*, 1971, 55, 324–330.

Swinth, R. L., & Tuggle, F. D. A complete dyadic process model of four man group problem-solving. *Organizational Behavior and Human Performance*, 1971, 6, 517–549.

Talland, G. A. Task and interaction process: Some characteristics of therapeutic group discussion. *Journal of Abnormal and Social Psychology*, 1955, 50, 105–109.

Taylor, D. W., Berry, P. C., & Block, C. H. Does group participation when using brainstorming facilitate or inhibit creative thinking? *Administrative Science Quarterly*, 1958, 3, 23–47.

Taylor, D. W., & Faust, W. L. Twenty questions: Efficiency in problem solving as a function of size of group. *Journal of Experimental Psychology*, 1952, 44, 360–368.

Varela, J. A. *Psychological solutions to social problems*. New York: Academic Press, 1971.

Vroom, V. H. Industrial social psychology. In G. Lindzey & E. Aronson (Eds.), *The handbook of social psychology*. (2nd ed.) Reading, Mass.: Addison-Wesley, 1969.

Wagner, A. B. The use of process analysis in business decision games. *Journal of Applied Behavioral Science*, 1964, 1, 387–408.

Weick, K. E. *The social psychology of organizing*. Reading, Mass.: Addison-Wesley, 1969.

Wolf, G. A model of conversation. *Comparative Group Studies*, 1970, 1, 275–304.

Wolf, G. *Recent history and social interaction*. Technical Report. New Haven, Conn.: Dept. of Administrative Sciences, Yale University, 1974.

Zajonc, R. B. The requirements and design of a standard group task. *Journal of Experimental Social Psychology*, 1965, 1, 71–88.

Group Process and Group Effectiveness: A Reappraisal[1]

J. Richard Hackman
YALE UNIVERSITY

Charles G. Morris
UNIVERSITY OF MICHIGAN

At about the same time that we were putting the finishing touches on the review chapter reprinted here, Ivan Steiner (1974) gave the first annual Katz–Newcomb lecture at the University of Michigan, entitled "Whatever Happened to the Group in Social Psychology?" Steiner noted the demise of interest during the 1960s in many of the truly social aspects of social psychology and the increased attention of social psychologists to individual and cognitive phenomena. He suggested that interest in groups may recede in time of relative serenity in society and then (with perhaps a decade lag) reassert itself when social turbulence reappears. According to Steiner's argument, we should be seeing about now a reemergence of research on groups, in response to the social unrest of the mid- to late 1960s. Or, in Steiner's concluding words: "If the tea leaves tell me true, social psychology in the late 70's is going to look a lot like social psychology in the late 40's, better, of course, but groupy once more [p. 106]."

I. What's Been Happening

It is now late in the 1970s, and indeed there are some signs of renewed interest in group and interpersonal phenomena. Yet if

[1] Preparation of this essay was supported in part by the Organizational Effectiveness Research Program, Psychological Sciences Division, Office of Naval Research, under Contract No. N00014-75-C-0269, NR 170-744. Reproduction in whole or in part for any purposes of the U.S. Government is permitted.

Group Processes

studies of the *performance effectiveness* of small groups are part of this movement, they clearly are not on the cutting edge of it: In the few years since our review chapter appeared, there has hardly been a pouring forth of conceptual and empirical attacks on the group effectiveness problem. Even eminently researchable conceptualizations such as Steiner's (1972) models of group process and productivity have generated relatively little research activity.

The work on small groups that has been published in recent years has been well-summarized in an *Annual Review of Psychology* chapter by Davis, Laughlin, and Komorita (1976) and in the second edition of Hare's *Handbook of Small Group Research* (1976). Rather than attempt here a systematic update of published material relevant to our chapter, we refer readers to these comprehensive reviews. There are, however, two recent contributions to research and theory on group process and group effectiveness that strike us as worthy of special mention.

The first is the publication of two books dealing with the use of structured intervention techniques to improve the effectiveness of interacting groups. Stein (1975) has compiled a diverse set of group procedures that are oriented specifically toward increasing group creativity, including brainstorming, synectics, creative problem-solving techniques, and so on. For each technique reviewed, practical procedures for using the technique and the theoretical basis of the approach are described, and the present state of evidence regarding its efficacy is assessed. Delbecq, Van de Ven, and Gustafson (1975) provide a practitioner-oriented guide to group techniques for program-planning activities, with detailed explication of procedures for using nominal group techniques (NGT) and Delphi procedures to enhance group decision-making effectiveness.

In terms of the framework provided in our chapter, both the creative techniques and the nominal group techniques can be viewed as devices for improving the quality of the *performance strategies* used by the group. Moreover, both attempt to improve performance strategies by altering the dynamics of the interaction process that occurs among group members. The nature and impact of those alterations are quite different for the two approaches, however. The creative techniques foster new, nonobvious patterns of interaction, with the intent of generating what we called in our chapter "process gains." The NGT and Delphi techniques, on the other hand, severely limit the amount of spontaneous interaction that can occur among group members and constrain the interaction that does take place in order to minimize the possibility of dysfunctional "process losses." Indeed, in the Delphi technique, communication among group members occurs only via summaries of member inputs compiled by a coordinator, thus negating the possibility of *any* spontaneous member-to-member interaction.

The development of sophisticated interventions such as those cited above adds to the pool of techniques available to those who would improve group effectiveness by altering what happens among group members. And when they are used competently in appropriate task and organizational circumstances, such interventions apparently can raise group effectiveness beyond the level that otherwise would have occurred.

The problem is that we have little in the way of systematic evidence (or even well-tested heuristics) to use in deciding just what "appropriate task and organizational circumstances" are for the various techniques that have been developed and espoused. Indeed, according to one of the developers of NGT, one problem with the intervention is that after group members have learned the technique they may spontaneously apply it to all manner of tasks and problems—including those for which nominal group procedures are inappropriate (A. H. Van de Ven, personal communication, 1977).[2]

The second recent contribution on which we wish to comment, an article by Herold (in press), specifically addresses the "under what circumstances" question. Herold begins by assuming that what will turn out to be helpful to a group has perhaps as much to do with the *demands of the group task* as with any other aspect of the group or its environment—a view with which we obviously are sympathetic. He then differentiates between the technical and social demands made by group tasks and shows how these demands vary on a dimension of complexity. For tasks with simple technical demands, such as pushing a stalled car, the procedures to be used in executing the task are obvious and available to group members; for tasks with complex technical demands, the data, solutions, or procedures needed are not readily available. For tasks with simple social demands, little or no interaction is required to complete the task successfully (e.g., a tug-of-war); for tasks with complex social demands, the success of a group is powerfully determined, perhaps in subtle ways, by how members work together and relate to one another. The result is a 2 × 2 matrix of task demands: simple/complex by technical/social.

---

[2] This phenomenon provides evidence contrary to the argument in our chapter that structured techniques for improving group performance strategies are rarely incorporated into the on-going process of the group and used for new, unfamiliar tasks.

When will a group be effective? Only, Herold argues, when its performance processes fit with the demands of the task. Thus, if the technical demands of a task are complex but the group has only simple technical skills in its repertoire, then the group cannot be effective. It also works the other way: If a group uses highly sophisticated technical procedures (such as those provided by many interventions that provide members with task-relevant training) when the task has objectively simple technical demands, performance will also be impaired.

This framework sets the stage for Herold's analysis of the efficacy of various group process interventions. An intervention that replaces routine group processes with nonobvious technical or social processes will improve effectiveness *only if it brings group process into closer congruence with the technical and social demands of the group task.* When both the social and the technical demands of a task are quite straightforward, no intervention is required—the routine, obvious performance processes of the group will suffice. When technical demands are complex and social demands simple, then an intervention that increases the sophistication of the technical procedures of the group will enhance effectiveness. When social (but not technical) demands are complex, an intervention into the social processes of the group is indicated. And when the task requires both complex technical and complex social processes for effective performance, then a sociotechnical intervention is required.

Herold tests his ideas by examining seven published experiments for which the group tasks varied widely in technical and social complexity. In three experimental conditions in which intervention was not indicated by the theory (e.g., sensitivity training for group members working on a task with little social complexity), performance was found to *decrease* as a result of the intervention. And in the four experimental conditions in which the intervention was congruent with the task demands (that is, technical assistance on tasks with high technical complexity and interpersonal assistance on tasks with high social complexity), the interventions enhanced measured performance effectiveness.

Although the tests reported are post hoc and are therefore open to alternative interpretations (especially regarding the measurement of the technical and social complexity of the tasks), the rather simple conceptual framework that Herold has developed appears to have a good deal of power and may turn out to be of considerable use both in research on small group effectiveness and in the development of interventions intended to improve it.

II. The Shape of Groups to Come

What then of research and theory in the next decade? In part it will be shaped by the methodologies of the past decade—although we may have learned from past approaches more about what we should *not* do than about what we *should* do in studying the effectiveness of small groups. We have found, for example, that research on behavior in minimal social situations in the experimental laboratory is relatively unlikely to generate the kinds of understanding that can be generalized to apply to group members working interdependently on complex tasks—especially when (as is usually the case) the group is located in an organizational milieu. We have learned that simply counting the frequency of various kinds of behavior in groups is pretty unlikely to lead to significant understanding of why some groups are effective and others are not. (Now we must hope, with Weick [1976], that the popularity of videotape technology does not rekindle the fantasy that more and more observation of the details of group interaction will somehow unravel the group effectiveness knot.) And, on the action side of the theory/practice coin, we have learned that running group members through various kinds of sensitivity training or team-building activities does not necessarily improve the effectiveness of work groups and, in some circumstances, may even impair it.

Now may be the time when we see not only a resurgence of interest in small groups but also some innovations in how we go about studying them. Here, then, are some predictions about future trends in research on group processes and group effectiveness, trends that may generate some new directions in both theory and practice having to do with task-oriented small groups:

1. *There will be an increase in clinical, idiographic, ethnographic, and sit-and-stare approaches to understanding small groups, at the expense of highly controlled laboratory studies with large sample sizes.* In a provocative paper written for a political science audience, Weick (1976) outlines what he would do—and what he wouldn't do—if he were starting out fresh to study small groups. Weick concludes that he would not attempt to understand what happens in groups using highly controlled experimental tasks (such as the prisoner's dilemma game), large sample sizes, elaborate interaction coding systems, and the like. Nor would he construe groups as rational performing units with clearly demarked boundaries specifying who is and who is not a member. Going about it this way, Weick argues, has not provided very clear views of what actually goes on in groups—

especially when the researcher, in the interest of scientific "objectiv- ity," keeps distant and detached from the actual goings-on of the group.

Maybe, Weick suggests, we should go about studying groups in a wholly different way. Maybe we should start by viewing participation in groups as the fluid, sometimes, now-and-then, only-part-of-me phe- nomenon that group membership actually is for most of us. Then we might focus our attention on what is really going on in most groups, such as stratification, ethnocentrism, and dealing with minority influ- ences, rather than persist in searches for underlying structures that seem always to elude us. And, perhaps most important of all, maybe we can unhook ourselves from our traditional conceptions of methodology and theory so as to experience and understand group phenomena in their own terms rather than within the sometimes rigid frameworks we have been trained to use.

Many social psychologists will find Weick's views a refreshing alternative to traditional strategies for gaining knowledge about groups. Moreover, his piece may provide both the impetus and the legitimacy needed for some researchers, frustrated for years by the constraints of traditional empirical methodology, to do what they have always *wanted* to do in studying groups—but never felt they *should* do. The result may be a moderate pouring forth of ethnographic, idiographic, and clinical studies of groups in all their looseness and irrationality—which is very much to the good in that an increased diversity of research approaches can only help the field, and new ideas and insights about what groups are and what happens within them are almost sure to appear. The challenge is to use such knowl- edge in developing systematic knowledge about groups at the concep- tual level—and, for those interested in the group effectiveness ques- tion, to build the extremely important bridge between understanding what is going on in a group and knowing what to do about it to make things better.

2. *There will be more and better models of group decision- making processes, as well as better-informed practical applications of group decision-making techniques to real life groups.* The litera- ture on group decision-making is mushrooming. We are seeing in- creasing numbers of structured techniques for improving group decision-making processes, such as those described by Delbecq, Van de Ven, and Gustafson (1975), mentioned earlier. At the same time, a number of researchers are generating and testing new (and often quite sophisticated) models of how individual preferences are translated via group interaction into group decisions. The theory of social decision

schemes developed by Davis (1973), for example, allows assumptions about the decision-relevant effects of group interaction to be precisely and formally specified and tested. Still another approach to the study of group decision-making effectiveness is that of Janis (Janis, 1972; Janis & Mann, 1977). Janis and his colleagues have identified a number of social and emotional barriers to effective decision-making, and set forth a number of specific intervention strategies that are posited to improve decision-making processes.

We expect that these various lines of research and thought about group decision-making will increasingly converge and complement one another in the years to come. The more normative and applied approaches, for example, could usefully be informed by the theorizing about social decision processes that has been central in research on the more formal models of group decision-making. Moreover, if there is to be real movement toward an answer to the "under what circumstances" question for the normative approaches, then surely there will have to be increased attention to the ways that task and situational factors affect group decision processes. The more formal and abstract models of group decision-making, on the other hand, may be limited by their emphasis on pure description and prediction. That is, if typical group decision processes are substantially sub-optimal, then descriptive models of these processes necessarily characterize sub-par processes—and they may not be of much help in the development of strategies for *improving* group decision-making effectiveness. We expect some significant and interesting conceptual sparks to fly in the near future as the several group decision-making paradigms come increasingly into contact with one another in research and in practice.

3. *There will be an increasing tendency for work to be designed for interacting teams in organizations, resulting in large numbers of case studies of work team effectiveness.* This prediction may seem a bit bizarre, given how inefficient we know groups can be in performing certain kinds of tasks and how little we know for sure about helping groups perform tasks more effectively. But the work group idea clearly is in good currency among those who study organizations and those who manage them. Cautious suggestions that "we don't know enough yet" are not likely to counter significantly the present movement to use small work groups as design devices in organizations.

Indeed, at a symposium commemorating the fiftieth anniversary of the well-known Hawthorne studies at Western Electric, Leavitt (1975) recently gave a paper entitled "Suppose We Took Groups Seriously . . . ," in which he persuasively suggests that organizations might function better if groups, rather than individuals, were used as the basic organizational building blocks. Leavitt speaks optimistically about both the efficacy of using groups as primary organizational units and our present level of understanding of groups, concluding his paper as follows:

> The major unanswered questions in my mind are not in the understanding of groups, nor in the potential utility of the group as a building block. The more difficult . . . question is whether or not the approaching era is one in which Americans would willingly work in such apparently contra-individualistic units [Leavitt, 1975, p. 77].

The idea of designing work for groups has grown largely out of the sociotechnical systems tradition in England, and the device of the "autonomous work group" is increasingly being employed in Europe, in the Far East, and (more recently) in this country (Davis & Cherns, 1975). Probably the best-known application in a U.S. organization is the Topeka pet food plant of General Foods, where an entirely new manufacturing organization was designed around the concept of the semiautonomous work group (Walton, 1972, 1975). The early reports of success at Topeka have spawned numerous similar experiments in both public and private organizations in this country. Though a quick look at case reports from these innovations would suggest that many or most of the experiments are quite successful, a closer review (including projects that do *not* get written up and published as case reports) suggests somewhat the contrary—namely, that many work teams in organizations are not functioning as effectively as had been hoped. Even the much touted Topeka experiment (which in fact involved many innovations in addition to the creation of work teams) is now being viewed with a good deal of skepticism by some commentators (see, for example, "Stonewalling Plant Democracy," *Business Week,* March 28, 1977, pp. 78–82).

Part of the reason why some work teams in organizations are not very effective is that existing knowledge about group processes and group performance has not been well-used in the design and management of the groups. But another reason, and one that is cause for considerable concern, is that some innovators have bought the work team idea to such an extent that they simply throw a number of workers together, give them a task that requires interdependent work, and assume that the group will "work things out."

If we know anything at all about group behavior, it is that groups rarely "work things out" automatically. There are many things that can go wrong within a group and in the relationship between a group and its systemic environment. Indeed, it may be harder, not easier, to effectively design and manage work done by interacting teams than work done by coacting individuals. And it probably requires more, not less, behavioral science sophistication.

That sophistication is unlikely to develop simply by accumulating additional case reports of "work team successes" in organizations (especially given that reports of "work team failures" rarely see the light of publication). Nor is it likely to emerge from a further accumulation of basic research on small group behavior in the social psychological laboratory. As Hare's *Handbook* (1976) documents, we have many thousands of these studies already, and usable knowledge about the design and management of small work groups continues to elude us. Rather, the sophistication necessary to make work teams successful is most likely to arise from better theory about the factors that actually *make differences* in the process and performance of work groups in organizational settings and from experiments that test these ideas *in situ*.

The conceptual work of Herold (in press; discussed earlier in this essay) may provide some help in developing theory and designing research to move toward this goal, as may some of the perspectives provided in our review chapter reprinted here. One of us recently has prepared a paper on "The Design of Self-Managing Work Groups" (Hackman, 1977) that attempts specifically to apply the ideas in our review chapter to the design and management of work teams in organizations. The paper gives special attention to four questions. First is the old "individual versus group performance" question in new, organizational clothes: What are the task and organizational circumstances under which it is more advantageous to design work for interacting teams than for coacting individuals? Second, if work is to be done by a team, how should that work be designed, and how should the team be composed? Third, after the basic design of a group has been set, what kinds of process interventions or managerial behaviors are most appropriate in maintaining the group and helping it increase its task effectiveness? And, finally, what are the organizational and technological factors that affect the process and performance of work teams, and how should such factors be dealt with in the design and management of work groups in organizations?

Unfortunately, little research evidence presently is available for questions such as these, either in the social psychology or in the organizational psychology literature. Moreover, in the short list of emerging research trends set forth in this essay, we have *not* listed such research and theory among the likely "hot" topics in research in the near future on group effectiveness. The fact is that we see much more interest and activity in proceeding to *install* work groups in organizations than we do in attempting to *understand* what affects their process and their performance. Especially helpful, in our view, would be undertakings in which the process of creating effective work groups and the process of understanding them proceed simultaneously—that is, research (whether in the laboratory or the field) that allows the testing of ideas about what most makes a difference in work group effectiveness even as attempts are made to create those conditions. Only through such activities, we believe, are we likely to advance simultaneously our conceptual understanding of groups in social systems and our supply of usable knowledge about how to design and maintain them at high levels of effectiveness.

REFERENCES

Davis, J. H. Group decision and social interaction: A theory of social decision schemes. *Psychological Review*, 1973, **80**, 97–125.

Davis, J. H., Laughlin, R. P., & Komorita, S. S. The social psychology of small groups. *Annual Review of Psychology*, 1976, **27**, 501–541.

Davis, L. E., & Cherns, A. B. (Eds.), *The quality of working life* (Vol. I: *Problems, prospects and the state of the art*. Vol. II: *Cases and commentary*). New York: Free Press, 1975.

Delbecq, A. L., Van de Ven, A. H., & Gustafson, D. H. *Group techniques for program planning*. Glenview, Ill.: Scott, Foresman, 1975.

Hackman, J. R. The design of self-managing work groups. In B. T. King, S. S. Streufert, & F. E. Fiedler (Eds.), *Managerial control and organizational democracy*. Washington, D.C.: Winston & Sons, 1977.

Hare, A. P. *Handbook of small group research* (2nd ed.). New York: Free Press, 1976.

Herold, D. M. Improving the performance effectiveness of groups through a task-contingent selection of intervention strategies. *Academy of Management Review*, in press.

Janis, I. L. *Victims of groupthink: A psychological study of foreign-policy decisions and fiascos*. New York: Houghton-Mifflin, 1972.

Janis, I. L., & Mann, L. *Decision making*. New York: Free Press, 1977.

Leavitt, H. J. Suppose we took groups seriously . . . In E. L. Cass & F. G. Zimmer (Eds.), *Man and work in society*. New York: Van Nostrand Reinhold, 1975.

Stein, M. K. *Stimulating creativity* (Vol. 2). New York: Academic Press, 1975.

Steiner, I. D. *Group process and productivity*. New York: Academic Press, 1972.

Steiner, I. D. Whatever happened to the group in social psychology? *Journal of Experimental Social Psychology*, 1974, **10**, 93–108.

Walton, R. E. How to counter alienation in the plant. *Harvard Business Review*, November–December 1972, **50**, 70–81.

Walton, R. E. From Hawthorne to Topeka and Kalmar. In E. L. Cass & F. G. Zimmer (Eds.), *Man and work in society*. New York: Van Nostrand Reinhold, 1975.

Weick, K. E. Some challenges for future group research: Reflections on the experience in psychology. In R. Golembiewski (Ed.), *The small group in political science: The last two decades of development*. Athens, Georgia: University of Georgia Press, 1976.

GROUP PROBLEM SOLVING[1]

L. Richard Hoffman
GRADUATE SCHOOL OF BUSINESS
UNIVERSITY OF CHICAGO
CHICAGO, ILLINOIS

I. Introduction

The basic question to be answered in the study of group problem solving is: How best can a group use the resources of its members in solving a given problem? This phrasing of the question assumes that the necessary resources do lie in the group. Furthermore, most experimenters have used problems whose difficulty is presumably commensurate with the level of ability of the group members.

[1] This chapter was prepared in conjunction with United States Public Health Service Grant MH-2704, N.R.F. Maier, principal investigator. The author was then with the Department of Psychology, University of Michigan.

Reprinted from *Advances in Experimental Social Psychology*, Volume 2, 99–132.
Copyright © 1965 by Academic Press, Inc.

From a broad point of view, however, this assumption avoids several interesting questions which could easily arise in naturalistic settings. How does a group know when its members have the resources which are needed to solve a problem? How long will the group persist in attempting to solve the problem if it does not possess the necessary resources? What actions does a group take when it decides it is unable to solve the problem? To my knowledge, there have been no experimental attempts to answer these questions. In fact, Feather's (1961) analysis of the motivational and expectational bases for persistence in individual problem solving is almost unique in its concern with these questions.

Even if the group members have the capability of solving the problems assigned to them, other questions remain. What factors tend to prevent the effective utilization of the group's resources, and what conditions promote problem-solving efficiency? This separation of inhibiting and facilitating factors suggests that the removal of the impediments to problem solution is often insufficient to produce effective group problem solving; facilitating factors must be brought to bear before the group's resources are mobilized effectively.

This chapter will review the experimental evidence which bears on these questions. The discussion will focus primarily on group problem solving and will bring in the classical group-versus-individual controversy only where it seems relevant. We assume here that it is helpful to conceive of groups as groups without necessarily falling into the trap of reifying the group and its processes independently of the members (Allport, 1924). Furthermore, since the management of large organizations often requires that various committees rather than single individuals must do the work, the question of how groups can solve their problems most effectively is of considerable practical importance.[2]

II. Factors Inhibiting Effective Problem Solving

The major barriers to effective problem solving are those conditions which prevent the free expression of ideas in a group. Restraints can decrease the likelihood that the correct solution or the elements of such a solution will be made available to the group. Factors both obvious and subtle can work against the group's use of its resources.

[2] The organizational requirement for group problem solving rests primarily on two assumptions: (1) the information needed for most management decisions must come from a variety of sources whose functional interdependence requires its simultaneous consideration and evaluation by all concerned; and (2) the acceptance of such decisions by the persons affected is often more important than the objective quality of the decision, and acceptance is promoted by participation in decision making.

A. Pressures Toward Uniformity

1. The Value of Unanimity

The continued existence of the group is itself a subtle barrier against the free expression of ideas. In laboratory groups there is a strong tendency for the members to agree on a single solution to a problem, even when instructed to ignore the prior discussion in reporting their own conclusions. It is as if the members did not want the group to be divided at the end of the session. We have often had subjects who violently opposed the majority's solution announce their capitulation with, "I thought we were all supposed to agree!"

The experimental evidence is quite clear on this point. Groups tend to produce unanimous decisions, and discussions tend to increase the uniformity of their members' individual judgments. Thomas and Fink (1961) gave Maier's Horse Trading Problem to groups varying in size from two to five people. Each member was asked for his individual answer after the discussion. Sixty-four percent of the groups regardless of size gave unanimous answers. Despite the instructions to the members to disregard the group's discussion if they wished this degree of uniformity was obtained on a relatively simple, if tricky, arithmetic reasoning problem. Almost half of these groups, moreover, were unanimously incorrect. Interestingly enough, the members of unanimous groups were more satisfied with their decisions than were members of split groups, whether they were correct or incorrect. We shall return to this phenomenon later.

Pennington et al. (1958), in a design similar to Edith Bennett's (1955), showed that groups which were required either just to discuss the problems assigned to them or to reach decisions on these problems (or to do both) agreed more following the group sessions than before. On the other hand, members of groups which merely reconsidered their judgments showed practically no change. However, Bass (1963) demonstrated recently that coalescence (increased agreement) has only a modest correlation with the effectiveness of the decisions ($r = .29$). Even on these objective problems there is no guarantee that unanimity produces truth; in fact, there often is very little relationship between the two. The more complex the problems, then, the more detrimental will this tendency be for the group members to promote consensus at the possible expense of arriving at a more effective solution.

2. Majority Rule

The majority's ability to coerce the people holding the minority view was recognized early by Thorndike (1938) and Timmons (1942). They showed that the majority generally produced only a slightly better solution than the average group member. When it is wrong, moreover, the majority frequently suppresses the expression of the correct solution. This was demonstrated by Maier and Solem (1952). Where the majority was initially correct, 60% of the persons who were initially incorrect adopted the correct answer following discussion. If a minority of the members was initially correct, however, and there was no appointed leader in the group, only 46% of the incorrect subjects adopted the correct answer. When a leader was appointed, however, 72% of the incorrect subjects changed to correct answers. The effect of the leader's presence was most apparent in those groups where only one of the five or six members was initially correct, thus supporting the investigators' argument that the leader served to protect the minority opinion from majority suppression.

Pressures toward uniformity may be merely the result of the typical experimental situation. The experimental conditions may have made the subjects think that proper performance as subjects required them to arrive at a single solution to the problem (Orne, 1962). In real life, on the other hand, there is often very little motivation to arrive at a unanimous decision, but just how often is hard to tell and worthy of study. No matter how frequently the phenomenon arises in naturalistic settings, pressures toward opinion uniformity may be detrimental to the group's effectiveness if it prevents search for and discussion of alternative possibilities.

B. Sources of Pressure Toward Uniformity

1. Personal Characteristics

Whatever the source of the uniformity pressures, these conditions cause many subjects to refrain from solving the assigned problem themselves and to depend on particular members of the group for the correct answer. The cited effects of majority influence—right or, especially, wrong—exemplify this abdication of individual problem solving (Maier and Solem, 1952; Thomas and Fink, 1961). As one might expect, then, Hochbaum (1954) found that subjects who were induced to lack confidence in their ability to do the problem given them were more inclined to conform to a fictitious group norm than were subjects who were made to be high in self confidence. Furthermore, the subjects' confidence in their own judgment was increased by conformity to the group norm. Thus, a person who feels uncertain of his own abilities can feel secure and successful by adhering to others' suggestions.

Many people also exhibit a tendency to agree with the other group members if they believe that, by agreeing with them, they might improve their popularity or standing with these others. Jackson and Saltzstein (1958) found that subjects who were highly attracted to the group they were in were likely to conform to the group norm. But in addition, subjects who previously had not been accepted by their groups conformed to the group judgment as much on an individual task as on a problem for which the group as a whole was to be evaluated. Thus, the insecurity generated by possible rejection by the group provides a strong incentive to prove oneself worthy by agreeing with the other members.[3] If the subject holds an opinion which deviates from that of the majority, he is presented with a difficult dilemma. Since deviation can lead to ultimate rejection by the group, he should conform in order to preserve his standing in the group (Schachter, 1951). On the other hand, if he fails to contribute the correct answer to the group, he has done the group a disservice. Experimental evidence suggests that he chooses the first alternative over the second to the extent that he values his membership in the group and feels threatened by rejection.

2. Group Members' Perceived Ability

The tendency to depend on other group members can also be rational, i.e., rational to the degree that the subject regards the other members as being able to solve the assigned problem. Mausner (1954) paired his subjects with a confederate who was either successful or unsuccessful on a previous task similar to the test task. Subjects paired with successful confederates tended to conform to the confederate's erroneous judgments in the test situation more than did subjects paired with unsuccessful confederates. In a somewhat similar investigation, Berkowitz (1957) showed that persons who were induced to like their ficticious partners rated them as more proficient and conformed more to their purported judgments than did subjects who were made to like their partners less. Thus, not only will subjects conform to the opinions of people they like, but they will judge these people to be more proficient. Believing the liked person to be proficient, of course, also heightens adherence to his views.

3. Prior Experience

An essentially similar phenomenon has been observed at the group level in the subjects' tendency to rely on the total group's ability to solve their problems. Kidd and Campbell (1955) gave three-person groups differential success experiences on an anagram task. They then had the

---

[3] The extensive literature on conformity pressures is clearly relevant here (e.g., Asch, 1956; Crutchfield, 1959; Tuddenham, 1959) but, since it has been reviewed elsewhere, I have limited this discussion to conformity in group problem solving.

group members make individual judgments about the number of flickers of a light, gave the subjects a purported group average, and then permitted them to change their judgments. There was an increase in the subjects' degree of conformity to the group average on the light judgments which corresponded to the degree of prior group success on the anagram task.

This generalization of group influence from one type of task to another can clearly be a detriment to an effective attack on some problems. Problems vary considerably in their solution requirements. Prior success may promote adoption of a readily available but poor solution (Hoffman *et al.*, 1963). Berkowitz and Levy (1956) suggested that success generates both pride in the work group and more positive motivation to work on group problems. While these may be salutary effects, the omniscient and omnipotent powers thus given to the group remind one of those questionable personnel selection decisions which assume that success in one situation predicts success in another: great generals make great executives; the best graduate students make the best professors.

It may well be that the more successes a group has had on one type of problem, the more inflexible it will be in attacking other types of problems. Thus, there must be careful feedback. A group must have information providing for continued motivation, but, at the same time, the members must never lose the ability to search for varied alternatives in attempting to solve new and different problems. The question of how to separate these two functions of feedback is important for research.

C. PARTICIPATION BIASES

1. Feedback

Unfortunately, it is often difficult to control the type and amount of feedback given to group members. People have a natural tendency to participate more actively in group discussion the more favorably their contributions are evaluated. Oakes and his colleagues (Oakes *et al.*, 1960) increased the participation of certain arbitrarily chosen members by indicating to them that their contributions were "insightful" and decreased the participation of other members by indicating that their contributions were "not insightful." Furthermore, this difference was maintained, even though not significantly, during a subsequent period in which feedback was not provided and the authors expected participation to "extinguish." Thus, the person whose ideas are initially accepted by the group is likely to increase his participation, while one whose suggestions are rejected will withdraw from the discussion (Pepinsky *et al.*, 1958).

Oakes (1962) later showed that the effectiveness of the feedback he provided in enhancing participation varied with the relevance of the purported source (professional psychologists versus peers of the subjects versus laymen) for the problem at hand. Nevertheless, since the feedback that one usually receives is typically relevant to the problem, the recipient of this information is likely to be influenced by it.

2. Influence on Solution Adoption

Riecken's study (1958) of the impact of the talkative member on the group's decision-making process identified a major barrier to successful problem solving. In groups which had solved two previous problems he determined who were the most talkative and least talkative members. He then gave a "hint" [the elegant solution to Maier's Parasol Assembly Problem (Maier, 1952)] to one member of each group: to the most talkative member in half the groups and to the least talkative member in the other half. Groups with the most talkative hint-holders generally adopted the elegant solution, while those with the least talkative hint-holders rejected it for the most part.

Riecken attributed this difference to the supporting remarks made by the highly talkative hint-holders. More recently, our own research has indicated that the solution which receives the highest number of favorable comments becomes the group's solution, but that most of the comments often come from a single member (Hoffman and Maier, 1964; cf. also Thomas and Fink, 1961). Thus, he who talks the most is likely to promote his solution to the group successfully.

There is a complementary aspect to this relationship. The person who talks the most is also most likely to be selected as a leader by the other group members. Slater (1955), Kirscht et al. (1959), and Blake and Mouton (1961) all found that leaders selected by their groups were generally the most talkative members. Even in the Shaw and Gilchrist study (1956), where subjects "interacted" only through notes placed in mailboxes at various times, the leaders elected in almost all of the groups had sent the most number of items, especially items relating to the means of organizing the group and to information about the problem.

More definitively, Mann (1961) factor analyzed members' behaviors as rated by peers and by observers and as coded by observers (Bales' system) on two different problems. He found a Task Prominence factor in both analyses. This factor included peer ratings of leadership and talkativeness, observer ratings of leadership, and observer coding of talkativeness and making suggestions, thus supporting the relationship between talkativeness and rated leadership.

This relationship would not interfere with group performance if all or most of the talk contributed to effective solutions to the problems. In Mann's study, however, as in most investigations, the relationship holds regardless of the quality of the solutions obtained from the group. As was noted earlier, Riecken (1958) demonstrated experimentally that the elegant solution was rarely adopted when its proponent was the least talkative member of the group, but was almost always adopted when he was the most talkative member.

3. Personal Characteristics

Undue personal influence is considered a barrier to effective problem solving. The reason becomes clear when we examine the personality characteristics of the more talkative members. Mann (1959) reported significant correlations on two problems of .25 and .36 between the Task Prominence behavioral factor referred to above and a Social Extroversion personality factor, marked most clearly by the General Activity, Ascendance, and Sociability scales of the Guilford-Zimmerman Temperament Survey. These scales also were positively correlated with leadership nominations in groups studied by Kaess et al. (1961). Reference to this phenomenon as the "GAS" syndrome—referring to the initial letters of the three traits—is probably apt in view of the amount of hot air which emerges. Unfortunately, these three variables have generally been found to be unrelated to any measures of cognitive ability (Joan S. Guilford, 1952; Mann, 1959).

Shaw's Individual Prominence scale (1959) shows the same relationship to talkativeness and influence and gives us some clue to the type of person identified by the GAS syndrome. He is the outgoing, domineering, yet friendly person who invests tremendous energy in all his activities. He takes the initiative in social situations, enjoys being with other people, and is able to mobilize others to gain ends that he sets.

Another kind of emergent leader is the one highly motivated to find a good solution. The member with a high stake in achieving a successful outcome is likely to enter the discussion early and to attempt to organize and discuss the problem his way (Hemphill et al. 1957). The persons selected to discuss their groups' solutions with representatives of other groups in the Blake and Mouton (1961) and the Kirscht et al. (1959) studies were identified as dominating the group and engaging in more directive and influencing acts than other members. Crockett's (1955) observations of management conferences revealed that, even in formal groups, leaders emerged who had higher stakes in the issues than the other members and who were generally higher in organizational rank, rated expertise, and rated expression of self-oriented needs. Unfortunately, the expression of self-oriented needs was judged to be related to ineffective group performance in these management conferences (Fouriezos et al., 1950).

The ability of these extroverted and highly motivated people to influence the other group members suggests that, when faced with complex problems, people are willing to rely on those who sound as if they know the answer. Thus, the self-confident manner in which the socially aggressive person presents his opinion may enhance its merit to others. Such a mechanism would account for the surprising findings by Shaw (1963) and Shaw and Penrod (1962). These investigators obtained a negative relationship between the amount of additional information about the problem they gave to certain members and the amount of influence those members had over the group's decision. Shaw's suggested explanation, consistent with the present hypothesis, was that the subjects with the least amount of additional information were able to promote a single solution, whereas those with even more information had several equally good alternatives. Thus, the first group could self-confidently influence the other members to adopt their proposed solutions, while the latter subjects were less confident in supporting several possibilities. A corollary hypothesis, of course, is that the more self-confident members prevent a group from utilizing all of the relevant information available to it.

D. GROUP STRUCTURE

1. Group Size

The disproportionate influence of the self-confident member on the group's decision would seem to grow with increases in group size. As groups grow the distribution of participation among the members becomes severely skewed, with one or two people doing most of the talking (Bales and Borgatta, 1955; Stephan and Mishler, 1952). The larger the group the more inhibited the introverted people are, unless they have a strong stake in the outcome. In such cases those who are willing to speak will railroad their ideas through the meeting. Potential dissenters are often reluctant to voice their opinion for fear of being thought deviant.

2. Formal Structure

These personality characteristics which often lead to an undue influence upon group members typically have their greatest effect in formally leaderless groups. These traits become somewhat less important in groups with some type of formal structure. Shaw (1959), for example, found a positive relationship between scores on his Individual Prominence scale and a person's influence over the decision in groups in which no decision power structure existed. No relationship was found, however, where some members were given more decision-making authority. Similarly, Crockett (1955) found that leaders emerged in management conferences more frequently when the formal leaders were less active in setting goals and proposing problems or their solutions and, thus, when the group structure was relatively ambiguous. A clear-cut status structure in the group seems to inhibit the tendency of the GAS personalities to dominate the discussion.

3. Power Structures

Power structures in groups, while possibly mitigating the negative effects of personality factors, may also create new barriers to effective problem solving. Torrance's series of studies (1955) of problem solving in Air Force crews demonstrated that the lowest-ranked member was least likely to influence the group's decision, even when he was correct. There are probably several reasons for this finding. For one, high-status members expect to exercise influence and require little support from the other group members in order to have their ideas adopted. (This explanation seems consistent with the self-confidence hypothesis stated above. Expecting to influence, the high-status person may assert his ideas more forcefully than the other group members.) Then, too, Torrance suggested that a low-status member may go along with the person in authority even when he knows his personal opinion is better. This notion is supported by the findings of Maier and Hoffman (1961) that people who are strongly identified with their organizations are especially likely to accept the suggestions of the authority person.

The presence of authority relations in a group seems to change the character of the discussion. There is a greater concentration on the ideas of the high-status person, and the group has to spend considerable time either supporting or rejecting his views rather than searching for alternatives (Maier and Hoffman, 1960b). Bass (1963) provided evidence supporting this characterization. He noted that in unstructured groups increases in group participation enhanced the quality of group judgments, whereas in highly structured groups, extent of participation was negatively correlated with the profitability of the discussions and the increases in agreement among the group members. Thus, leaders in formal authority positions should either refrain from suggesting solutions to a group or should be very certain that they are right if they want to avoid introducing error into the problem-solving process.

The inhibiting effect of an authority figure in the group may also be due to the threat he implies to other group members. We are all familiar with the insecure person who is afraid to contradict the boss. The study of bargaining in two-man groups by Deutsch and Krauss (1960) suggests that even implied threats can produce reactions which are harmful to the threatener. Only when neither member of the dyad could threaten the other were they able to develop a cooperative solution to the bargaining problem so that both could make a profit. Neither member was able to make a profit when either or both were able to threaten the other.

Communication between the people involved in the bargaining situation may help overcome some of the negative consequences of implied threat. Loomis (1959) demonstrated that permitting communication in varying degrees between the partners increased the members' perception of the others' trustworthiness and also increased trustworthy behavior on the same problem. Thus, some of the adverse effects produced by the presence of high status group members may be traced to restraints upon communication. Authority figures may be seen as potential threats and inhibit the free expression of ideas, as Mellinger (1956) and Read (1962) demonstrated. But whatever the reason for the negative consequences, the implied threat created by the presence of authority may lessen the members' motivation to arrive at the best possible decision (Hoffman, 1961) in favor of what is apparently acceptable to the boss.

Status differences affect the behavior of those who are high in authority as well as those with little power. Placing a person in a leadership position generally influences his actions in the group. Even in laboratory groups where one person is told he is the "leader" of a group, but is without any real power, he tends to participate more actively and to direct and organize the group's effort, more than the other group members. Berkowitz (1956) showed that persons occupying central positions in a communication net, whether they were strongly ascendant or weakly submissive personalities, tended to dominate the problem-solving process of the group. In a recent unpublished study of our own, we found that subjects who were randomly selected to be leaders were almost invariably the most active members of the group, often engaging in as many problem-directed acts as the other three members combined. Furthermore these acts were critical in determining which of a variety of solutions being discussed by the group was finally selected. The appelation "leader" seems to legitimize the high volume of attempted direction giving and also serves to focus the control of the group's operation in his hands. This control may or may not enhance the effectiveness of the group, depending on the leader's ability to solve the problem (Fiedler and Meuwese, 1963). When power to reward and punish is added to the title, the focus seems to become even sharper. An apparent dilemma is thereby created. The more power a leader has the more effectively he can control the group's procedure, but the more likely he is to become a barrier to the free exchange of ideas.

It is important to repeat here that the barriers to problem solving that I have just discussed have their negative impact on the creative aspects of group problem solving. In terms of other criteria of group effectiveness, however, these conditions may be beneficial. For example, the pressures toward uniformity may be useful when time is an important factor or where solution alternatives are of approximately equal quality. Where, however, an exchange of ideas is needed or only one member has the necessary information, these barriers seem to prevent effective group problem solving.

E. Failure to Search for Problems

Another contributor to ineffective problem solving is the failure of most groups to organize or plan their attack on the problem. Shure et al. (1962) showed that groups tend not to use the opportunity to organize for effective problem solving. The pressure to reach a solution (cf. Maier and Solem, 1962) probably inhibits discussion of the problem or of how to attack it, thus preventing the group from systematically reviewing the problem requirements or their approach to it.

This lack of planning is most clearly seen in the Cohen and Bennis study (1962) of changes in communication networks. Groups which remained in circle networks persisted in a circuit problem-solving system, in which each member received all information and solved the problem himself, rather than developing the more efficient hierarchical system. Groups which shifted from wheel to circle networks, on the other hand, almost invariably developed a hierarchical communication flow. Apparently the necessity for concentrating on the problem activities in the all-circle groups prevented their consideration of problem-solving organization. But the wheel-to-circle groups were able to benefit from their previous experience and explore the possibilities of organizing within the circle network. Even in the original study by Bales and Strodtbeck (1951) on phases in group problem solving, only 6 of the 22 groups actually conformed to the orientation-evaluation-control sequence postulated by the researchers. Whether their definition of orientation is equivalent to "defining the problem" or not, it is clear that comparatively few groups spend much time on that activity.

III. Factors Promoting Effective Problem Solving

A. Group Composition

1. Ability

Several theories have been offered to relate the abilities of the group members to the outcome of the group's problem-solving efforts (e.g., Ekman, 1955; Lorge and Solomon, 1955; Thomas and Fink, 1961).

These have generally applied to so-called "insight" problems in which there is a single key to obtaining the correct answer to the problem. The Lorge and Solomon and the Ekman accounts assume that the group product reflects the ability of the most able member. Thus, they claim, if one member of a group can solve the problem, the group as a whole will solve it. They explain the typical superiority of groups over individuals on these problems as the greater probability of having at least one able person in a group. By extension, groups should be even more superior than individuals on more complex problems, since the probability of having one person who can solve one part of the problem and another who can solve another part, etc., is much greater than having individuals who can solve all parts.

Thomas and Fink (1961) showed, however, that such a model was grossly inadequate for the Horse Trading Problem which requires only a single insight. On the basis of solutions collected from individual members following group discussion, they found a much smaller proportion of unanimously correct groups and a much larger proportion of unanimously incorrect groups than expected on the basis of the "rational" probabilistic model. Thomas and Fink proposed a "consensus model" which predicts that group members will tend to converge on a single answer. This solution is likely to be correct or incorrect according to the difficulty and "verifiability" of the problem.

These theoretical formulations dealing with ability in group problem solving face the difficult task of measuring the relevant abilities. While mathematical reasoning scores on general intelligence tests are usually positively correlated with individual success on deductive reasoning problems, they are completely uncorrelated with success on the Horse Trading Problem and other similar insight problems. Thus, for any particular problem it is difficult to determine the ability level of the group or its most able member other than by having subjects solve the problem itself. Such a procedure obviously introduces new complexities (e.g., practice effects) into the research.

Furthermore, having examined the barriers of personality and social structure to effective group problem solving in previous sections, we can understand why the few studies which varied the general level and distribution of abilities of the group members produced inconsistent findings. In some instances the group product reflected the most able member's ability (Palmer, 1962; Comrey and Staats, 1955; Wiest et al., 1961). In other cases, however, group performance was either uncorrelated with the ability of the most able member (Roby and Lanzetta, 1961), or the member with the right answer was actually suppressed (Maier and Solem, 1952).

Tuckman and Lorge (1962) tested the hypothesis that the group product reflects the best member's ability. Using the relatively complex Mined Road Problem, they had groups of five solve the problem first individually and then as groups. Although they concluded that the group product was no better than the best individual's, their data show that in some groups the group product was considerably better than the best individual's, in others considerably worse, and in a few exactly the same. There is also some suggestion in the group problem-solving literature that the less straightforward the problem, the less the group product reflects the single, most able individual. The evidence presented in succeeding sections suggests that the group product is rarely a simple extension of the combinational model which Lorge and Solomon (1955) offered to account for insight problems. Rather, the group answer typically represents a utilization of abilities in identifying deficiencies, recommending solutions, and applying the appropriate evaluative criteria to select the correct one. Occasionally it means skillfully combining several alternatives into a single elegant package.

2. Other Factors

Characteristics other than the intellectual ability of the group members have been shown to affect a group's problem-solving performance. Two factors appear to stand out as promoting creative problem solving: (1) the members' motivation to work cooperatively on the probem, and (2) the diversity of the points of view and information relevant to the problem within the group.

a. Motivation. In general, all-female groups tend to do less well than male groups. While this may reflect ability differences between the sexes, the equivalence and occasional superiority of mixed-sex groups to all-male groups suggests a motivational explanation (Hoffman and Maier, 1961a, 1961b; Hoffman *et al.*, 1962). Women in all-female groups probably reinforce each other's rejection of the problem-solving task, as when they solve problems individually (Carey, 1958). Thus, such groups presumably fail to put as much effort into the task as is needed. When the problem is more involving, as when they are role-playing with men or dealing with a problem of feelings, all-female groups seem able to function effectively. Men are generally relatively highly motivated to solve problems, although it is not clear how they would act if faced with problems more clearly feminine in content (Milton, 1958). In an unpublished study, we have found that the same male groups which were substantially superior to female groups on the Horse Trading Problem were slightly inferior to them on a human relations problem. Whether these results are a product of the men's reduced motivation on the human relations problem or of the women's increased motivation is not known at this time.

The work of Stock and Thelen (1958) also suggests that personality factors in the group members may promote or inhibit effective group interaction. Working from Bion's framework (1959), they define people's orientations to groups in terms of tendencies toward fight, flight, dependency, and counter-dependency. People with flight or dependency tendencies avoid conflict by denying it or surrendering. They supposedly are less likely than the other two types to engage in the expression and adjustment of different points of view necessary for creative problem solving. J. C. Glidewell's study (reported in Stock and Thelen, 1958) compared the problem-solving effectiveness of groups which he characterized as having one of these four orientations. The fight groups which were able to focus their conflicts on the task produced the most effective solutions. The use of the Reactions to Group Situations Test, an incomplete-sentence, projective device developed by Stock and Thelen, should permit the identification of group members with productive or inhibiting personalities and provide further insight into the emotional factors in group problem solving.

b. Diversity of viewpoints. The superiority of mixed-sex groups even to all-male groups in certain instances may be attributable, in part, to the divergent approaches and orientations of men and women to the assigned problem (Hoffman and Maier, 1961a; Hoffman et al., 1962). The seemingly greater sensitivity of women to feelings and interpersonal relations combined with men's ability to find the facts provides a potent combination for many complex human relations problems.

Other evidence of the value of diverse approaches to problems was supplied by Hoffman (1959) and Hoffman and Maier (1961a) with heterogeneous personalities, Ghiselli and Lodahl (1958) with heterogeneous decision-making approaches, Ziller and Exline (1958) with heterogeneous age groupings, and Triandis et al. (1962) with heterogeneous religious and political attitudes. In all of these studies the more heterogeneous groups were more effective and/or more creative. The usual explanation for these successes is that in heterogeneous groups there are more different kinds of ideas or different possible directions available for approaching the problem (Hoffman, 1961).

It is not clear, however, how one determines whether a group will be sufficiently heterogeneous or not heterogeneous enough to be effective. On what dimensions should group members be heterogeneous? Shaw (1960) showed that group problem-solving performance was either uncorrelated or negatively correlated with the variance (heterogeneity) of member characteristics on such scales, taken singly, as Acceptance of

Authority and Individual Prominence. Heterogeneity clearly has to be relevant to the problem requirements, yet we rarely know what these requirements are. The successes of the heterogeneous groups cited earlier probably stem from the wide variety of characteristics on which the members differed.

The two exceptions to the positive relationship between heterogeneity and effectiveness occur in the all-female groups of the Ziller and Exline study (1958) and in the comparison of homogeneous and heterogeneous religious groups in Holland done by Fiedler *et al.* (1961). Both results are probably due to the lack of appropriate motivation in the group. The poor performance of the all-female groups is consistent with the explanation, given above, that women tend to lack interest in problem-solving tasks. As for the other study, the religious groups (Calvinist and Catholic) have a long history of antagonism in Holland which, Fielder *et al.* implied, restricted communication among the group members. It would seem that a diversity of viewpoints must be accompanied by a tolerance for differences of opinion if a group is to exploit its potential creativity.

B. GROUP PROCESS

1. Idea Generation

It is apparent, however, that even when diversity exists in a group, the varied viewpoints are not always heard. The many barriers listed above can operate successfully to restrict the benefits of diversity. As Osborn (1953) pointed out, and as Hoffman and Maier (1964) have confirmed quantitatively, one of the most effective barriers is the tendency for group members to evaluate suggested solutions as they appear, instead of waiting until all suggestions are in and then making their choice. This tendency to evaluate suggestions one at a time may promote a mediocre solution or it may kill off a good one early. Frequently groups will evaluate the pros and cons of one solution for a considerable length of time before discarding it. During this period the suggestions made prior to the discarded solution have been forgotten and may never be recovered (Hoffman and Maier, 1964).

As a cure for these difficulties Osborn (1953) offered the brainstorming technique, in which an evaluation-free period of idea production permits the exploration of even the most harebrained solutions. By delaying the evaluation and by systematically recording all solutions offered, brainstorming prevents the amassing of support for a single solution, as well as the slaughter and forgetting of solutions which occurs in ordinary discussion (Hoffman and Maier, 1964).

Experimental evaluations of the usefulness of brainstorming have produced generally favorable results when individuals with training in the technique are compared to those without training (Parnes and Meadow, 1959). But despite the introduction of brainstorming as a group problem-solving technique (Osborn, 1953), we have not located any experimental tests of its superiority to free discussion. The technique is, in any case, difficult to evaluate. The difficulty can be illustrated in a study by Weisskopf-Joelson and Eliseo (1961) in which brainstorming groups were compared with groups instructed to be "critical" in developing possible brand names for three different products: a cigar, a deodorant, and an automobile. (Note that this was an advertising-like task for which Mr. Osborn originally developed the brainstorming technique.) The brainstorming groups produced significantly more possible names than did the critical groups, as expected. Next, 150 students at the same university rated each of the 902 different brand names suggested on a five-point scale of attractiveness, with a high degree of agreement. The superior production of the brainstorming groups apparently resulted primarily from a higher number of poor quality suggestions. The same number of high quality suggestions was produced by groups with both types of instructions. We cannot question the validity of the student ratings for judging the effectiveness of these groups since the groups' task was to produce brand names which would "attract the Purdue male." The results do not, however, refute the creative advantages of the brainstorming technique. As the authors indicate, "such conventional names as 'Sportsmen,' 'Esquire,' or 'Century' were among the three most highly rated names [which] suggests that the imaginative ideas of the brainstormers were wasted on the conservative taste of the judges."

A period of idea-production free from evaluation can lead to a solution of high quality—if the members can refrain from being self-critical and can avoid staying in particular directions. The studies by Taylor *et al.* (1958) and by Dunnette *et al.* (1963) suggest that brainstorming groups tend to follow a particular train of thought and fail to use all the members' ideas.

2. Solution Evaluation and Selection

Brainstorming is not, however, a panacea. While the brainstorming technique may produce a large number of possible solutions to a problem, the increase in number does not guarantee that the group will adopt the best or most creative solution. The process of evaluating and selecting from the enlarged pool of ideas creates new opportunities for the old biases and individual personality and motivational factors to operate again.

In a recent study in our own laboratory, Springborn (1963) found a significant *negative* correlation between the number of alternative solutions suggested by a group and the quality of its final solution. The great number of proposed solutions may increase the ambiguity of the group situation and thus heighten the extent to which sources of error can operate. Little attention has been given and much invention is needed to develop effective techniques for evaluating solution suggestions.

Maier's Screening Principles (1960) appear to offer a highly sophisticated means for forcing a group to attend more carefully to the facts given in a problem. The four principles he describes are designed to eliminate solution possibilities which cannot be supported by the facts in the case, or whose support cannot be agreed upon by the group members. Solutions selected in accord with the Maier principles are those which have the largest number of facts from different sources of information. While it is analytically meaningful, however, this technique has not been subjected to experimental test. The difficulties in validating this procedure are similar to the problems of evaluating the brainstorming technique. In any case, if ideas are to be produced in a way which avoids the motivational and social problems common to group problem solving, the evaluation and selection of final solutions must be governed by the realities of the problem situation.

3. Identifying the Problem

Both of these phases (idea-generation and solution-evaluation) will be assisted immeasurably if the group first identifies the problem requirements. More often than not, a group will begin to offer solutions to the immediately perceived problem rather than exploring the facts to define the real problem. As Maier pointed out early with respect to individual problem solving, the direction which the person initially takes in solving the problem may prevent him from finding the correct answer (Maier, 1930). In the same way a group may fall into a rut and produce an inferior solution by agreeing early on an inadequate definition of the problem requirements.

Maier and Solem (1962) have shown that when a group first explores the circumstances surrounding a problem before attempting to solve it, they are more likely to produce a creative solution to the problem. They used a three-step procedure. After the problem was presented by the foreman, there was: (1) a brief airing of everyone's view; (2) a listing of the important factors in the problem by the group; and (3) a use of the list as the basis for the final solution. The groups which followed this procedure produced significantly more creative solutions than groups which used the usual free discussion, presumably because the former groups were more likely to consider all facets of the problem before they offered possible solutions.

4. Exploiting Conflict

Hoffman (1961) suggested that ideational conflicts can be conducive to creative problem solving. Faced with differing alternatives, none of which is acceptable to the entire group, the members may be encouraged either to search for new solutions or to integrate the alternative suggestions into a more complete and more effective single possibility. Experimental support for this idea was provided (Hoffman et al., 1962) by encouraging low-status members, via the wording of their role instructions, to oppose the suggestion of a high-status member to change their work method. The resulting conflict produced more creative solutions than did the comparable, less conflictful situations.

One has to be sure, however, that the conflict is based principally on different facts or on different ways of interpreting the facts, rather than on the likes and dislikes the group members have for each other. Guetzkow and Gyr (1954) made this distinction in their comparisons of effective and ineffective management conferences. Successful conferences were marked by substantive conflict, while unsuccessful ones suffered from emotional difficulties.

It is extremely difficult, too, to dissociate your feelings about another persons from your feelings about his ideas (Hoffman, 1961). When someone disagrees with you, you dislike him, yet from disagreement can come creativity.

Furthermore, when differences are resolved creatively, positive feelings are generated. Maier and Hoffman (1964) found that group members were more satisfied with their group's solutions to a role-playing problem where creative solutions emerged than where their initially preferred, but objectively less adequate, solution was adopted. The successful resolution of conflict appears to be a satisfying experience.

C. LEADERSHIP IN GROUP PROBLEM SOLVING

1. The Function of a Leader

Is a leader necessary for the effective functioning of a problem-solving group? In a now classical article, Benne and Sheats (1948) suggested that the important consideration was whether the functions they listed—energizing, recording, harmonizing, etc.—were performed, not who performed them. This proposition has never been tested empirically, so the question remains unanswered.

As pointed out earlier, those who are usually called the group leaders, i.e., those who emerge to direct the operations of the group, do not necessarily help the group reach effective solutions. Furthermore, formally appointed leaders (managers, deans, etc.) tend to dominate and limit the range of discussion. Maier and Solem (1952) showed, however, that a formally appointed leader can enhance a group's effectiveness by ensuring the hearing of a minority, but correct, viewpoint. Without the leader the incorrect viewpoint favored by the majority usually prevailed. Thus, when the leader acts to ensure the free expression of ideas, he enhances the group's effectiveness. If, on the other hand, he tends to dominate the discussion (Section II, C, 2), he hinders the group's operation. He is especially likely to dominate if he holds a position of authority in the group, as was pointed out earlier (Torrance, 1955).

2. Skills and Techniques

Struck by the effectiveness of the "democratic" leader in the studies of children's groups (Lewin et al., 1939; Lippitt, 1940), and seeing the relevance of the non-directive method developed by Rogers (1942, 1951), Maier trained foremen in permissive conference leadership (Maier, 1952, Chapter 2). He evaluated the training by having the trainees and an untrained control group lead role-playing problem-solving conferences. While the control groups produced no creative solutions to the problem, almost 40% of the trained leaders' groups did so (Maier, 1953). Since then Maier has had a continuing interest in developing more refined leadership skills based on empirical research for formal leaders, usually in hierarchial settings (Maier, 1963). While the principles and techniques enunciated thus far require considerable refinement and further development, they seem promising as ways of overcoming some of the barriers to effective problem solving noted earlier and stimulating group members to more creative thought.

These principles and techniques provided a loose framework for a series of experiments concerned with the quality of problem solving by groups. For example, the principle that groups should be *problem-minded* rather than *solution-minded* (Maier, 1958) was tested by Maier and Solem (1962) by having experimental groups review the facts of the problem under the leader's direction, before they tried to solve the problem. In contrast to control groups, these experimental groups produced more than three and a half times as many inventive solutions.

Maier also demonstrated that the quality of a group's decision could be improved when the leader asked stimulating questions causing the group to question its current approach or to consider other aspects of the problem (Maier, 1950). Asking stimulating questions must still be done within the context of a permissive atmosphere which encourages free expression of divergent viewpoints. But this procedure faces difficulties. What the leader considers to be a "stimulating question" may be rejected by the group as irrelevant. When his questions point to facts which merely support his position, they may serve as a barrier to effectiveness. Nevertheless, to the extent that his questions call the group's attention to previously neglected facts, their potential for improving quality is great.

The "developmental discussion" technique (Maier, 1952, p. 53 ff.) was introduced both to separate the influence of feelings and facts on decisions and to ensure the systematic attention of the group to the several facets of the problem. The technique in its most advanced form consists first of analyzing the total problem into sub-parts, then discussing and solving these sub-problems in turn before reaching a final total decision. In this way the group members can avoid the difficulties in communication which arise when two people disagree with each other because they are talking about different facets of a complex problem, not because they truly disagree.

The usefulness of the developmental discussion technique in promoting good solutions to a complex problem has been demonstrated experimentally (Maier and Maier, 1957), especially when the leaders have been trained in its use (Maier and Hoffman, 1960a). Maier succeeded in reducing the proportion of poor quality decisions (in comparison to that produced by permissively led groups) merely by asking group leaders to follow written descriptions of the developmental technique. The majority of decisions were still, however, of poor quality. Maier and Hoffman then trained leaders in the use of the technique and succeeded in producing good decisions in more than 60% of their groups.

The screening principles, noted earlier (Maier, 1960), are the most recent contribution to the discussion leader's skill repertoire. These techniques were designed to depersonalize the evaluation process and focus the group more closely on the facts of a complex, ambiguous, and uncertain problem situation.

Besides these techniques for improving the use of information by a group, leaders in positions of authority have been advised to have their groups come up with two different solutions to a problem from which a choice might be made. An experimental test of this technique on groups of students and nursing administrators showed the second solution to be an inventive one three and a half to five times as often as their first solutions (Maier and Hoffman, 1960b). This process of resolving a problem is thought to have two advantages: (1) it restores the group to a state of *problem-mindedness* (cf. Maier and Solem, 1962) and starts them thinking again about the problem requirements and the character of the information available; and (2) it permits the leader, the authority figure, to pay more attention to the members' views, since his need to arrive at a solution was met by the first solution.

3. Attitudes

The leader's attitude toward the group and toward his proper function as leader also has come under investigation. It is clear that his group's problem solving effectiveness is greater: (1) the more the leader sees his role as stimulating the most creative thoughts in the group and encouraging the members to air their ideas and make their own decisions, and (2) the less he feels it necessary to be the final arbiter of the decision. In one role-playing problem, Solem (1958) instructed half the foremen to decide on the best solution to the problem before they started the meeting. The other half were told merely to present the problem to the group members and to accept their solution. The groups whose foremen had not thought of a solution before the meeting were the more creative. Solem suggests that the group leaders who came into the meeting with a formed opinion were even less open to other ideas than were the "bosses" in the control groups and, therefore, less effective in utilizing the creative potential in the group.

The leader is often not completely free to accept whatever the group's decision is. He is often limited, or believes himself to be limited, by the desires and orders of people above him in the organization to which he and the other group members belong. Epstein (1955) reported an ingenious attempt to introduce the leader's organizational superiors into the group problem-solving discussion by means of instructions. As part of the leader's role instructions, he was told that his boss had asked him to solve his problem and would either (1) support whatever he came up with, or (2) support actions only within a limited realm. The clarity of each of these admonitions was also varied. The major differences in quality of solutions favored the clearly free leaders over the clearly restricted leaders, with the leaders given vague instructions of both types producing solutions of intermediate quality. The clearly restricted foremen tended to be more concerned with management's goals and less with the workers', to restrict discussion, and generally to reject the various ideas brought forth by the group members. Thus, the process of effective group problem solving may be hampered by the leader's attitude, which is created by his perception of higher management's expectations.

D. ACCEPTANCE

These leadership principles and techniques have been developed to improve the quality of problem solving in groups. They have grown, however, within the context of a free and permissive discussion atmosphere, which reflects Lewin's concern (cf. K. Lewin, 1947) about the members' identification with the group's decision. In keeping with this concern, Maier early distinguished between two dimensions of a solution to a problem—*quality* and *acceptance* (Maier, 1952). Quality "refers to the objective features of a decision" while acceptance "refers to the degree to which the group that must execute the decision accepts it" (Maier, 1963, p. 253). The theoretical usefulness of this distinction lies in the fact that the quality of a decision reflects the group's ability to produce and utilize information effectively, while acceptance reflects the members' feelings about the solution and about the way it was reached. This is a helpful distinction since effective real-life decisions require both high quality and high acceptance, else they fail from lack of attention to the facts or from the members' unwillingness to carry out the decision. Yet the two concepts merge, as Hoffman (1961) demonstrated. Our ideas at times are so affect-laden that we feel proud when they are accepted by the group. If they are rejected, on the other hand, we may feel rejected also and may lower our regard for the group.

A number of experiments have investigated the relationship between acceptance and quality. Acceptance has been measured most often by the subjects' responses to a five- or six-point, Likert-type question, "How satisfied are you with the group's solution?" While such an item can have but limited reliability, it has been employed successfully and certainly has face validity as a measure of acceptance.

In the studies noted earlier, which compared the problem-solving effectiveness of groups composed of heterogeneous and homogeneous personalities (Hoffman, 1959; Hoffman and Maier, 1961a), the heterogeneous groups produced higher quality solutions than did the homogeneous groups, but the general level of members' satisfaction with the solutions hardly differed. Correlations between the quality score and members' satisfaction ratings further confirmed the lack of relationship between the quality and acceptance of solutions in these groups. Furthermore, members' ratings of satisfaction with the solutions were highly correlated with the subjects' own ratings of satisfaction with their influence over the decision, although they were uncorrelated with the other members' ratings of each member's influence over the decision. Thus, acceptance seems to be related to the freedom of the problem solving process, rather than to the objective quality of the decision itself. A group member will feel satisfied to the extent that he feels he has influenced the group decision appropriately.

The importance of the leader in promoting this feeling of shared influence and, thereby, encouraging member acceptance is illustrated in studies by Maier and Hoffman (1960b; 1962). When members were asked whether their leader used group decision making or whether he made the final decision himself, members of groups characterized by unilateral decisions were least satisfied with the decision, while members of group-decision groups were most satisfied. The proportion of satisfied members in a group had no relationship to the character of the solution produced.

More recent research, using a newly developed method for quantifying the problem-solving interaction in a group, provides additional evidence. These findings suggest that the most important factor in determining the members' acceptance of a solution to a complex problem is the amount of influence they actually exerted over the final decision (Hoffman *et al.*, 1963). The rated satisfaction with the solution was correlated .59 with the amount of verbal support given by the member for the solution finally adopted by the group. Considering the modest reliability of the satisfaction measure, this value probably accounts for a considerable portion of the common variance. Acceptance, again, was completely unrelated to the quality of the solution.

If this relationship is confirmed in future studies, it suggests that acceptance can be increased if everybody is encouraged to express his feelings about the solution to be adopted. If the feelings are favorable, their expression will presumably promote the member's commitment to the decision. If unfavorable feelings are exhibited, however, they may be emotionally based and their mere expression may relieve the resistance to the proposed solution. Voicing negative feelings which are based upon information can promote a search for new and perhaps superior alternatives. High acceptance and high quality, while not necessarily positively correlated, may be associated when the problem-solving process considers both aspects of the solution. The leader's attitudes and skills will be of paramount importance in determining their joint consideration.

IV. The Present State of Group Problem-Solving Research

Although the history of the experimental study of group problem solving is a long one, it is replete with discontinuities. A survey of the literature reveals numerous difficulties which have been relatively ignored over the years, but which have retarded the advance of knowledge in this area. An enumeration of these difficulties may help in defining problems for needed research and analysis.

A. THE NATURE OF PROBLEM-SOLVING TASKS

1. Problem Content

One of the foremost difficulties concerns the term "problem solving" itself. Problem solving has been used with reference to tasks as varied as judging the number of dots briefly displayed on a large card, to providing answers to arithmetic reasoning problems, to solving the complex problems faced by the managements of large business organizations. On a priori grounds one might expect that the factors producing effective performance should vary greatly in these different types of problems, just as qualitatively different abilities are required for individuals to solve simple addition problems as against problems in topology, even though both are "mathematics" problems. This rather obvious point has generally been neglected. It calls for the systematic development of a taxonomy of problems.[4]

The few experiments in which the same groups solved more than one problem have almost invariably produced discrepant or ambiguous conclusions. For example, in our study of the effects on group problem solving of homogeneous personality composition (Hoffman and Maier, 1961a) the same groups were compared on four different problems. The quality of the solutions produced varied considerably from problem to problem, as did the magnitudes of the differences between the two types of groups. At our present state of knowledge it is impossible to identify the causes of these variations in results. In some general sense all of the problems are somewhat similar; they are all complex and deal with relations among people in quasi-real situations. There were many specific differences, however, in the character of the problems and in the circumstances surrounding their administration and scoring which could account, singly or in combination, for the differences obtained. As one question we might ask, what different requirements are placed on groups when all the members share all the information, as compared to problems where only part of the information is generally known?

The data obtained from this study show clearly that the assigned problems elicited substantially different amounts of problem-related activity among the group members. In addition, we found consistent differences among groups as well as significant and varied interactions between groups and problems (Hoffman and Smith, 1960). In other words, each problem produced a particular reaction from the group members, which was modified somewhat by the idiosyncratic nature of the groups to which they belonged. Our inability to identify these behavioral differences in relation to the quality of the group products limits our ability to theorize about the determinants of effective problem solving.

2. Grading the Quality of Solutions

Related to these difficulties is the lack of suitable methods for assessing the quality of solutions to complex problems. How can we compare,

---

[4] This need for a taxonomy of problems, as I have indicated elsewhere (Hoffman, 1961), is also pressing in the study of individual problem solving, but Guilford's work (J. P. Guilford, 1956) in this area has provided at least one approach to its satisfaction.

using an appropriate metric, the adequacy of the solutions to altogether different problems? Thus, on one problem used in the previously cited study of homogeneous and heterogeneous groups the range of possible scores went from 0 to 100, while on another problem there were only three qualitatively different categories of solutions. If we try to answer the question, "Did the homogeneous groups do relatively worse on the one problem than the other?" we are stymied by a lack of comparable scales for measuring solution quality. If, as was done, we compare the percentage of groups of each type whose solutions were above the median score on each problem, we ignore possible differences in the difficulty level of the two problems. To the extent that the problems are either very easy or very difficult, the effects of the experimental variation may be obscured. I believe we will not develop adequate, comparative scoring systems until a method for identifying task requirements is obtained. Without a good theoretical reason for making particular distinctions among solutions to problems, scores assigned to represent differences in quality must remain arbitrary and limit the generality of most experimental results (Coombs, 1963).

3. Analysis of the Process

Roby and Lanzetta (1958) made the most systematic attempt to provide an adequate framework for a taxonomy of group problems. Although their paradigm was designed for use with all group tasks and populations of groups, they illustrated its application to the "common symbol" task, which is used so frequently in studies of communication nets. They first recommend that the properties of the problem and the distribution of these properties among the group members be described in elemental terms. There then can be an analysis of the ways in which these properties are transformed during the group interaction into some output (the group's solution). Recognizing that we are unable to describe the elements of the task and the group members so specifically at present, Roby and Lanzetta suggested a more macroscopic analysis of the task and the group process in terms of the "critical demands" required to produce an effective solution.

Unfortunately, little effort has been directed toward applying this paradigm to group problem solving. Most systems for observing group interaction (e.g., Bales, 1950; Heyns and Lippitt, 1954) ignore the relevance of what is being said for the solutions to the problem. These systems, while useful for determining the general socioemotional versus task atmosphere of the group interaction, yield little information about the quality of such interaction for effective group problem solving. The assumption, frequently made, "that the process would surely be self-defeating and self-limiting if there were more . . . negative reactions than positive" (Bales, 1953) must be open to question in view of our findings of the creative possibilities in exploiting conflict among group members (Hoffman *et al.*, 1962).

A few attempts have been made to relate the activities of the group to the solutions produced. Oakes *et al.* (1960), in reinforcing suggestions for particular solutions, recorded the number of suggestions made for each of three solutions. They found an increase in the number of favorable comments made for the arbitrarily reinforced solution, which led to its selection by the group in most cases. Hoffman and Maier (1964) developed a more elaborate system, based on some ideas proposed by Hoffman (1961), for recording the types of comments made about the suggested solutions offered in a group. They distinguish among several different types of remarks, e.g., descriptions of the solution, justifications for it, criticisms of it, both favorable and unfavorable. With this technique they were able to follow the course of a potential solution through the group discussion and found consistent relationships with the group's selection of a final solution and with the members' satisfaction with the decision.

Such recording systems, however, represent only a beginning in the analysis of the cognitive aspects of group problem solving. There also should be a systematic examination of the way the information available to the group is offered and used. As Roby and Lanzetta (1958) implied, giving more attention to the cognitive requirements of group problem-solving tasks may greatly increase our understanding of the group problem-solving process.

B. POPULATION CHARACTERISTICS

1. Group Composition

A collateral difficulty arises from the fact that the particular characteristics of the groups studied may severely restrict the generality of the results obtained. While most experiments have been conducted on that favorite subject, the college freshman or sophomore, the intellectual quality of students at different universities varies considerably and may often be the principal source of discrepant conclusions in similar experiments (e.g., Shaw, 1961). Calvin *et al.* (1957) compared groups of "low" I.Q. and "high" I.Q. students in "authoritarian" and "permissive" experimental atmospheres on Twenty Questions problems at Michigan State and Hollins Colleges. In two sessions run in Michigan, the permissive, bright subjects used fewer questions per problem and solved a higher percentage of problems, while at Hollins the results showed a slightly reversed trend.

Similarly, the oft-noted description of subjects as "members of a class in introductory psychology" neglects the sex composition of the group (a group characteristic whose importance has already been discussed; in Section III, A, 2, a), as well as other factors such as age, personality, socio-economic status, etc. The unknown characteristics of groups formed from other populations, of course, also should limit statements regarding the generality of the experimental results, but rarely do. The approaches taken by Cattell and Wispe (1948) and by Hemphill and Westie (1950) in identifying the dimensions of groups on a variety of tasks could be followed profitably in more specific studies of group problem solving. If such an effort sampled groups from many diverse populations, it would provide a framework of norms into which specific experiments could be classified. Until the characteristics of different populations relevant to problem solving have been systematically studied and identified, the comparative interpretation of different experimental results will remain obscure.

2. Real and Ad Hoc Groups

In the same vein, the criticism by Lorge *et al.* (1958) that the groups studied in comparison with individuals are almost invariably ad hoc, temporary groups without history, tradition or norms, holds almost equally true for studies of group problem solving per se. Unfortunately, these authors provided no directions or clues as to the differences one might expect to find between these two types of groups. Their criticism, while potentially valid, thus serves principally as a reminder that generalizations from experiments on ad hoc groups may have only questionable validity for "traditioned" groups.[5]

Probably the major difficulties which experimenters avoid by remaining with the study of laboratory groups are: (1) the lack of comparability of problems, and (2) the members' involvement in the problems, which one meets in studying the problem solving of real groups. As mentioned earlier, even in comparing groups of men and women on laboratory problems, we find that the females' reduced motivation often results in poor performance. To try to find a problem which might be of considerable interest to a sample of executives from different companies is clearly an almost insuperable obstacle.

The Michigan Conference Research Project (Guetzkow and Gyr, 1954) encountered a major difficulty in trying to study the problem solving of executive groups in business and government. Their decision to use

---

[5] Before I am accused of taking a "holier-than-thou" position about these difficulties, I should like to admit my own contributions to these problems and recognize the considerable difficulty in overcoming them.

the number of agenda items completed as an index of the effectiveness of the meetings observed seems a far cry from even the arbitrary systems discussed earlier. In a number of cases an agenda item was "completed" by tabling it for the future, presumably when the conflict it generated in the group could not be resolved. An ability to classify different tasks would probably have helped immeasurably.

Another difficulty in studying real groups stems from the experimenter's usual lack of knowledge of the past history of the group. In a typical existing group the members have developed a shorthand language through their shared experiences which permits them to refer to complex issues in terms foreign to the outside observer. To study a group solving its real problems, the experimenter must familiarize himself with the group's history and environment. This often requires more time than we care to invest. Unless we are willing to spend the time required to study groups outside University and college settings, we must settle for a very restricted theory of group problem solving.

These difficulties—the absence of taxonomies for types of problems and for populations of groups and of systems for describing the cognitive aspects of the problem-solving process—have provided major barriers to any attempt to integrate the literature on group problem solving. With rare exceptions there has been a notable lack of a continued, consistent, and additive effort in this area. The typical experimenter does one or two studies on a single facet of the topic, with a problem (described in too general and vague terms) which nobody else has ever used. He produces suggestive, but inconclusive results, and is never heard from again. This practice has left the literature on group problem solving a large conglomeration of unrelated experiments, with only the faintest suggestion of commonality. Also, most of the experiments to date have concentrated on identifying the barriers to effective problem solving, rather than on discovering means to stimulate group creativity. Admittedly, the barriers discovered so far occur so ubiquitously that, unless they are overcome, the chance of promoting creative problem solving is rather small. Nevertheless, effort directed to inventing and testing new ways of encouraging creative group problem solving should advance our understanding of the problem-solving process and, when successful, would have practical value for society as well.

REFERENCES

Allport, F. H. (1924). *Social psychology.* Boston, Mass.: Houghton Mifflin.
Asch, S. E. (1956). Studies of independence and conformity. A minority of one against a unanimous majority. *Psychol. Monogr.* **70**, No. 9 (Whole No. 416).
Bales, R. F. (1950). *Interaction process analysis: a method for the study of small groups.* Reading, Mass.: Addison-Wesley.

Bales, R. F. (1953). The equilibrium problem in small groups. In T. Parsons, R. F. Bales, and E. A. Shils (Eds.), *Working papers in the theory of action.* New York: Free Press, 111–161.

Bales, R. F., and Borgatta, E. F. (1955). Size of group as a factor in the interaction profile. In A. P. Hare, E. F. Borgatta, and R. F. Bales (Eds.), *Small groups: studies in social interaction.* New York: Knopf, pp. 396–413.

Bales, R. F., and Strodtbeck, F. L. (1951). Phases in group problem solving. *J. abnorm. soc. Psychol.* 46, 485–495.

Bass, B. M. (1963). Amount of participation, coalescence and profitability of decision-making discussions. *J. abnorm. soc. Psychol.* 67, 92–94.

Benne, K. D., and Sheats, P. (1948). Functional roles of group members. *J. soc. Issues* 4, 41–49.

Bennett, Edith (1955). Discussion, decision, commitment and consensus in "group decision." *Hum. Relat.* 8, 251–274.

Berkowitz, L. (1956). Personality and group position. *Sociometry* 19, 210–222.

Berkowitz, L. (1957). Liking for the group and the perceived merit of the group's behavior. *J. abnorm. soc. Psychol.* 54, 353–357.

Berkowitz, L., and Levy, B. I. (1956). Pride in group performance and group task motivation. *J. abnorm. soc. Psychol.* 53, 300–306.

Bion, W. R. (1959). *Experiences in Groups.* New York: Basic Books.

Blake, R. R., and Mouton, Jane S. (1961). *Group dynamics—key to decision making.* Houston, Texas: Gulf Publ. Co.

Calvin, A. D., Hoffmann, F. K., and Harden, E. L. (1957). The effect of intelligence and social atmosphere on group problem solving behavior. *J. soc. Psychol.* 45, 61–74.

Carey, Gloria J. (1958). Sex differences in problem-solving performance as a function of attitude differences. *J. abnorm. soc. Psychol.* 56, 256–260.

Cattell, R. B., and Wispé, L. G. (1948). The dimensions of syntality in small groups. *J. soc. Psychol.* 28, 57–78.

Cohen, A. M., and Bennis, W. G. (1962). Predicting organization in changed communication networks. *J. Psychol.* 54, 391–416.

Comrey, A. L., and Staats, Carolyn K. (1955). Group performance in a cognitive task. *J. appl. Psychol.* 39, 354–356.

Coombs, C. H. (1963). *A theory of data.* New York: Wiley.

Crockett, W. H. (1955). Emergent leadership in small, decision-making groups. *J. abnorm. soc. Psychol.* 51, 378–383.

Crutchfield, R. S. (1959). Personal and situational factors in conformity to group pressure. *Acta Psychol.* 15, 386–388.

Deutsch, M. and Krauss, R. M. (1960). The effect of threat upon interpersonal bargaining. *J. abnorm. soc. Psychol.* 61, 181–189.

Dunnette, M. D., Campbell, J., and Jaastad, Kay (1963). The effect of group participation on brainstorming effectiveness for two industrial samples. *J. appl. Psychol.* 47, 30–37.

Ekman, G. (1955). The four effects of cooperation. *J. soc. Psychol.* 41, 149–162.

Epstein, S. (1955). An experimental study of some of the effect of variations in the clarity and extent of a supervisor's area of freedom upon his supervisory behavior. Unpublished Ph.D. thesis, Univer. of Michigan.

Feather, N. T. (1961). The relationship of persistence at a task to expectation of success and achievement related motives. *J. abnorm. soc. Psychol.* 63, 552–561.

Fiedler, F. E., and Meuwese, W. A. T. (1963). Leader's contribution to task performance in cohesive and uncohesive groups. *J. abnorm. soc. Psychol.* **67,** 83–87.

Fiedler, F. E., Meuwese, W., and Oonk, Sophie (1961). An exploratory study of group creativity in laboratory tasks. *Acta Psychol.* **18,** 100–119.

Fouriezos, N. T., Hutt, M. L., and Guetzkow, H. (1950). Measurement of self-oriented needs in discussion groups. *J. abnorm. soc. Psychol.* **45,** 682–690.

Ghiselli, E. E., and Lodahl, T. M. (1958). Patterns of managerial traits and group effectiveness. *J. abnorm. soc. Psychol.* **57,** 61–66.

Guetzkow, H., and Gyr, J. (1954). An analysis of conflict in decision-making groups. *Hum. Relat.* **7,** 367–382.

Guilford, Joan S. (1952). Temperament traits of executives and supervisors measured by the Guilford personality inventories. *J. appl. Psychol.* **36,** 228–233.

Guilford, J. P. (1956). The structure of intellect. *Psychol. Bull.* **53,** 267–293.

Hemphill, J. K., and Westie, C. M. (1950). The measurement of group dimensions. *J. Psychol.* **29,** 325–342.

Hemphill, J. K., Pepinsky, Pauline N., Kaufman, A. E., and Lipetz, M. E. (1957). Effects of task motivation and expectancy of accomplishment upon attempts to lead. *Psychol. Monogr.* **71,** No. 22 (Whole no. 451).

Heyns, R. W., and Lippitt, R. (1954). Systematic observational techniques. In G. Lindzey (Ed.), *Handbook of social psychology.* Reading, Mass.: Addison-Wesley, pp. 370–404.

Hochbaum, G. M. (1954). The relation between group members' selfconfidence and their reactions to group pressures to uniformity. *Amer. sociol. Rev.* **79,** 678–687.

Hoffman, L. R. (1959). Homogeneity of member personality and its effect on group problem-solving *J. abnorm. soc. Psychol.* **58,** 27–32.

Hoffman, L. R. (1961). Conditions for creative problem solving. *J. Psychol.* **52,** 429–444.

Hoffman, L. R., and Maier, N. R. F. (1961a). Quality and acceptance of problem solutions by members of homogeneous and heterogeneous groups. *J. abnorm. soc. Psychol.* **62,** 401–407.

Hoffman, L. R., and Maier, N. R. F. (1961b). Sex differences, sex composition, and group problem solving. *J. abnorm. soc. Psychol.* **63,** 453–456.

Hoffman, L. R., and Maier, N. R. F. (1964). Valence in the adoption of solutions by problem-solving groups: concept, method and results. *J. abnorm. soc. Psychol.* **69,** 264–271.

Hoffman, L. R., and Smith, C. G. (1960). Some factors affecting the behaviors of members of problem-solving groups. *Sociometry* **23,** 273–291.

Hoffman, L. R., Harburg, E., and Maier, N. R. F. (1962). Differences and disagreement as factors in creative group problem solving. *J. abnorm. soc. Psychol.* **64,** 206–214.

Hoffman, L. R., Burke, R. J., and Maier, N. R. F. (1963). Does training with differential reinforcement on similar problems help in solving a new problem? *Psychol. Rep.* **13,** 147–154.

Jackson, J. M., and Saltzstein, H. D. (1958). The effect of person-group relationships on conformity processes. *J. abnorm. soc. Psychol.* **57,** 17–24.

Kaess, W. A., Witryol, S. L., and Nolan, R. E. (1961). Reliability, sex differences, and validity in the leaderless group discussion technique. *J. appl. Psychol.* **45,** 345–350.

Kidd, J. S., and Campbell, D. T. (1955). Conformity to groups as a function of group success. *J. abnorm. soc. Psychol.* **51**, 390–393.

Kirscht, J. P., Lodahl, T. M., and Haire, M. (1959). Some factors in the selection of leaders by members of small groups. *J. abnorm. soc. Psychol.* **58**, 406–408.

Lewin, K. (1947). Group decision and social change. In T. M. Newcomb and E. L. Hartley (Eds.) *Readings in Social Psychology.* New York: Holt, pp. 330–344.

Lewin, K., Lippitt, R., and White, R. K. (1939). Patterns of aggressive behavior in experimentally created "social climates." *J. soc. Psychol.* **10**, 271–299.

Lippitt, R. (1940). An experimental study of the effect of democratic and authoritarian group atmosphere. *Univer. Iowa Stud. Child Welf.* No. 16.

Loomis, J. L. (1959). Communication, the development of trust, and cooperative behavior. *Hum. Relat.* **12**, 305–315.

Lorge, I., and Solomon, H. (1955). Two models of group behavior in the solution of Eureka-type problems. *Psychometrika* **20**, 139–148.

Lorge, I., Fox, D., Davitz, J., and Brenner, M. (1958). A survey of studies contrasting the quality of group performance and individual performance, 1920–1957. *Psychol. Bull.* **55**, 337–372.

Maier, N. R. F. (1930). Reasoning in humans. I. On direction. *J. comp. Psychol.* **10**, 115–143.

Maier, N. R. F. (1950). The quality of group decisions as influenced by the discussion leader. *Hum. Relat.* **3**, 155–174.

Maier, N. R. F. (1952). *Principles of Human Relations.* New York: Wiley.

Maier, N. R. F. (1953). An experimental test of the effect of training on discussion leadership. *Hum. Relat.* **6**, 161–173.

Maier, N. R. F. (1958). The appraisal interview: objectives, methods and skills. New York: Wiley.

Maier, N. R. F. (1960). Screening solutions to upgrade quality: A new approach to problem solving under conditions of uncertainty. *J. Psychol.* **49**, 217–231.

Maier, N. R. F. (1963). *Problem-solving discussions and conferences: leadership methods and skills.* New York: McGraw-Hill.

Maier, N. R. F., and Hoffman, L. R. (1960a). Using trained "developmental" discussion leaders to improve further the quality of group decisions. *J. appl. Psychol.* **44**, 247–251.

Maier, N. R. F., and Hoffman, L. R. (1960b). Quality of first and second solutions in group problem solving. *J. appl. Psychol.* **44**, 278–283.

Maier, N. R. F., and Hoffman, L. R. (1961). Organization and creative problem solving. *J. appl. Psychol.* **45**, 277–280.

Maier, N. R. F., and Hoffman, L. R. (1962). Group decision in England and the United States. *Personnel Psychol.* **15**, 75–87.

Maier, N. R. F., and Hoffman, L. R. (1964). Financial incentives and group decision in motivating change. *J. soc. Psychol.* **64**, 369–378.

Maier, N. R. F., and Maier, R. A. (1957). An experimental test of the effects of "developmental" vs. "free" discussions on the quality of group decisions. *J. appl. Psychol.* **41**, 320–323.

Maier, N. R. F., and Solem, A. R. (1952). The contribution of a discussion leader to the quality of group thinking: the effective use of minority opinions. *Hum. Relat.* **5**, 277–288.

Maier, N. R. F., and Solem, A. R. (1962). Improving solutions by turning choice situations into problems. *Personn. Psychol.* **15**, 151–157.

Mann, R. D. (1959). The relation between personality characteristics and individual performance in small groups. Unpublished Ph.D. thesis, Univer. of Michigan.

Mann, R. D. (1961). Dimensions of individual performance in small groups under task and social-emotional conditions. *J. abnorm. soc. Psychol.* **62,** 674–682.

Mausner, B. (1954). The effect of one partner's success in a relevant task on the interaction of observer pairs. *J. abnorm. soc. Psychol.* **49,** 557–560.

Mellinger, G. D. (1956). Interpersonal trust as a factor in communication. *J. abnorm. soc. Psychol.* **52,** 304–309.

Milton, G. A. (1958). Five studies of the relation between sex-role identification and achievement in problem solving. Tech. Rep. No. 3, Departments of Industrial Administration and Psychology, Yale University [Contract Nonr609 (20)].

Oakes, W. F. (1962). Effectiveness of signal light reinforcers given various meanings on participation in group discussion. *Psychol. Rep.* **11,** 469–470.

Oakes, W. F., Droge, A. E., and August, Barbara (1960). Reinforcement effects on participation in group discussion. *Psychol. Rep.* **7,** 503–514.

Orne, M. T. (1962). On the social psychology of the psychological experiment: with particular reference to demand characteristics and their implications. *Amer. Psychologist* **17,** 776–783.

Osborn, A. F. (1953). *Applied Imagination.* New York: Scribner's.

Palmer, G. J. Jr. (1962). Task ability and effective leadership. *Psychol. Rep.* **10,** 863–866.

Parnes, S. F., and Meadow, A. (1959). Effects of "brainstorming" instructions on creative problem solving by trained and untrained subjects. *J. educ. Psychol.* **50,** 171–176.

Pennington, D. F., Jr., Haravey, F., and Bass, B. M. (1958). Some effects of decision and discussion on coalescence, change, and effectiveness. *J. appl. Psychol.* **42,** 404- 408.

Pepinsky, Pauline, Hemphill, J. K., and Shevitz, R. N. (1958). Attempts to lead, group productivity, and morale under conditions of acceptance and rejection. *J. abnorm. soc. Psychol.* **57,** 47–54.

Read, W. H. (1962). Upward communication in industrial hierarchies. *Hum. Relat.* **15,** 3–15.

Riecken, H. W. (1958). The effect of talkativeness on ability to influence group solutions of problems. *Sociometry* **21,** 309–321.

Roby, T. B., and Lanzetta, J. T. (1958). Considerations in the analysis of group tasks. *Psychol. Bull.* **55,** 88–101.

Roby, T. B., and Lanzetta, J. T. (1961). A study of an "assembly effect" in small group task performance. *J. soc. Psychol.* **53,** 53–68.

Rogers, C. R., (1942). *Counseling and psychotherapy.* Boston, Mass.: Houghton-Mifflin.

Rogers, C. R. (1951). *Client-centered therapy.* Boston, Mass.: Houghton-Mifflin.

Schachter, S. (1951). Deviation, rejection, and communication. *J. abnorm. soc. Psychol.* **46,** 190–207.

Shaw, M. E. (1959). Some effects of individually prominent behavior upon group effectiveness and member satisfaction. *J. abnorm. soc. Psychol.* **59,** 382–386.

Shaw, M. E. (1960). A note concerning homogeneity of membership and group problem solving. *J. abnorm. soc. Psychol.* **60,** 448–450.

Shaw, M. E. (1961). Some factors influencing the use of information in small groups. *Psychol. Rep.* **8,** 187–198.

Shaw, M. E. (1963). Some effects of varying amounts of information exclusively possessed by a group member upon his behavior in the group. *J. gen. Psychol.* **68**, 71–79.

Shaw, M. E., and Gilchrist, J. C. (1956). Intra-group communication and leader choice. *J. soc. Psychol.* **43**, 133–138.

Shaw, M. E., and Penrod, W. T., Jr. (1962). Does more information available to a group always improve group performance? *Sociometry* **25**, 377–390.

Shure, G. H., Rogers, M. S., Larsen, Ida M., and Tassone, J. (1962). Group planning and task effectiveness. *Sociometry* **25**, 263–282.

Slater, P. E. (1955). Role differentiation in small groups. *Amer. sociol. Rev.* **20**, 300–310.

Solem, A. R. (1958). An evaluation of two attitudinal approaches to delegation. *J. appl. Psychol.* **42**, 36–39.

Springborn, B. A. (1963). Some determinants and consequences of the locus of evaluation in small group problem solving. Unpublished Ph.D. thesis, Univer. of Michigan.

Stephan, F. F., and Mishler, E. G. (1952). The distribution of participation in small groups: an exponential approximation. *Amer. sociol. Rev.* **17**, 598–608.

Stock, Dorothy, and Thelen, H. A. (1958). *Emotional dynamics and group culture.* Washington, D.C.: National Training Laboratory.

Taylor, D. W., Berry, P. C., and Block, C. H. (1958). Does group participation when using brainstorming facilitate or inhibit creative thinking? *Admin. sci. Quart.* **3**, 23–47.

Thomas, E. J., and Fink, C. F. (1961). Models of group problem solving. *J. abnorm. soc. Psychol.* **68**, 53–63.

Thorndike, R. L. (1938). The effect of discussion upon the correctness of group decisions, when the factor of majority influence is allowed for. *J. soc. Psychol.* **9**, 343–362.

Timmons, W. M. (1942). Can the product superiority of discussors be attributed to averaging or majority influences? *J. soc. Psychol.* **15**, 23–32.

Torrance, E. P. (1955). Some consequences of power differences on decision making in permanent and temporary three-man groups. In A. P. Hare, E. F. Borgatta, and R. F. Bales (Eds.), *Small groups: studies in social interaction.* New York: Knopf, pp. 482–492.

Triandis, H. C., Mikesell, Eleanor H., and Ewen, R. B. (1962). Task set and attitudinal heterogeneity as determinants of dyadic creativity. Tech. Rep. No. 8, Univer. Illinois.

Tuckman, J., and Lorge, I. (1962). Individual ability as a determinant of group superiority. *Hum. Relat.* **15**, 45–51.

Tuddenham, R. D. (1959). Correlates of yielding to a distorted group norm. *J. Pers.* **27**, 272–284.

Weisskopf-Joelson, Edith, and Eliseo, T. (1961). An experimental study of the effectiveness of brainstorming. *J. appl. Psychol.* **45**, 45–49.

Wiest, W., Porter, L. W., and Ghiselli, E. E. (1961). Relationships between individual proficiency and team performance and efficiency. *J. appl. Psychol.* **45**, 435–440.

Ziller, R. C., and Exline, R. V. (1958). Some consequences of age heterogeneity in decision-making groups. *Sociometry*, **21**, 198–211.

The Group Problem-Solving Process

L. Richard Hoffman
GRADUATE SCHOOL OF BUSINESS
THE UNIVERSITY OF CHICAGO

Of the several problems identified in the preceding chapter that need research, our attention has focused on illuminating the problem-solving process itself. This approach contrasts with the usual attempts to infer the process by manipulating independent variables and measuring outcomes (Davis, 1973). It recognizes the importance of the timing of events in the discussion and provides specific links between the independent variables and the outcomes. Greater understanding of the problem-solving process can lead to the invention of new techniques for improving it, as well as an analysis of the efficacy of currently available techniques (Bouchard, 1969; Dalkey, 1963, 1969; Osborn, 1957). I shall concentrate in this essay on our work on the problem-solving process and briefly note some of the research on "techniques" as they seem appropriate to that discussion.

Our research on the group problem-solving process has followed directly from the theory and results of two articles briefly mentioned in this chapter (Hoffman, 1961; Hoffman & Maier, 1964). In the first article I presented a model of group problem solving; it was based on an adaptation of Lewin's (1935) concept of valence as applied to the group's attraction toward or away from alternative solutions. Valence was conceived of as both a group concept and an individual one, and operationally the group measure was the sum of the individual measures. Valence for any particular solution was presumed to change during, and as a result of, the discussion, although members with prior

Group Processes

acquaintance with the problem would have some valence for particular solutions before the discussion began. The model of the adoption process assumed: (a) that one or more solutions accumulate more valence than a minimum value (the adoption threshold); (b) that beyond that threshold the amounts of valence accumulated determined the likelihood of the group's adoption of the solution; (c) that the group adopted the solution with the highest valence; and (d) that the members' commitment to the group's decision would be a positive function of their valence for the solution.

The second article reported the first empirical test of these basic propositions. The results confirmed that the valence of a solution, as measured merely by the *frequency* of favorable and unfavorable comments made about that solution, was strongly correlated with the probability of its adoption by the group. Table 1 shows these results for two different problems. A net surplus of 40 favorable over unfavorable comments for a solution practically guaranteed its adoption in most groups. Although it is not surprising that groups adopt the solution about which they say the most favorable things, that the frequency alone of favorable minus unfavorable comments is so strongly correlated with adoption is intriguing. Furthermore, it is not the amount of discussion about a solution that is important but the degree of favorability of that discussion for the solution. Moreover, that the numerical value of the index can be so strongly associated with the probability of adoption regardless of the group in which it occurs is an even more surprising result. The data shown in Table 1 were combined from more than 40 groups each, thus showing the independence of this valence–adoption relationship and of the associated numerical values from the size of the group (varying from 3 to 5 members) and the amount of total discussion (varying from 150 codeable comments to over 300).

In the first study (Hoffman & Maier, 1964), the valence index also predicted which solution would be adopted at a rather early stage of the process. Whereas the mean valence index for adopted solutions was approximately 40, the first solution to attain an index of 15 was adopted in about two-thirds of the groups. Since most groups suggested from 10 to 12 different solutions to the problem, the ability to choose the adopted solution on the basis of this frequency measure is a very significant finding. The original model posited a minimum threshold value necessary for a solution to qualify for adoption by the group. On the basis of the results of this first study the valence value of 15 seemed to be a good candidate for such a threshold.

Finally, with regard to the members' acceptance of the decision,

TABLE 1
VALENCE INDEX AND ADOPTION OF SOLUTIONS

Valence index	Generational problem (N = 45 groups)				Choice task (N = 43 groups)		
	Percentage of solutions				Percentage of solutions		
	Adopted		Not adopted	No. of solutions	Adopted	Not adopted	No. of solutions
	Principal	Subordinate					
≥40	81.8%	6.1	12.1	33	80.0%	20.0	35
30–39	50.0%	16.7	33.3	12	58.3%	41.7	12
15–29	26.2%	35.7	38.1	42	36.4%	63.6	22
1–14	.3%	2.9	96.8	312	.0%	100.0	28
–9–0	.0%	.0	100.0	137	.0%	100.0	52
≤–10	.0%	.0	100.0	1	.0%	100.0	66
				537			215

the valence index again showed strong predictive properties. Acceptance turns out to be an individual phenomenon. Although the original model had predicted that the members' acceptance of the decision would be a positive function of the group's valence for the adopted solution, the correlation between that index and the members' satisfaction with the decision was only .26. However, when each member's individual valence for the group's solution was correlated with his or her reported satisfaction with the decision, the correlation was .59, clearly significant. Thus, acceptance of the decision was increased by each member's public support of that decision.

Thus, the basic assumptions of the valence model of the process of group problem solving were confirmed, and the principal variable, valence of solutions, was shown to be reliably measurable. The process of arriving at a solution to a problem was shown to be a dynamic one of mutual persuasion in which the members' statements for and against different solutions move the group to a decision. The decision is acceptable to the members as a function of each member's personal contribution to that adoption process.

I. Elaborating the Valence Model

Extensions of this initial study have illuminated three important aspects of the group problem-solving process that have not been and could not be studied with conventional input–outcome designs. The focus on process highlights the importance of the timing of events for the ultimate decision, for its acceptance by the members, and for the influence exercised by different members in the group. By tracing the history of the extensions of the initial valence study, we can see the faults in the traditional views of these issues and point to directions for further enlightenment.

A. Groups with Leaders

The first question we raised was how replicable and generalizable the original results were. Three experiments have addressed this problem. In one (Hoffman & Maier, 1967), groups with appointed leaders solved the same problem used in the first study under four different motivational conditions. By instructing the leaders and members independently, we offered rewards to each for achieving a solution of high quality or one with which the other party was satisfied. The four goal conditions thus defined—both leaders and members striving for quality, both leaders and members attempting to satisfy each other, leader striving for quality and members for his satisfaction, and members striving for quality and the leader for their satisfaction—showed identical relationships between group valence and the adoption of solutions. Moreover, these values were identical to those obtained in the first study, despite substantial differences in the behavior of leaders and members observed among the four conditions. In brief, the stability of the valence–adoption relationships was maintained by differing contributions of leaders and members in different conditions.

B. Additional Problems and Prior Commitment

To determine whether these relationships were somehow artifactually associated with the one problem studied so far, we conducted another study (Hoffman & O'Day, unpublished) in which four different problems—two reasoning and two value problems—were solved by the same groups. In each case the problem was solved first by the members individually and then by them in groups. We anticipated that this latter modification in the procedure would cause the members to each develop prediscussion preferences for or commitments to (valence for) the solutions they had favored or rejected. Thus the numerical valence values obtained from the discussion would probably be smaller than those obtained in the previous studies.

Such was the case. The same valence–adoption relationships were obtained on all four problems, but the valence values needed for adoption were smaller than those obtained earlier. A crude exploratory attempt to measure the prediscussion valence members had for their individual solutions showed promise, when combined with the discussion-generated valence, in replicating the numerical values obtained in the earlier studies. Thus, the valence model not only was generalized to several different types of problems, but it also predicted accurately the effect of prior commitment on the adoption process.

Members' prior valence for solutions may facilitate the group's reaching consensus—as when all have positive valence for the same solution—or may delay consensus when members have strong valence for opposing solutions (cf. Hoffman, 1961). The specific ways in which an individual member's valence for solutions affects the group decision will be examined in future studies. Preliminary work on this topic has been limited to an insufficient number of groups, but it suggests that it is manifested at an early stage of discussion and persists as a function of the magnitude of prediscussion valence. Gary Coleite conducted a study on this topic in which three groups of three persons solved a problem at the same time. Following the initial problem-solving session, the groups were reconstituted into high, medium, and low valence groups to solve the problem again. The high group consisted of the persons with the highest valence for the solution adopted in their original group, the medium group of the next highest people, and the low group of the three lowest people. The extent to which people advocated their previous group's solution in the early part of their new group's discussion was a direct function of the amount of valence they had contributed in the initial group, not of their rank in that group.

C. EXPLICIT AND IMPLICIT PROCESSES

Our most recent attempt to extend our understanding of the valence–adoption process has reinforced our sense of the implicit nature of this process and has provided another significant numerical measure of valence (Hoffman, Friend, & Bond, 1978). In this study, groups solved a hypothetical personnel selection task in which they chose the best of five candidates for a job. The "solutions" to this problem were indivisible, mutually exclusive, and exhausted the possibilities. Most groups manifestly divided the process into two stages: (a) a rejection phase, in which unacceptable candidates were discarded, leaving the others for further discussion; and (b) the adoption phase, in which one candidate was chosen as the best. The division between the two stages was reliably identified in about three-quarters of the groups from such comments as "We have these two guys left." However, the valence analysis showed no such clearcut division. According to the valence analysis, solutions passed rejection and adoption thresholds (the former being the new numerical measure) at various times in the discussion, relatively independent of the formal announcement of the end of the rejection phase. That announcement caused the group to focus discussion only on the candidates identified. So it had a profound effect on the character of the discussion. Yet at the implicit level, as measured by the valence index, the group had often already "decided" to reject or adopt one or more of the remaining candidates by the time of the announcement. The relationship between events at the manifest and implicit levels will be the subject of our future work.

II. Participation, Influence, and Leadership

The subtlety of the implicit valence–adoption process has also shed new light on the question of who exercises influence over the group's decision. In the discussion in my original chapter of participation biases as inhibitors of effective problem solving (Section IIC), I reviewed the studies which supported the proposition that the highest participator in the group is the most influential. Acceptance of this proposition had lead to an ongoing controversy about the shape of the curve that best describes the skewed (from equal) participation rates in most groups so as to determine precisely the influence structure in the group (Fisek & Ofshe, 1970; Kandane & Lewis, 1969).

On the basis of two studies of the valence–adoption process, however, we are convinced that participation per se is not the important factor in influencing the group's decision. Rather, a high rate of participation gives a member a greater opportunity to contribute valence to the solution adopted by the group, but it is that contribution that is most critical.

The logic of our argument that leads to this conclusion is as follows. The group's task is to provide a solution to the problem presented. Since almost all groups adopt the solution which has the highest valence in that group, a person's influence in the group may be considered proportional to the amount of valence he/she contributed to that solution. Thus, by identifying the amount each member contributes to the valence for the adopted solution, we measure that person's influence over the decision.

Molly Clark and I (1976) examined the relationship between participation and influence by analyzing data on two problems solved by 3-, 4-, and 5-person groups. The results showed a modest relationship that decreased with increasing group size. Whereas in 3-person groups the same person was the highest participator *and* had the most influence in 73% of the groups, only 50% of them were identical in 5-person groups. Furthermore, although the proportion of participation of the most active members decreased in the larger groups, the proportion of valence contributed by the most influential member to the adopted solution remained the same. Other evidence derived from these analyses clearly supports the separateness of the total amount of discussion from the accumulation of valence for the adopted solution.

The highest participator seems to regulate the pace and length of the discussion but only influences that decision when he/she suggests solutions or evaluates them in ways acceptable to the other group members. Thus, though it was rare for the most active member to be the least influential or for the most influential to be the least active, the exercising of influence even in these ad hoc groups was more subtle than merely dominating the conversation. To be influential, a person must be effective in enunciating the beliefs of the other members or demonstrating his/her own expertness on the problem. In that way his/her comments add valence to his/her preferred solution, encourage others to agree, and stimulate little or no rebuttal from others, thereby cumulating valence toward that solution's adoption.

The sensitivity of the valence measure of influence also demonstrated the differential effects of the goals of acceptance versus quality of solutions set for appointed leaders and members in the study described earlier (Hoffman & Maier, 1967). I indicated then that, despite the regularities in the valence–adoption relationships across the four goal conditions, there were substantial differences in the leaders' and members' behaviors. The most striking of these was the relative contributions of each to the valence for the adopted solution. When both leaders and members strove for the quality of the solutions, the leaders were only slightly more influential than the average members and were the most influential at only a chance level. When the leader's goal was solution quality and the members wanted his/her satisfaction with the solution, the leader was substantially more influential, contributing the most valence for the adopted solution in three-fourths of the groups. The leaders' efforts to gain the members' acceptance for the group's solution in the other two conditions (about equally in both) also produced high levels of influence—the leader being the most influential in half the groups. These results suggest that in order to gain the members' satisfaction these leaders attempted to persuade them about the advantages of the leader's favored solution rather than to allow them to select freely their own solution and work through their differences to a successful and harmonious conclusion. In this way the leaders often failed to achieve their goal of members' satisfaction with the solution when they were unable to resolve conflicts among group members by taking one side or the other. Thus, not only does the group valence index reflect the movement of the group toward a decision, but the individual contributions to that index also measure each person's influence over the decision.

III. Conclusions

The principal significance of the valence approach to the group problem-solving process lies in its implications for improving that process.

The valence approach contrasts with and complements purely cognitive problem-solving perspectives (e.g., Arrow, 1963; Davis, 1973; Lorge & Solomon, 1955). All information is not processed equally. Whether the group uses the information available in its most able member is problematic (Lorge & Solomon, 1955). A solution offered before any other solution has passed the adoption threshold is more likely to be considered seriously by the group than if it is suggested afterward. A solution may be adopted by the group without general support from the members if some one person promotes it often enough to overwhelm the opposition. Yet the members are unlikely to be committed to a decision adopted in that way. In fact, when one member overcomes the resistance of the others and effects adoption of a solution, the others usually withdraw from or sabotage the implementation of that solution.

Possibly of greatest importance, the valence–adoption process is an implicit one in that groups seem unaware of its occurrence. There is nothing in the actual content of the discussion that reveals the valence status of a solution. What may appear to be a debate between two contending solutions is often the continued accumulation of positive valence for one solution accompanied by a balanced discussion of the merits and demerits of the other. In one group such a contender had 80 comments made about it, but its final valence value was zero. In contrast, the valence for the adopted solution was 26. Only by actually recording the valence values during the discussion is one able to know what the valences of different solutions are and, therefore, how the group is moving toward the final solution.

Since, of course, no groups record their valences during their discussion, they are subject to a process over which they have little control. The members often feel that they have discussed many alternative solutions to a problem and a simple count might indicate that they have. Yet the care and thoroughness with which each alternative had been discussed were probably quite uneven, depending on the valence of other alternatives at the time each one was suggested. Prior to one solution's achieving the adoption threshold, new solutions are welcomed and examined in detail for their advantages and weaknesses; following the achieving of the threshold, new solutions are often compared to the implicitly adopted one, often with an orientation toward finding the newcomer's weaknesses.

In leaderless groups the attention paid to different suggestions is often confounded with members' attempts to dominate the group. As shown in our review of studies of participation biases, personal factors unrelated to competence on the task facilitate the dominance of certain members whose suggested solutions are then more likely to accumulate valence beyond the adoption threshold at an early stage of discussion. Providing arguments in favor of his/her solution and eliciting support for it from others become the vehicle for this person's emergence as the group's leader. Solutions of greater merit suggested by others may thereby be ignored or treated in cursory manner by the group.

Even in cases in which the group's leader has been formally appointed, the tendency of such leaders to exercise undue influence in the group has often been noted (Hoffman, Harburg, & Maier, 1962; Hoffman & Maier, 1967; Maier & Hoffman, 1960). Since few leaders have been trained in techniques for effective group problem solving, they assume the right to propose solutions, to evaluate theirs and others, and to so control discussion that their preferences are realized. Members tend to legitimate these rights by responding appropriately. Since the valence–adoption process is so implicit, even when the leader attempts to elicit the members' views he exercises undue influence due to his anxiety to reach some acceptable decision (Maier & Hoffman, 1960).

An understanding of the subtleties of the valence–adoption process provides a basis for inventing techniques for improving the quality of group solutions without sacrificing their acceptance by the members. Many of the techniques that have been invented have attempted to overcome some of the barriers or to capitalize on the opportunities revealed by this process. Brainstorming (Osborn, 1957), for example, attempted to prevent the accumulation of valence for any one solution before others had been suggested, thus preventing any one from passing the adoption threshold before all suggestions could be considered thoroughly. Similarly, Bouchard's (1972) forced participation technique forestalls the dominance of any one member by invoking the rule that each member successively either offers a suggestion or passes his/her turn. The Delphi technique (Dalkey, 1969; Dalkey & Helmer, 1963), in which solutions are evaluated anonymously as a basis for reaching consensus, attempts to reduce the effects of status distinctions among the members. In this way, when status is independent of the task competence of the members, valence will not accumulate for incorrect solutions proposed by higher status members (Torrance, 1955). Chapanis' (1971) interesting demonstration that problems are solved better on telephone conference calls than in face-to-face interaction may have several valence-related effects. On the one hand the nonverbal cues for encouraging or resisting suggestions are reduced, thus possibly inhibiting longer justifications of proposed solutions. For the same reason, emotional contagion effects are less likely, and greater attention can be paid to the content of what is said.

Though these valence-related effects may be the reason why each of these methods has proved effective in certain contexts, the specificity of the effect must be kept in mind in understanding why none of them, by itself, is likely to be a panacea. Each of the techniques has proven effective upon introduction, but their generality has been rightly questioned (Meadow & Parnes, 1959). Brainstorming, for example, has not been a universally effective technique, even for generating a variety of solutions (Bouchard, 1969). Rather than discard the technique entirely, we need to specify the conditions for which it is most effective. Extensions of valence theory help us recognize that the valence of implicit definitions of a problem may result in constrained brainstorming (Hoffman, Bond, & Falk, unpublished). On the other hand, many evaluations of techniques rely on groups with no experience or very limited training with the method (Bouchard, 1969; Taylor, Berry, & Block, 1958), which is somewhat like evaluating the safety of automobile travel after watching new drivers take their first turns at the wheel.

The principal purpose of this essay has been to summarize the major findings in our study of the problem-solving process and thereby to stimulate more effort in that direction. We have been misled often by the conclusions of input–outcome experimental designs. It is still appropriate to raise questions about when to make decisions individually or in groups, as Vroom and Yetton (1973) have done in an elaborate model, or to suggest individual brainstorming for greater variety, followed by group decision for acceptance, as Bouchard (1969) implied. However, progress in promoting group problem-solving effectiveness is likely to come from a better understanding of the process itself.

REFERENCES

Arrow, K. *Social choice and individual values*. New York: Wiley, 1963.

Bouchard, T. J. Personality, problem-solving procedure, and performance in small groups. *Journal of Applied Psychology*, 1969, **53**, 1–29.

Bouchard, T. J. Training, motivation, and personality as determinants of the effectiveness of brainstorming groups and individuals. *Journal of Applied Psychology*, 1972, **56**, 324–331.

Chapanis, A. Prelude to 2001: Explorations in human communication. *American Psychologist*, 1971, **26**, 949–961.

Dalkey, N. C. *The Delphi method: An experimental study of group opinion*. Santa Monica, Ca.: The Rand Corporation, 1969.

Dalkey, N. C., & Helmer, O. An experimental application of the Delphi method to the use of experts. *Management Science*, 1963, **9**, 458–467.

Davis, J. H. Group decision and social interaction: A theory of social decision schemes. *Psychological Review*, 1973, **80**, 97–125.

Fisek, M. H., & Ofshe, R. The process of status evolution. *Sociometry*, 1970, **33**, 327–346.

Hoffman, L. R. Conditions for creative problem solving. *The Journal of Psychology*, 1961, **52**, 429–444.

Hoffman, L. R., Bond, G., & Falk, G. Implicit criteria in group decision making. Unpublished manuscript.

Hoffman, L. R., & Clark, M. M. Participation and influence in problem solving groups. Paper presented at the Eastern Psychological Association Meetings, New York, April, 1976.

Hoffman, L. R., Friend, K. E., & Bond, G. An analysis of the process of making choices in groups. Paper to be presented at the 49th Annual Meeting of the Eastern Psychologial Association, Inc., Washington, D.C., March 29–April 1, 1978.

Hoffman, L. R., Harburg, E., & Maier, N. R. F. Differences and disagreement as factors in creative group problem solving. *Journal of Abnormal and Social Psychology*, 1962, **64**, 206–214.

Hoffman, L. R., & Maier, N. R. F. Valence in the adoption of solutions by problem-solving groups: Concept, method, and results. *Journal of Abnormal and Social Psychology*, 1964, **69**, 264–271.

Hoffman, L. R., & Maier, N. R. F. Valence in the adoption of solutions by problem-solving groups, II: Quality and acceptance as goals of leaders and members. *Journal of Personality and Social Psychology*, 1967, **6**, 175–182.

Hoffman, L. R., & O'Day, R. Valence in the adoption of solutions by problem solving groups, III: Reasoning and value problems. Unpublished manuscript.

Kadane, J., & Lewis, G. The distribution of participation in group discussions: An empirical and theoretical reappraisal. *American Sociological Review*, 1969, **34**, 710–723.

Lewin, K. *A dynamic thoery of personality*. New York: McGraw-Hill, 1935.

Lorge, I., & Solomon, H. Two models of group behavior in the solution of Eureka-type problems. *Psychometrika*, 1955, **20**, 139-148.

Maier, N. R. F., & Hoffman, L. R. Quality of first and second solutions in group problem solving. *Journal of Applied Psychology*, 1960, **44**, 278–283.

Meadow, A., & Parnes, S. J. Evaluation of training in creative problem solving. *Journal of Applied Psychology*, 1959, **13**, 189–194.

Osborn, A. F. *Applied imagination*. (Rev. ed.) New York: Scribner, 1957.

Taylor, D., Berry, P. C., & Block, C. H. Does group participation when using brainstorming facilitate or inhibit creative thinking? *Administrative Science Quarterly*, 1958, **3**, 23–47.

Torrance, E. P. Some consequences of power differences on decision making in permanent and temporary three-man groups. In A. P. Hare, E. F. Borgatta, & R. F. Bales (Eds.), *Small groups: Studies in social interaction*. New York: Knopf, 1955. Pp. 482–492.

Vroom, V. H., & Yetton, P. W. *Leadership and decision-making*. Pittsburgh, Pa.: University of Pittsburgh Press, 1973.

STUDIES IN LEADER LEGITIMACY, INFLUENCE, AND INNOVATION[1]

Edwin P. Hollander and James W. Julian
STATE UNIVERSITY OF NEW YORK AT BUFFALO

[1]The studies reported here are part of a program of research supported under ONR Contract 4679 from the Group Psychology Branch, Office of Naval Research. We are indebted to Lanning J. Beckman, Lorraine Heilbrun, John B. Morganti, Franklyn A. Perry, C. Robert Regula, Richard M. Ryckman, Richard M. Sorrentino, and David Wiesenthal for their various kinds of assistance in this research. Experiment one appeared originally as Julian, J. W. & Hollander, E. P. A study of some role dimensions of leader-follower relations. Tech. Report No. 3, ONR Contract 4679, State Univ. of New York, Buffalo, N. Y., 1966; experiment two appeared as Hollander, E. P., Julian, J. W. and Perry, F. A. Leader style, competence, and source of authority as determinants of actual and perceived influ-

Reprinted from *Advances in Experimental Social Psychology,* Volume 5, 33–69.

I. Introduction to the Contemporary Study of Leadership

Leadership affords a rich field for the study of a wide range of sociopsychological phenomena. Among these are role behavior, interpersonal influence, attitude change, conformity, socialization, and intergroup relations. However, even with the broad range of interests inherent in its study, leadership continues to be approached in relatively limited ways with essentially static conceptions. An especially major limitation is the continuing emphasis on leaders as managers, without reference to the wider ramifications of the leadership enterprise. Side by side with this narrow emphasis is the still prevailing view of leaders as occupiers of a fixed position, rather than, in more dynamic terms, as attainers or maintainers of their standing with followers. Together, these emphases have tended to slight the influence process which is basic to leadership phenomena.

While the field of leadership research has made a number of forward strides, it has been encumbered by the lingering effects of earlier fads. The dominant approach for some time was to focus on the leader and his attributes; in this view, the leader was taken to be someone with qualities which would make him achieve leadership in any situation. This approach not only lacked a conception of leadership as a process, but suffered as well from the implicit assumption of a certain homogeneity in the leader's role. Reviews of the leadership literature by Stogdill (1948), Hemphill (1949), and Mann (1959), amply revealed these failings, and led the way for the more recent situational approach to leadership. Among the distinctions which came with this approach was the recognition that some leaders are appointed as leaders, and have their authority vested in them from above, while others are chosen by followers through a literal election or through some less formal, emergent process.

Within the situational framework, leaders were seen to fulfill various role expectancies depending upon the way the group tasks and group structures emphasized particular requirements for leadership. A variety of experiments essentially supported the consideration that who became a leader depended upon the nature of the task and its setting (e.g., Carter, Haythorn, Meirowitz, & Lanzetta, 1951; Carter & Nixon, 1949; Gibb, 1947).

In the 1950's, when it had its greatest impact, the situational ap-

ence. Tech. Report No. 5, ONR Contract 4679, State Univ. of New York, Buffalo, N. Y., 1966; experiment three appeared as Julian, J. W., Hollander, E. P., and Regula, C. R. Endorsement of the group spokesman as a function of his source of authority, competence, and success. *Journal of Personality and Social Psychology*, 1969, **11**, 42–49; and experiment four appeared as Hollander, E. P., Julian, J. W., and Sorrentino, R. M. The leader's sense of legitimacy as a source of his constructive deviation. Tech. Report No. 12, ONR Contract 4679, State Univ. of New York, Buffalo, N. Y., 1969.

proach went a long way toward rectifying the balance which had been overweighted for so long toward leader traits. The research generated by this approach indicated that different situations demanded different capabilities from the person called a "leader." Yet this situational focus, by itself, failed to give sufficient attention to the process of leadership. For the most part, the group's task was regarded as the key element differentiating situations, and the leader was considered to be someone who occupied his position almost in a fixed sense as a consequence of his competency with respect to that task. Less attention was given to followers, their perceptions of the leader, and reactions to his assertions of influence over time. Furthermore, the situational approach encouraged a separation of the leader from the group's situation when, in fact, the leader is an important element in the situation from the followers' vantage point.

All in all, it suffices to say, as we have recently stated elsewhere, that ". . . the two research emphases represented by the trait and situational approaches afforded a far too glib view of reality. Indeed, in a true sense, neither approach ever represented its own philosophical underpinning very well, and each resulted in a caricature" [Hollander & Julian, 1969, p. 388].

A. THE TRANSACTIONAL APPROACH

Given the limitations of these earlier emphases, some new formulations of leadership phenomena were required. Leadership is now coming to be seen as a transaction between leaders and followers which implies a reciprocal influence process. To be influential, as Homans (1961) has observed, the leader must be willing to be influenced by others and, in effect, to exchange rewards with them. Thus, Katz and Kahn (1966) view leadership functions in system terms as an interchange of inputs for outputs. Accordingly, the leader provides a resource for the group by facilitating leadership functions, among which is the direction of the enterprise. Although the leader's contributions and their consequences vary with system demands, the leader is perceived by the other group members as providing the more valued resources needed for the attainment of their common goal (Hollander & Julian, 1968, p. 891).

Put in transactional terms, the leader who fulfills expectations and helps to achieve group goals provides a rewarding resource for others which is exchanged for status, esteem, and greater influence. Thus, he gives something and gets something. And what he gets contributes to his legitimacy insofar as he is "validated" in his role by followers. It is the leader's sense of this legitimacy which then serves as the base on which he may operate to exert influence.

B.　Idiosyncrasy Credit and Legitimacy

Before elaborating further the nature and function of legitimacy, it is useful to observe that the functional effectiveness of an individual's status intimately depends upon other persons' perceptions. The "idiosyncrasy credit" model (Hollander, 1958, 1964) deals in particular with the processes affecting acceptance as a feature of status. Briefly, the essential point of this model is that the leader's influence depends upon how competent others in the group believe he is in helping the group achieve its goals, and his conformity to the group's normative expectations as a sign of his motivation to belong to the group. If seen favorably in these respects, the leader's subsequent assertions of influence are then more readily accepted, even though they may represent deviations from group patterns. Thus, a person gains credits, in terms of the positive impressions held by relevant others, which he may then draw on in exerting influence, particularly regarding deviations from normative expectancies.

The process orientation of this model has produced research with a distinctive sequential feature. An early exemplification of this is revealed in an experiment by Hollander (1960) which showed that a group member with high competence on the task could be highly influential, both with regard to the task and the norms of the group, if he conformed earlier to those norms, but not if he failed to do so. Under those conditions, where such a group member manifested early non-conformity, his influence was markedly curtailed later. A reversal of this sequence, however, produced high influence. Other research by Berkowitz and Macaulay (1961), Harvey and Consalvi (1960), Hollander (1961b), Julian and Steiner (1961), Sabath (1964), and Wiggins, Dill, and Schwartz (1965), support the essential proposition that higher status members could deviate from the group norms with greater impunity. A recent experiment by Alvarez (1968) on deviance in simulated work organizations, provides an interesting confirmation of this. He found that for the same acts of deviance, the higher status person lost esteem at a slower rate, but only in "successful" organizations; where the organizations were "unsuccessful," the higher status person lost esteem at a faster rate than his lower status counterpart.

II. The Study of Legitimacy

Prominent among the several elements making up a leader's role are the manner by which the leader achieves his position, what he is perceived to be doing, and how his actions and motivations are seen to contribute to the group's task. Depending upon the expectations and perceptions held by the others involved, these elements may be interrelated in various ways, thereby determining the leader's ability to influence the others in working effectively toward common goals. This is the essence of the leader's legitimacy. In Goffman's (1959) terms, the impressions he "gives off" will alter the balance of favorability for the leader's success. Indeed, Fiedler (1967) attached particular significance in his "contingency model" to the favorability factor in predicting outcomes of the leadership process.

As we have suggested, one way to approach legitimacy is to understand it as a process involving an exchange of rewards which provides the leader with greater sway for asserting influence. Thus, the person in the role of leader who is able to achieve what is expected of him comes to have greater credits at his disposal. In broadest terms, his legitimacy grows out of the authority vested in his role, the source of that authority, and a set of impressions which followers hold regarding his competence and his motivations. In signalizing the acceptance of a leader's legitimacy, two steps appear to be involved: first, the acceptance of the leader in his role as an authority; and, second, the willingness to respond affirmatively to his assertions of influence over time. Ultimately, the latter responses are the more crucial in establishing leader-follower relationships of a sustained sort.

A. VARIABLES AFFECTING LEGITIMACY

In framing the program of research to be reported here, three leader characteristics were chosen as determinants of legitimacy which could be experimentally manipulated. These are: (a) the source of the leader's authority, whether the origins are internal or external to the group; (b) the leader's competence on a task which facilitates the attainment of group goals; and (c) his motivations with respect to the task itself and the needs of his followers. How these factors are perceived by followers may be hypothesized to determine the favorability with which they respond to the leader, as an indication of his legitimacy.

Apart from the perceived competence of the leader, few investigations have explored the impact of these variables in eliciting favorable follower reactions. A number of studies (e.g., Croner & Willis, 1961; Hollander, 1960; Mausner, 1954) have demonstrated that individuals seen to be more highly competent exert greater influence on others' task behavior. By contrast, the effects of perceived leader motivations have remained relatively unexplored, although the early study by Fouriezos, Hutt, and Guetzkow (1950) presented some suggestive leads in this vein. Similarly, the effects of the source of the leader's authority have mainly gone unattended, with several exceptions. Cohen and Bennis (1961) found that the continuity of leadership was maintained more by allowing groups the option of electing their own leaders. Goldman and Fraas (1965) have demonstrated differences in the productivity of groups established under four procedures of leader selection, including appointment and election. Also noteworthy are the findings of Raven and French (1958) which indicated that the election of a supervisor produced private as well as public compliance in groups while its absence only produced conformity when it was observable by the supervisor. Taken together, the results of these studies point to the potentialities of further research on the effects of the leader's source of authority.

B. APPROACH TO THE PROBLEM

In the present set of experiments, we employed a factorial design within which were manipulated several variables. At the outset, we considered the effects on followers of the leader's perceived source of authority, task competence, and relevant motivations. We selected as major dependent variables the followers' acceptance of the leader in his position, and his influence, in terms of their responses and perceptions. In the first experiment, we explored all three of these variables as inputs; in the second, we were concerned with their influence effects, both in behavioral and perceptual terms; for the third experiment, we returned to a focus on the acceptance of the leader where success and failure were introduced subsequent to the followers' initial perceptions of the leader's capability to serve as the group's spokesman; finally, the fourth experiment looked at the phenomenon from a different perspective, that of the leader's sense of his own legitimacy as a determinant of his assertions of influence.

Across conditions, we predicted in general that the leader's perceived competence should be positively related to his acceptability and influence. We also expected that his perceived motivations toward the task and toward others in the group should be positively related to these measures. We recognized, however, that perceived competence and motivation might interact statistically to determine the outcome produced by each. With reference to the "idiosyncrasy credit" concept, it might be predicted for instance that a more competent, task-oriented leader would be reacted to more positively than one of less competence who is group-oriented, even though a simple reward model might favor the latter.

In addition to this possibility of statistical interaction, there were also effects predictable from interactions with source of authority. For the present studies, two sources of leader authority were explored: appointment and election. While clearly not representative of all sources of authority, they are two characteristic and readily understandable ways by which individuals are legitimized in the status of group leader. And they can readily be seen to shape other relationships in a differential way.

III. Experiment One: An Exploration into the Phenomenology of Leadership

The first experiment sought to explore the phenomenology of leadership from the vantage point of followers. Following a procedure employed by Hollander (1961b), information was presented about a leader through a simple description technique. Subjects could then rate the acceptability of this stimulus person and thereby provide a view of the way in which the three kinds of leader characteristics already noted, alone and in various combinations, differentially affected these ratings. This approach was considered to be preparatory to the later experiments presented in sections IV, V, and VI.

A. METHOD

1. Subjects and Procedure

Six hundred thirty-three undergraduate men and women at the State University of New York at Buffalo participated as subjects. All were students enrolled in introductory psychology and took part during a special one-hour session of the course. The problem was presented as a study of the personal attitudes of people in groups. Salient attributes of a hypothetical leader were described and subjects were then asked to indicate their responses to him. A brief questionnaire booklet included instructions to "Think of a group to which you belong or did belong and imagine an elected leader of your own sex who is described as follows: good performer in the group activity, interested in the group activity, and interested in the members of the group." This exemplifies the several characteristics of the leader which were manipulated by systematically varying the content of the information in this description. Subjects then completed a number of ratings of the hypothetical leader.

2. Leader Characteristics

Source of leader authority was varied in three ways, by describing the leader as "elected," "appointed," or by not mentioning the source of his authority. Leader competence was also varied in three ways by describing him as a "good performer in the group activity," a "poor performer in the group activity," or by not mentioning his competence. The two aspects of leader motivation were each varied in two ways, either by describing the leader as "interested in the members of the group" or not mentioning his motivation, and by describing him as "interested in the group activity" or not mentioning his activity motivation.

3. Design

Combining all levels of these four leader characteristics, 36 hypothetical leaders were described to yield a 3 × 3 × 2 × 2 factorial design. An analysis of variance was then conducted to discern significant differences in the ratings of these leaders by cell. To compensate for inequalities of N within cells, 36 cases were deleted at random from particular cells and 15 artificial mean cases were added to other cells thus yielding 17 subjects per cell. Given the substantial size of our sample, this procedure provided computational ease without introducing between-cell variability. Degrees of freedom for the estimate of within-cell variability were, of course, appropriately reduced. Each cell of the design represented both men and women in a ratio of approximately one to two.

4. Ratings of the Leader

Reactions of subjects to the group leaders were obtained by ratings along four basic role dimensions: 1. "How willing would you be to have this person continue as the leader of the group?" (leader); 2. "How willing would you be to have this person serve as a spokesman for the group in dealing with other group leaders where important issues are at stake?" (spokesman); 3. "How willing would you be to have this person as a member of the group if you were leader?" (follower); 4. "How willing would you be to have this person as a close friend?" (friend). Each rating response was obtained on a graphic scale with six points ranging from "extremely willing" to "definitely not willing."

B. Results and Discussion

Table I presents average ratings of the leader as a function of each level of the leader characteristics, collapsed across the four ratings. An overall analysis of variance of these ratings yielded significance beyond the .01 level for the factors of leader competence, interest in group members, and interest in group activity, but not for source of authority. Thus, we have confirmation of the major prediction that the leader's acceptance varies as a function of these perceived attributes.

TABLE I

MEAN RATING OF LEADER FOR EACH INDEPENDENT VARIABLE
ACROSS THE FOUR RATING SCALES
IN EXPERIMENT ONE

	Source of authority	Leader competence[a]	Member motivation[a]	Activity motivation[a]
Elected	4.53			
Appointed	4.45			
Good performer		5.28		
Poor performer		3.19		
Interested in group members			4.65	
Interested in group activities				4.69
Variable not mentioned	4.51	5.02	4.35	4.30

[a]Mean ratings in these columns were significantly different beyond the .01 level.

An additional result from this analysis, in line with prediction, is the significant interaction between the competence of the leader and his interest in the focal activity of the group ($p < .01$). Table II presents the mean ratings displaying this interaction. The contrast between ratings of leaders who were seen to perform well or poorly at the task was most striking. We note that where level of competence was not indicated, favorable endorsement of the leader was quite similar to that for "good performer." It is apparent in Table II that although the leader's interest in group activities had a significant positive impact on member reactions, its major contribution was to the status of the "poor performing" leader. A poor performer could apparently retain considerable status by evidencing a sincere interest in "playing the game." A similar trend was observed for the contribution of "interest in group members," although this failed to reach significance.

TABLE II

INTERACTION BETWEEN COMPETENCE OF LEADER AND HIS INTEREST
IN GROUP ACTIVITY ACROSS THE FOUR RATING SCALES
IN EXPERIMENT ONE[a]

	Competence of leader		
	Good	Poor	Not mentioned
Leader interested in group activity	5.34	3.58	5.15
Not mentioned	5.22	2.80	4.88

[a]The interaction between leader competence and his interest in the group activity was significant beyond the .01 level.

Before we examine the relative importance of these leader characteristics for each of the four role ratings, it is worthwhile to consider the overall average rating for each role. These means are averages taken across all cells of the design, and may therefore be viewed as indicating the relative level of acceptance of a group leader in each of the four roles. They are shown in the bottom row of Table III. In general, subjects were relatively more willing to accept these "leaders" as followers or friends than to accept them as leaders or spokesmen. These results support the obvious contention that group members will tend to be wary in endorsing someone for a position of authority over them.

The overall analysis of variance also indicated significant interactions between levels of leader competence, member motivation, and activity motivation with the four types of roles. The mean ratings showing these interactions are given in Table III. All three interactions took similar form, with levels of independent leader characteristics having a greater effect on the two more relevant roles, i.e., leader and spokesman. Thus, being a "poor performer on the task" resulted in a dramatic loss of status in terms of endorsement *as a leader or spokesman*, but had less effect on endorsement as a follower or friend.

For the most part, the findings of this first experiment confirmed our expectations. The perceived competence of the leader and his motivations relevant to the group setting had a significant and strong impact on

TABLE III

MEAN RATING OF LEADER IN EACH ROLE FOR ALL LEVELS OF COMPETENCE, MEMBER MOTIVATION, AND ACTIVITY MOTIVATION IN EXPERIMENT ONE[a]

Condition		Role			
		Leader	Spokesman	Follower	Friend
Competence:	Good	5.20	5.22	5.49	5.20
	Poor	2.41	2.20	3.84	4.32
	Not mentioned	4.93	4.92	5.27	4.95
Interested in	Yes	4.38	4.35	4.90	4.95
group members:	Not mentioned	3.98	3.88	4.84	4.69
Interested in	Yes	4.39	4.39	5.05	4.95
group activity:	Not mentioned	3.97	3.83	4.70	4.70
Overall average rating[b]		4.18	4.11	4.87	4.82

[a]The interaction of each leader characteristic and the four roles was significant beyond the .01 level.

[b]All pair comparisons here were significant beyond the .01 level (by Duncan Range Test) except between "follower" and "friend." Degrees of freedom for error equalled 1658.

group members' willingness to accept and endorse him in four distinct role relationships. The inference that these characteristics are salient features in the phenomenology of leader-follower relations is therefore well sustained.

The major variable which failed to produce significant effects, either overall or in interaction with other characteristics, was source of authority. As shown in Table I, levels of response for election and appointment were nearly identical, and certainly not different from the "blank" source of authority condition. Although the null hypothesis remains unassailable from these data, a possible methodological parameter might have rendered the source of authority unimportant in our experimental procedure. As noted above, subjects were instructed to "think of a group to which you belong or did belong and imagine an elected (appointed or not mentioned) leader of your own sex who is described as follows:." There followed a schematic box in which were presented the leader characteristics which comprised the manipulations of the competence and motivational variables. Hence, the information as to source of leader authority was less prominently displayed and may not have been "received" by our subjects.

This interpretation suggested the desirability of an early replication in which all leader characteristics were given equal prominence. Such a replication was conducted, although on a more limited scale than in the original study. The procedure was precisely the same, except that all the independent leader characteristics, including source of authority, were presented within a schematic box. The specific characteristics replicated were: 1. source of authority, *elected, appointed,* and *not mentioned*; 2. competence, *good performer* and *poor performer*; and 3. member motivation, *interested in other group members* and *not mentioned*. Combinations of levels of these three variables yielded 12 hypothetical leaders in a 3 × 2 × 2 design. Ratings were again obtained for each of the four roles. An overall analysis of variance produced results paralleling those of the main study.

Of particular interest again was the failure of source of leader authority to influence the endorsement of the leader, even though care had been taken to give this factor equal weight in the leader descriptions. In view of the size of our samples, the homogeneity of response within the studies, and the replication of results, it could be concluded that for these subjects election or appointment of the group leader was not a salient aspect of the leader-follower relationship. However, due to the artificial nature of the leadership situation with which they were confronted, it seemed best to reserve judgment and pursue the effects of this distinction under more concrete conditions in the second experiment.

IV. Experiment Two: Leader Style, Competence, and Source of Authority

In the next experiment, we extended our investigation to a laboratory situation in which these leader characteristics could be varied under conditions permitting greater control and immediacy. Once again we were interested in the source of the leader's authority, his task competence, and his motivations. In examining the last variable, we employed a manipulation that more closely approximated leader "style." We distinguished between a leader who appeared to be group-oriented as against one who appeared to be self-oriented. The former is a direct parallel to the description of the leader in the first experiment as someone "interested in group members," while the latter is its obverse. This time our major dependent variables were the leader's actual and perceived influence. A multifactor design was again employed, with two sources of leader authority, i.e., election vs. appointment, two levels of leader competence, i. e., high vs. low, and two kinds of leader style, group-oriented vs. self-oriented, thus yielding eight experimental treatments.

A. METHOD

1. Subjects and Procedure

Eighty male undergraduates enrolled in introductory psychology at the State University of New York at Buffalo participated as subjects in the experiment. They were organized into 20 groups, each composed of four true subjects plus a mock subject who in each case became the leader and accordingly served as the stimulus person for the experiment. Each of the 20 groups followed an identical procedure but for one major difference regarding either election or appointment of the leader.

In each group, subjects were seated at a table with partitions that did not allow them to see one another once they had taken their seats. The task was explained as one requiring judgments of which one of three stimulus lights on the wall went off first. It was presented as a matter requiring the recognition and communication of a sequential pattern of information to achieve a common group judgment. The actual sequence with which the stimulus lights extinguished was randomly scheduled using intervals of about .05 seconds. This differential was previously found to be unambiguous.

Following this presentation of the task, subjects filled out forms giving information about themselves and indicating their preference for appointment or election of a group leader. The leader was described as having the following functions: He would report his judgment first on any trial to the group via a signal panel; he would then take account of the other members' judgments communicated to him and would report to the experimenter what he took to be the "group judgment"; and finally, he would decide the distribution of winnings within the group. The personal information sheets were then collected by the experimenter and fictitious carbon copies of each group member's sheet were distributed to the other group members, with the explanation that they could be "read over in order to get better acquainted." Each group member actually received the same set of information about his fellows.

There were 50 trials in all. The first ten trials were described as individual test trials. For these, all members responded to the lights at the same time and there was no "communication" among group members. Performance on these trials was used to manipulate the perceived competence of the group members. For the second set of ten trials, the procedure was as described above with the leader communicating his judgment first to the other group members, and then the members presumably communicating back to the leader. At the end of this first set of *group* trials, group performance was scored and points based on the performance were ostensibly allocated to the members by the group leader. His distribution of the points was varied to manipulate the leader style variable. Then a final set of thirty group trials was conducted following the procedure described. The matching by members of the leader's initial erroneous judgments was the basis for the primary influence measure. It should be noted that the leader always made 20 incorrect judgments distributed across the last 30 trials.

At the end of the session, all members rated the leader in terms of his competence, fairness, and perceived influence, and in addition, reacted to the experiment generally in terms of their satisfaction and enjoyment. Ratings of the leaders were also obtained on ten bipolar scales of the semantic differential sort. The scales were chosen because of their previous loadings on the evaluative, potency, and activity dimensions of connotative meaning (see Osgood, Suci, & Tannenbaum, 1957).

2. Experimental Inductions

To manipulate the source of leader authority, the experimenter first ostensibly studied the forms on which group members indicated their preference for election or appointment of the group leader. In ten groups, taken on an alternate basis, the experimenter then announced that a majority favored election, and a contrived election was held. In the other ten groups, the experimenter appointed the leader on what appeared to be an arbitrary basis, after informing the group that there was no clear preference for either election or appointment.

Leader competence was manipulated within each group by early feedback to members regarding their ability at the task relative to the leader and the other group members. This information was communicated following the individual test trials so that two subjects in *each group* perceived the leader to be relatively more competent than they were at the task, while two perceived him to be less competent than they were. This was done with a single feedback score sheet which always showed two members distinctly superior to the others and two members inferior, with the leader always approximately in the middle.

Style of the leader was manipulated across levels of competence after the first ten group trials by having the leader appear either to divide the money earned by the group heavily in his favor (60% to himself and 10% to each of the others) or to divide it on an equal basis (20% to each). One of the two subjects in each of the pairs, who saw themselves as either more competent or less competent than the others, believed the leader to be relatively "self-oriented," while the other members of these pairs saw the leader as "group-oriented."

In any group, therefore, each subject represented one of four treatments in a two-by-two design. Depending on the cell, the leader was variously perceived to be: relatively more competent and self-oriented, more competent and group-oriented, less competent and self-oriented, or less competent and group-oriented. As previously noted, then, the design was a $2 \times 2 \times 2$ factorial, including the variables source of leader authority (election or appointment), leader style, and relative competence of the leader.

B. RESULTS AND DISCUSSION

Data reflecting the extent to which group members were influenced by the leader are displayed in Fig. 1. An analysis of variance testing these means revealed a significant main effect for leader competence only. Replicating the first experiment and previous findings, leaders seen as relatively more competent at the task were significantly more influential than were leaders seen as less competent ($F = 4.08$; $p < .05$ for 1 and 72 degrees of freedom). Effects of other leader characteristics, although not significant, are shown in trends toward greater influence exerted by the elected leader as compared with the appointed leader (8.3 *vs.* 7.1), and the greater apparent influence of the self-oriented leader as compared with the group-oriented leader (8.2 *vs.* 7.1).

In addition to the primary measure of acceptance of influence, the postsession reactions to the leader in terms of his manipulated characteristics lend further insight into the operation of these variables. Although

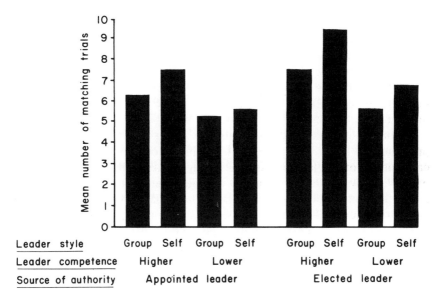

Fig. 1.   Mean number of twenty trials in Experiment Two on which subjects gave leader's erroneous judgments as their own for each treatment. $N = 10$ for each treatment.

we did not find strong effects of the two leader styles on influence, the mean postsession ratings of the fairness of the leader testified to the successful induction of the leader-style variable. Table IV shows the average ratings of leader "fairness" for each experimental condition. Leaders who were group-oriented in distributing points were rated as significantly "fairer" than those who were self-oriented ($F = 160.3$; $p <$

TABLE IV

MEAN POSTSESSION RATINGS OF LEADER FAIRNESS AND COMPETENCE
ON FIVE-POINT GRAPHIC SCALES
IN EXPERIMENT TWO[a]

	Leader style			
	Group-oriented		Self-oriented	
Competence of leader:	High	Low	High	Low
Fairness of leader	4.3	4.6	2.3	2.0
Competence of leader[b]	3.3	3.2	3.3	2.8

[a]No average differences were found as a function of source of authority. Therefore, this breakdown has been omitted.

[b]No comparisons were significant for the rated competence of the leader.

.01). As previously indicated, leaders presented as more competent exerted significantly greater influence during the group judgment task. However, they were not rated more competent following the task. This unexpected result suggested a shift in the evaluation of the leader's competence during the course of the group judgments. Postsession ratings bear the effects of the subjects having seen the group leader make a number of "questionable" judgments.

Ratings of task difficulty and of the leader's perceived influence are presented in Table V and are also pertinent to the issue of the relative competence of the leader. For the rating of task difficulty, group members who saw the leader as of low competence rated the task as significantly less difficult (3.4 vs. 4.1; $F = 7.9$; $p < .05$) and reported the leader as exerting less influence on their task judgments (1.5 vs. 2.3; $F = 15.0$; $p < .01$). These latter perceptions were, of course, quite veridical, supporting the efficacy of the induction of relative competence on the task, and the validity of the task measure of influence.

Table V also presents the subjects' postsession reactions to the experiment generally. In terms of "enjoyment of the task," no differences were found; however, significant differences emerged on the measure of "satisfaction with one's performance," as a function of both relative competence and an interaction between competence and the style of the leader. Group members who saw the leader as relatively low on competence were significantly better satisfied with their performance (3.1 vs. 2.1; $F = 17$; $p < .01$), and further, the orientation of the leader appeared to have a subtle influence on the degree of member satisfaction. For the group members presented with leaders of low competence, leader style had no effect, but for the opposite treatment, leader style was important. Also, group members with leaders of high competence were significantly less satisfied with their *own* performance when the

TABLE V
MEAN POSTSESSION REACTIONS TO EXPERIMENT TWO[a]

	Leader style			
	Group-oriented		Self-oriented	
Competence of leader:	High	Low	High	Low
1. Difficulty of task	4.1	3.3	4.0	3.5
2. Influence of leader	2.3	1.5	2.2	1.4
3. Enjoyment of task	3.5	3.3	3.1	3.4
4. Satisfaction with performance	1.8	3.2	2.5	3.0

[a]No average differences were found as a function of source of authority. Therefore, this breakdown has been omitted.

TABLE VI

MEAN RATINGS OF THE LEADER BY CONDITION IN EXPERIMENT TWO FOR THE
EVALUATIVE, ACTIVITY, AND POTENCY DIMENSIONS[a]

| | Leader style | | | |
| | Group-oriented | | Self-oriented | |
Competence of leader:	High	Low	High	Low
1. Evaluative	4.3	4.5	3.7	4.0
2. Activity	2.9	2.6	2.5	2.8
3. Potency	4.1	4.0	4.3	4.1

[a]No average differences were found as a function of source of authority. Therefore, this breakdown has been omitted.

leader had acted in a group-oriented, unselfish way than when he had acted in a self-oriented way.

While source of leader authority was not by itself significant in producing leader influence, it did affect the relationship between actual and perceived influence significantly. The correlation between actual and perceived influence was .57 for the elected leader condition but only .21 for the appointed condition. This difference is significant at the .05 level and raises the interesting point that with elected leadership there exists a greater willingness to acknowledge having been influenced by the leader in one's task performance.

Supplemental information characterizing group member reactions to the leader was available from postsession semantic differential descriptions of the leader. These ratings were scored in terms of the previously identified dimensions of connotation (see Osgood *et al.*, 1957). The mean levels of these ratings by condition are shown in Table VI. The group-oriented leaders were rated more positively on the evaluative dimension ($F = 27.7$; $p < .01$), but in terms of "activity," leader style and competence interacted, with the self-oriented, low-competence leader being seen as distinctly more active than his group-oriented counterpart ($F = 4.02$; $p < .05$). In addition, the "potency" of the leader also varied as a function of his relative competence, with the high-competence leader being described as more potent ($F = 4.44$; $p < .05$). From these descriptions it can be readily inferred that the leader's style had greater impact on the affective relation between the leader and member, but that under the short-term conditions of the group task these differences in style did not affect his ability to exert an influence on the group. Power or potency was tied exclusively to the relative competence of the leader.

In sum, we find again that the single most important variable in determining the leader's successful influence on the group was his perceived competence. Source of the leader's authority, while not a significant factor in itself, did however qualify the relationship between the group members' actual and perceived acceptance of influence. Also, leader style or orientation showed its effect in the more positive evaluations given to group-oriented than to self-oriented leaders following the completion of the task.

V. Experiment Three: Support for the Leader as Group Spokesman

We had indications in the last experiment that source of the leader's authority had relevance in an intriguing, second-order fashion to the subjects' willingness to acknowledge being influenced by the leader. But since our group task was constrained by the absence of face-to-face interaction, we believed there had been only a limited prospect for generating much feeling about the leadership process and the leader. Accordingly, we carried out a third experiment, making a distinction once again between election and appointment as bases for legitimizing leadership. This time we created a task with face-to-face contact under more naturalistic conditions, and also introduced the variable of group "success" or "failure."

Essentially, the effects of three factors were studied: (a) the leader's source of authority in either election or appointment; (b) his competence in the sense of his perceived capability on the task; and (c) his subsequent task success. The main dependent variable was acceptance by group members of the leader in his role as spokesman for the group. Broadly speaking, it was hypothesized that competence, along with success, would be significant determiners of the group's endorsement and that the elected spokesman would be more strongly endorsed than the appointed one, recognizing the distinct prospect of certain statistical interactions that would shape these relationships differentially.

A. METHOD

1. Subjects and Procedure

One hundred thirty-six undergraduate men at the State University of New York at Buffalo volunteered to participate as members of four-man discussion groups. At their appointed hour, an average of 20 men reported to a central screening point where they were assigned to individual rooms, four men to a room. Group discussions began with a review of a fictitious case of cheating on an exam by a hypothetical friend of the group members. Groups were told that as part of the experiment, "Roger's case" was to be presented before a board of inquiry and that their immediate task was to develop arguments for his defense. After a 20-minute discussion, an experimenter indicated that it was time to choose a spokesman to represent the group before the board in Roger's behalf. The experimenter then arranged the procedure so as to manipulate the source of authority (either election or appointment) and the perceived competence of the spokesman (either high or low).

After the spokesman was chosen, he was asked to report to another room where the inquiry would take place. The spokesman was allowed to take any notes or information from the group's discussion which he thought would be helpful. When the spokesman had left, the experimenter distributed a brief questionnaire asking for evaluations of the spokesman and the group's discussion. After a 10- to 15-minute delay the spokesman returned to his group with a sealed envelope containing the board's verdict. After opening the envelope and reading the verdict aloud, the experimenter asked the spokesman again to step outside the room while the group evaluated the verdict. Group members then filled out a second questionnaire responding to the verdict and evaluating the spokesman's performance. It was at this point that the group members indicated the strength of their endorsement of the spokesman.

2. Experimental Inductions

To manipulate the perceived competence of the spokesman, the experimenter for each group made tallies during the discussion of the frequency of participation of each member. The most frequent contributor to the discussion was chosen as the spokesman in the "high perceived competence" condition, while in the "low perceived competence" condition the member who ranked next to lowest in participation was chosen. The purpose was to identify individual group members who were perceived by their peers as more or less able to perform well as the spokesman for the group.

To manipulate his source of authority, the spokesman was either chosen by mock election or by the experimenter's appointing him. For election conditions, ballots were secret, and members rank-ordered their choices, with the winner announced after an ostensible calculation of the highest average rank. Appointment of the spokesman was made arbitrarily by the experimenter after an obvious inspection of the notations he had made during the discussion.

The manipulation of the spokesman's success or failure was distributed randomly across groups. Verdicts of "acquitted" or of "guilty," were simply handed to the spokesman before they returned to their groups. This constituted the sole basis for the group's knowledge of his success or failure.

3. Variables and Measures

The independent variables of source of authority and perceived competence formed a 2 × 2 factorial design for the first questionnaire, and then, with the addition of the success variable, they formed a 2 × 2 × 2 design for the analysis of the second questionnaire. Since there were slight discrepancies in the number of groups per cell, an unweighted means analysis was used. All scores were group averages of individual member ratings.

The major dependent measures were the ratings completed by the discussants immediately following the discussion and then following the announcement of the verdict. These questionnaires were comprised of 6-point graphic rating items. Endorsement of the group spokesman was assessed in the second questionnaire by asking: "How willing would you be to have the spokesman represent the group again?" (Table VII, item 4).

B. RESULTS AND DISCUSSION

The main findings of this experiment concern group members' reactions to and perceptions of the spokesman. In general, these indicate that the perception of the spokesman's task competence or his subsequent success led to greater endorsement by group members. While there were no average differences in the endorsement of elected versus

TABLE VII
MEAN RATINGS OF GROUP SPOKESMAN IN EXPERIMENT THREE

Variable	Competence:	Appointed		Elected	
		High (N = 8)	Low (N = 8)	High (N = 9)	Low (N = 9)
1. How much did spokesman contribute to discussion?		4.7	3.1	5.1	3.7
2. How well qualified is spokesman?		4.0	3.3	4.4	3.7
3. How satisfied with choice of spokesman?		4.4	3.7	4.6	3.8
4. How willing would you be to have spokesman represent group again?		5.1	4.4	4.9	4.2

appointed spokesmen, the impact of this source of authority became apparent in the complex and significant interaction between this variable and perceived competence and success. The nature of the interaction between success and perceived competence took quite a different form for spokesmen who had been elected than for those appointed.

1. Effects of Competence

Reactions to the spokesman gave strong testimony to the important influence of his perceived competence. Variable *1* was most similar to the criterion of competence used for the selection of the spokesman: "How much did the spokesman contribute to the discussion?" Here, differences as a function of competence were highly significant, with the more talkative spokesman rated as having contributed more to the discussion (14.7 and 10.5 for high and low competence conditions, respectively; $F = 40.1, p < .01$).

Confirming our prediction, group members were also better satisfied with the more competent spokesman (Variable *3*; $F = 4.13, p < .05$) and more willing to have him represent the group again (Variable *4*; $F = 11.2, p < .01$). These judgments were consistent with the perceived qualifications of the spokesman (Variable *2*; $F = 8.18, p < .01$). Hence, we find that the criterion of competence fits the group members' perceptions well, and that they evaluated the spokesman on this basis, with stronger endorsement given to the spokesman under those conditions in which a more competent man was chosen.

2. Effects of Source of Authority

As anticipated, there were no overall differences in reactions to the spokesman as a function of his source of authority; however, there was a tendency to evaluate the elected spokesman more positively. He was judged as having contributed more (Variable *1*; $F = 2.88, p < .10$) and as being better qualified (Variable *2*; $F = 3.99, p < .10$).

Although, as seen in Table VII, endorsement of the incompetent spokesman (Variable *4*) was low regardless of whether he was elected or appointed, the significance of source of authority did become apparent in its qualification of the effects of the spokesman's success.

3. Effects of Success

Success was defined as a verdict of acquittal, while failure was a verdict of guilty. The appropriateness of these definitions is seen in a final questionnaire item which asked: "How satisfied were you with the outcome?" Satisfaction with the acquittal verdict was rated 15.7, with the guilty verdict receiving a mean satisfaction rating of 6.2 ($F = 123.1, p <$

.01). The spokesman's success in representing the group also influenced member endorsement. Endorsement of the successful spokesman was 14.8 while the unsuccessful spokesman was rated 13.0 ($F = 7.04$, $p <$ .05).

In addition to these findings, Fig. 2 displays the interactive effects of the spokesman's success with his perceived competence and source of authority. Although the level of endorsement did not differ overall for the elected or appointed spokesman, the nature of the interaction between success and perceived competence took quite a different form for the elected and appointed conditions. For the elected spokesman, an incompetent man was rejected regardless of his success at the task, whereas success increased the endorsement of a competent man. For the appointed spokesman, however, it was the competent man who was relatively immune to the effects of success or failure, while his incompetent counterpart suffered a dramatically lowered endorsement if he failed.

Returning to our original conception, we had proposed that endorsement of the group spokesman would vary significantly as a function of his perceived competence and the source of his authority. In addition, we hypothesized that the spokesman's success in representing the group

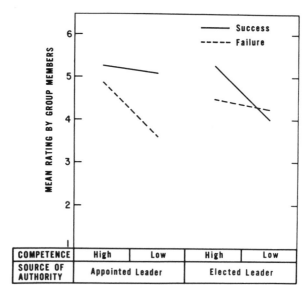

Fig. 2.   Mean member willingness to have spokesman represent group again, presented for each condition, in Experiment Three.

would influence member support. These hypotheses received substantial confirmation, although there is clearly a need for clarification of the effects of election and appointment. The more competent and successful spokesmen were endorsed more strongly than less competent and unsuccessful spokesmen, respectively.

The major question remains of how election and appointment qualify endorsement of the spokesman. As noted in the results, there was a tendency to perceive the elected spokesman more positively immediately following the group discussion, although this "halo" did not carry over to the final assessments of member endorsement. In addition, the complex interaction which emerged among competence, source of authority, and the spokesman's success appeared to indicate that there were important differences in member expectations as a function of election or appointment. For the appointed spokesman, endorsement depended merely on some indication that the "system worked," i.e., that it resulted in *either* a competent choice or a successful outcome. For the elected spokesman, however, endorsement depended upon *both* a competent choice and a successful outcome. This result may be interpreted as supporting the general position that *election builds higher expectations for success or higher demands by group members on the leadership role.* When the spokesman was appointed, members were more readily satisfied and responded positively to any sign of a positive result. When he was elected, members failed to support him if he was incompetent or unsuccessful.

If anything, this interaction indicates that election, rather than making the spokesman more secure and immune from task effects, creates a situation in which the spokesman is more vulnerable. This vulnerability may result from the greater involvement and responsibility which group members feel under election conditions, producing a stronger sense of control over the authority structure of the group. Hence, they show a greater willingness to consider alternatives. If their first choice of spokesman did not work out, they could try someone else next time around. Under appointment conditions, however, there was no assurance that members would be involved in any subsequent choice of the spokesman.

VI. Experiment Four: Leader Legitimacy and Attempted Innovation

From the earlier work presented here, it is clear that the source of the leader's authority does have an effect on followers, even in subtle ways. To review briefly, we found in Experiment Two that for elected leaders, in comparison to appointed leaders, there was a significantly stronger relationship between the followers' actual acceptance of the leader's influence attempts and their reports of having been influenced by him. In Experiment Three we saw evidence that the "success" or "failure" of an elected or appointed leader was differentially reacted to by followers, especially in regard to their continued endorsement of him in his role as leader. Thus, depending upon the followers' perception of the leader's initial competence, elected leaders were more vulnerable to a withdrawal of endorsement when they had been unsuccessful in representing the group. This finding lends support to the proposition that election, at least in this context, builds higher demands by group members on the leader's role.

As we have suggested, the group is likely to feel a greater investment in elected leaders, especially when such leadership is both desired and possible to attain. Along with this investment, the elected leader may also be someone to whom the group gives greater latitude to act in behalf of group goals. This interpretation is in line with the "idiosyncrasy credit" concept that a person gains credits which he may then draw on in exerting influence, particularly through innovations from normative expectancies. An experiment by Hollander (1960) revealed that such assertions were likely to have an effect on both the procedures for guiding the group's activity and the actual content of the group's task. For this process to work, however, the person who is a leader must be aware that these credits are at his disposal; he must, in short, sense his legitimacy. Accordingly, our emphasis in this next experiment turned to how the leader's perception of his own legitimacy determines his expressions of deviation. The term "deviation" refers here to an attempt to redirect the group's effort at a task solution. Using the idiosyncrasy credit concept of nonconformity as a feature of higher status, we may consider a willingness to deviate from the group as an indicator of attempted influence. It may also be that deviation reveals resistance to conformity pressures from the group. Thus, the leader may manifest his sense of legitimacy and higher status by assertions of influence or by his resistance to pressures from the others in the group.

It follows from the concept of gaining credits that the elected leader, in light of his supposed choice by the group, will already sense support from the others; for the appointed leader, such support may still need to be gained. The latter may indeed have to overcome the prevailing disposition that a leader should be elected. Therefore, when team members come together as a new group, election should provide a greater sense of credit than does appointment. This ought to be so in many circumstances in which groups, such as committees, are newly organizing for mutual activity. On the other hand, where appointed leadership is characteristic, as in many institutional structures, the legitimacy which goes with the role normally vests the incumbent with the expectancy of having influence. Even with such backing, however, the need exists for him to gain credits through appropriate actions as a leader.

With these considerations in mind, we predicted that where group members expect that leaders may be elected as well as appointed, elected leaders should display greater constructive deviation, as acts of influence assertion, than appointed leaders. We also predicted that a sense of strong endorsement should heighten this deviation. Thus, both source of authority and strength of endorsement as bases for legitimacy should contribute to the leader's willingness to attempt influence or counter-influence. In addition, we expected that, when they are deviating, elected and appointed leaders should differ in the amount and kind of communication they direct to their followers. With a greater sense of legitimacy, the leader's need to justify and conciliate should be lessened. Thus, his messages to the group should be less elaborate in these respects.

A. METHOD

1. Task and Setting

The task was specifically selected to engage interest and to permit the free flow of discussion among subjects in the first phase of an essentially quasinaturalistic experiment. Subsequently, it was to serve as a reasonable vehicle for the clash of ideas so as to assess the willingness of subjects placed in the position of "team leader" to overturn proposals from their team members.

For the first phase of the experiment, 52 male students from introductory psychology classes at the State University of New York at Buffalo took part in one of four discussion sessions. Each of these sessions drew at least 12 subjects, and all four were treated in as comparable a way as possible. Students had volunteered for this study of "group decision making in urban planning" from among other alternative studies available to them. Participation in research was part of their course requirement.

Once assembled for the discussion session, the subjects were provided with name tags and seated in a circle with an identifying number before them. A faculty member was introduced to them as an "expert in urban problems" who would serve as the discussion leader. Orientation materials were then distributed in which they were told: "As we all know, our cities are faced now with a multitude of problems ranging from financial shortages to major social ills. You are here today to help decide which of the problems a city faces appear to be most critical, and what actions can be taken to help alleviate these problems. Our procedure today also will help in looking at the decision-making processes involved in coping with these problems." They were then asked to read a description of about 700 words which had been developed by the Western Behavioral Sciences Institute to cover the problems of an imaginary city dubbed "Colossus." Subjects were then asked to write down the problems they individually considered to be most important, after which each would have a chance to voice his views on these.

The discussion period lasted approximately 40 minutes. Subjects were told that this was a time for them to shape their ideas and to get to know the opinions of others. The level of evident involvement was high. At the close of the discussion, the group was told that the discussion group members would now be separated and reconstituted into three teams, each consisting of a task leader and his staff. Subjects were further informed that the leaders would be separated from the others in their group, although they could pass messages to each other, largely to determine how communication processes affect decisions about ways to deal with big city problems.

2. Design and Induction of Treatments

From each discussion session, ten subjects were isolated in rooms for the second phase of the experiment, the induction of the experimental conditions. Two raters observing the group discussions in the first phase had been tallying interactions to ascertain which subjects among the 12 or more participated least; these subjects remained behind in the discussion room while the other 10 went to their assigned rooms. On arriving there, half of the subjects found an election ballot on which to nominate three people for team leader, by name or number, whichever they recalled; these subjects were in an election condition.

After a few minutes, each subject in the election condition was informed by an experimenter that he had been elected to lead a team, while each person in the appointed condition learned that the expert who had served as discussion leader had chosen him as a team leader. Cutting across these treatments, half of the subjects were told that they were the "top choice" for the leader position and would lead "Team A," and the other half were told they were the "third choice" and would lead "Team C." The result was a 2×2 factorial design with two sources of authority, and two levels of strength of endorsement. Given the five subjects each in the election and appointment conditions, following each discussion session, a counterbalanced number of subjects (2-3, 3-2) was assigned to the strong and not strong endorsement conditions. Furthermore, with four sessions it was possible to have each subject assigned to a particular room following each discussion session to be representative of a different cell in the design. Thus, each room was used only once for each of the treatments, thereby reducing any possible position effects.

3. Dependent Measures

Once isolated, all of the subjects were provided with a sheet listing various urban problems and four action programs for each problem area. In order, the ten problem areas were: Education, Welfare, Culture and Recreation, Housing, Urban Renewal and Beautification, Fiscal Affairs, Race Relations, Police and Riot Control, Industrial and Economic Development, and Transportation.

When each subject had been told he was to be a team leader, he was also provided with a sheet entitled "Task-Group Procedure." This gave him full instructions for the third phase of the experiment. It indicated that the team would be discussing each of the ten problem areas and would consider the four action programs proposed to alleviate each. On a "communication form" the team would then be ranking all four action programs within a problem area by placing a "1" in front of the program they favored most, a "2" before the one they favored next, and so on through "3" to "4," the one they favored least of all.

Each communication form was delivered to the team leader by an experimenter; the leader's task was to decide which of the four action programs should be put into effect. He did this by sending his own rankings back to the team on the same form with any comments he wished to make as a message to the team. In actuality, all of these communication forms had been prepared in advance with the alleged team rankings. The first and third, dealing respectively with the problem areas of Education, and Culture and Recreation, were presented with the actual preference rankings obtained from a pilot study with 17 male subjects drawn from the same population of introductory psychology students. The forms for the other eight problem areas were contrived to present the "team leaders" with precisely the reverse ranking from the true preference order obtained from the subjects in the pilot study. The last seven presentations of these erroneous rankings, beginning with the fourth problem area, "Housing," constituted the seven "critical trials" in sequence. The format for this form is shown below as an illustration of the general pattern employed.

The major dependent variable was the number of critical trials on

COMMUNICATION FORM – 4

Indicate your ranking of these action programs in the column marked "Leader's Ranking." Your first choice should be marked "1," your second "2," and so on. Whatever you choose as "1" will be recorded as your team's recommendation.

HOUSING

Team Ranking	Leader's Ranking	
————	————	Construct a "satellite city" offering housing for families at every income level.
————	————	Convert public-housing projects into cooperatives — owned and run by the residents.
————	————	Create a city-housing authority to own and rent housing in the city.
————	————	Sponsor low-cost private housing for poor families, with no down payments and long-term mortgages.

Any Comment:

Fig. 3. Example of the form used for communicating the team's decisions to their leader.

which the subjects reversed the team's first choice. A reversal would mean assigning a "4" to the action program the team had supposedly marked "1," and a "1" to the action program marked "4." Other dependent measures were obtained through a postdecision questionnaire, which contained five scales, and an analysis of the volume and content of the "comments" the subjects sent back to their teams on the communication forms.

Following consideration of the last of the ten problem areas, subjects were asked to fill out the postdecision questionnaire and were then brought together for a briefing about the nature of the experiment and the inductions employed. The responses to two open-ended items on that questionnaire ("Why were you chosen the leader of one of the three urban planning teams?" and "Would you please take a moment and comment on the nature of your experience as leader of a decision team and your ideas about the hypotheses under investigation?") revealed relatively few signs of suspiciousness about the procedure employed. Indeed, the bulk of subjects found no difficulty whatever in giving a response to the first question in line with either their election or appointment, and with evident satisfaction at having been selected for the role.

B. RESULTS AND DISCUSSION

1. Constructive Deviation

Figure 4 shows the major finding regarding the differential deviation of appointed and elected leaders under the two conditions of endorsement, strong or not strong. The index used in this figure is the mean number of trials, out of the seven critical trials, when the leader totally reversed the team's first-rank choice. Recall that four options are provided to be rank-ordered, and that these are presented in a preference order precisely opposite to the actual preference ranking of similar subjects. As will be seen in this figure, elected leaders deviated from their teams considerably more than appointed leaders, and in each case the presence of strong endorsement tended to increase this deviation. The highest reading is for elected leaders under conditions of strong endorsement, a value of 3.4, which indicates that on approximately half of the

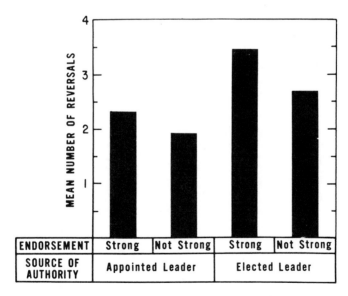

Fig. 4. Mean number of seven critical trials on which leaders reversed team's first rank choice in Experiment Four. $N = 10$ Ss for each treatment.

seven critical trials these leaders totally overturned the team's judgment. An analysis of variance of these data yielded a significant effect for source of authority ($F = 5.11$, $df = 1/36$, $p < .05$).

This finding represents a highly specific confirmation of our main hypothesis. More generally, however, we find that the sum of all deviations from the group rankings on seven trials is significantly higher for elected than appointed leaders. A chi-square analysis shows elected leaders to be significantly inclined toward higher deviation and appointed leaders toward lower deviation ($X^2 = 5.18$, $df = 1$, $p < .05$).

2. Postdecision Ratings

With regard to the success of the inductions, Table VIII provides a number of relevant findings. In response to the first item, as expected, elected leaders reported themselves as having more support. Here again, a chi-square analysis yielded a significant difference between the elected and appointed leaders ($X^2 = 3.8$, $df = 1$, $p < .05$).

TABLE VIII

MEAN RATINGS ON QUESTIONNAIRE ITEMS FOR EXPERIMENT FOUR[a]

Items	Source of authority and strength of endorsement			
	Appointed leader		Elected leader	
	Strong ($N = 8$)	Not strong ($N = 10$)	Strong ($N = 10$)	Not strong ($N = 10$)
1. To what extent team members supported you as their choice for task leader.	3.1	2.7	3.7	3.3
2. To what extent you agreed with planning judgments recommended by your team.	2.8	2.5	2.8	2.5
3. How competent you think you are to recommend policy on urban problems.	3.9	3.8	4.0	4.6
4. How much you enjoyed participating in this study.	4.9	4.9	4.5	4.8
5. How much restrictions on communication affected your final decisions.	4.0	3.6	3.7	4.7

[a] Ratings were on six-point scales, positively oriented.

In connection with the report of agreement with the judgments recommended by the members of the team, the second item of Table VIII reveals no significant difference by treatment. Accordingly, a further analysis was done, employing tetrachoric correlation, to test the association between the deviation reported and the actual sum of a subject's deviations on the seven critical items. The coefficient obtained was .78, with $p < .01$.

Regarding perceived competence in the handling of urban problems, the third item in Table VIII reveals differences which produced significant main effects for both source of authority and strength of endorsement, as well as a significant interaction term. The highest F obtained was for source of authority ($F = 14.70$, $df = 1/36$, $p < .001$), with elected leaders significantly more inclined to see themselves as competent. The main effect for endorsement was less pronounced ($F = 4.41$, $df = 1/36$, $p < .05$). The interaction term, as already noted, was also significant ($F = 9.55$, $df = 1/36$, $p < .01$), and is attributable to the high mean of 4.6 for elected leaders without strong endorsement. This finding is anomalous and will be discussed subsequently.

For the fourth item in Table VIII, enjoyment in participating in the study, no significant difference was found between the treatments, in line with expectation. In general, however, it is noteworthy that all of the means for this item approximate a value of 5 on the 6-point scales, thus revealing a uniformly high degree of enjoyment. For the fifth item, regarding the effect of the restrictions on communication in reaching decisions, a significant interaction was obtained by analysis of variance ($F = 4.50$, $df = 1/36$, $p < .05$). Here again, elected leaders with less strong endorsement showed the high value.

3. Leader Communications

Figure 5 summarizes the results dealing with the number of words used in communications to the team when the leader totally reverses the team's first choice. An analysis of variance revealed a nearly significant main effect for strength of endorsement ($F = 3.01$, $df = 1/36$, $p < .10 > .05$). The strongly endorsed leaders used fewer words when reversing their team's rankings.

Pursuing a related consideration regarding the nature of communications, a content analysis was undertaken of the messages directed to the teams by each of the subjects as leaders. The messages from each were rated independently by the three investigators without knowledge of condition on a scale that ranged from 0 to 3. The quality which proved to have the greatest reliability of ratings, and the greatest relevance to the issues under study here, was referred to as "group orienta-

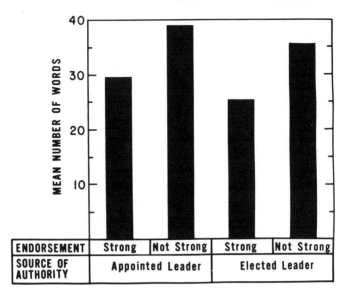

Fig. 5. Mean number of words contained in written messages from leaders to teams on trials where team's first rank choice is reversed in Experiment Four. $N = 10$ Ss for each treatment.

tion." This is taken to be a willingness to show a recognition of the viewpoint of the team and to conciliate differences. The corrected reliability of this measure, for three raters, was .89. Figure 6 reports the average ratings obtained, on the rating scale of 0 to 3, for leaders in the four treatment conditions. As can be seen, those *least* likely to reveal such an orientation in their messages to members were the elected leaders with strong endorsement. The interaction term, in an analysis of variance, was significant ($F = 6.95$, $df = 1/36$, $p < .01$).

Overall, these findings provide substantial additional evidence of the effects of election or appointment on leader behavior. The leader's sense of legitimacy does appear to be a compelling factor in determining his willingness to assert influence through constructive deviation. Evidently, the leader acts with an awareness of his source of authority, and this has consequences in his responses to the team.

Specific aspects of the results are also in accord with our expectations. Thus, the strongly endorsed elected leader not only deviates significantly more, but he also uses fewer words in his messages to the team when deviating. Furthermore, his messages in general reveal less conciliation and, presumably, less need to justify his position. Alternatively, the appointed leader without strong endorsement appears by comparison to be far weaker as a source of influence, in terms of these measures. Both aspects of legitimacy are therefore found to be effective in producing the leader's response to followers.

Among the results from the postdecision questionnaire are two unexpectedly high means for elected leaders without strong endorsement (see Table VIII, items 3 and 5). The first of these is not in accord with prediction. One may conjecture that some kind of compensatory reaction is being expressed to reveal a greater sense of competence in the absence of strong endorsement. In the case of the effect of restriction of communication, shown in item 5, again the elected leaders without strong endorsement appear to be affected most. All in all, these results suggest that the elected leader expects strong endorsement, as a function of the very process by which he was cast in the leader's role, and that its absence may prove disquieting.

VII. Conclusions and Implications

The findings from these experiments point to several conclusions, both general and specific. On the most general level, it seems eminently clear that the study of leadership can be fruitfully pursued in terms of

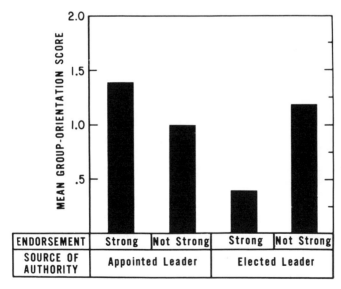

Fig. 6.   Mean Group-Orientation Score from content analysis of written messages from leaders to teams in Experiment Four. $N = 10$ Ss for each treatment.

interpersonal processes, notably the perceptions and expectations of followers. The observed attributes and actions of the leader have a discernible role in creating the impressions which contribute to his legitimacy. Furthermore, as the last experiment indicated, the leader's sense of his legitimacy has a direct impact on his willingness to assert influence.

There is also substantial support in these findings for the efficacy of multifactor designs in the study of leadership phenomena. Several variables may be looked at in combination to reveal their interactions. Indeed, among the more repetitive findings is the extent to which these interactions govern outcomes. Thus, while we had consistent confirmation of the leader's perceived competence as a significant determinant of his acceptability and influence, we also found that this relationship was affected by the followers' perceptions of the leader's motivation. In the first experiment, for example, there was a significant interaction between the leader's perceived competence and his perceived interest in the group's activity. The consequence of this interaction effect was most evident for the allegedly "poor performing" leader who apparently could retain his status when seen as having a sincere interest in that activity.

These relationships are further qualified by the contextual effects of the leader's source of authority. Whether he is elected or appointed also matters, especially under conditions of strain. In the third experiment, for example, the members' support for the elected leader depended both on his initially perceived competence for the role of spokesman and a successful outcome of his activity in behalf of the group; on the other hand, support for the appointed leader appeared to depend on either the perception of his competence or a successful outcome. This suggests, as we mentioned earlier, that election builds higher demands on the leadership role, at least by group members in this situation. It is by no means clear, however, whether this finding will hold where different kinds of group tasks and settings are involved. It is no great surmise to suggest that such differences should shape expectations regarding the leader's source of authority and what its payoffs ought to be. For instance, the relatively stronger position of the appointed leader in the third experiment does not jibe with at least one other relevant finding from the research of Carter, Haythorn, Shriver, and Lanzetta (1951): Their data indicated that the emergent leader was in a more favorable position. It is noteworthy that the dependent measures used in that study were somewhat different from those employed in our work. Yet the point is worth making that characteristics of the setting itself may shade the factor of favorability for the leader, an influence on which Fiedler (1967) has laid particular stress.

In this same vein, the results of the last experiment indicated broadly that the elected leader felt more confident and was more willing to exert influence through constructive deviation from the group's judgments than was the appointed leader. Unlike the third experiment, we did not introduce the variable of success or failure, nor did we examine the reactions of followers to the leader, since neither was a primary focus. Further research can deal with these relationships more fully. In any event, we have confirmation in this last experiment that, given a situation in which it is appropriate to have an elected leader, the incumbent of that role senses the idiosyncrasy credits at his disposal and can expend them through deviation from the group. Within a situation laden with formal organizational constraints, the expectation that there even could be an elected leader would be severely limited and this finding would be unlikely.

Another point that requires attention is the confinement imposed by research over a limited time span in the laboratory. As we moved toward increasingly naturalistic settings in the third and fourth experiments, we found it increasingly possible to study the dynamics of leadership attainment and maintenance. Though difficult to manage at times, the introduction of more reality in the task heightened the involvement of the participants and the scope of our findings. In addition to the often heard injunction to extend research into the field, which seems reasonable enough, it may also prove helpful to create "quasinaturalistic" experiments along the lines of our later studies.

Since leadership implicates various leader roles, beyond the traditional function summed up by Hemphill (1961) as "initiating structure," there is considerable room for studying these functions. In the third experiment we considered the spokesman as an occupant of a leader role, and in the fourth experiment we ventured into decision-making and innovating functions. These distinctions deserve more attention in research on leadership than they have thus far received. Also, additional study might be turned to the sequence implied by attaining a leader role and then maintaining it successfully. Those starting mechanisms which operate in emergent leadership (cf. Hollander, 1961a) may be quite critical in determining the leader's subsequent acceptance and influence, but this process needs to be more fully detailed through research.

We embarked upon this set of experiments within a process-oriented, transactional framework. It has afforded the basis for looking into interpersonal processes of perception and exchange in leadership. On balance, that framework appears to be well sustained as an avenue by which we can understand the workings of these phenomena better.

REFERENCES

Alvarez, R. Informal reactions to deviance in simulated work organizations: A laboratory experiment. *American Sociological Review*, 1968, **33**, 895-912.

Berkowitz, L., & Macaulay, J. R. Some effects of differences in status level and status stability. *Human Relations*, 1961, **14**, 135-148.

Carter, L. F., Haythorn, W., Meirowitz, B., & Lanzetta, J. The relation of categorizations and ratings in the observation of group behavior. *Human Relations*, 1951, **4**, 239-254.

Carter, L. F., Haythorn, W., Shriver, E., & Lanzetta, J. The behavior of leaders and other group members. *Journal of Abnormal and Social Psychology*, 1951, **46**, 589-595.

Carter, L. F., & Nixon, M. An investigation of the relationship between four criteria of leadership ability for three different tasks. *Journal of Psychology*, 1949, **27**, 245-261.

Cohen, A. M., & Bennis, W. G. Continuity of leadership in communication networks. *Human Relations*, 1961, **14**, 351-367.

Croner, M. D., & Willis, R. H. Perceived differences in task competence and asymmetry of dyadic influence. *Journal of Abnormal and Social Psychology*, 1961, **62**, 705-708.

Fiedler, F. E. *A theory of leadership effectiveness*. New York: McGraw-Hill, 1967.

Fouriezos, N. T., Hutt, M. L., and Guetzkow, H. Measurement of self-oriented needs in discussion groups. *Journal of Abnormal and Social Psychology*, 1950, **45**, 682-690.

Gibb, C. A. The principles and traits of leadership. *Journal of Abnormal and Social Psychology*, 1947, **42**, 267-284.

Goffman, E. *The presentation of self in everyday life*. Garden City, N. Y.: Doubleday Anchor, 1959.

Goldman, M., & Fraas, L. A. The effects of leader selection on group performance. *Sociometry*, 1965, **28**, 82-88.

Harvey, O. J., & Consalvi, C. Status and conformity to pressures in informal groups. *Journal of Abnormal and Social Psychology*, 1960, **60**, 182-187.

Hemphill, J. K. The leader and his group. *Education Research Bulletin*, 1949, **28**, 225-229, 245-246.

Hemphill, J. K. Why people attempt to lead. In L. Petrullo & B. M. Bass (Eds.), *Leadership and interpersonal behavior*. New York: Holt, Rinehart & Winston, 1961. Pp. 201-215.

Hollander, E. P. Conformity, status, and idiosyncrasy credit. *Psychological Review*, 1958, **65**, 117-127.

Hollander, E. P. Competence and conformity in the acceptance of influence. *Journal of Abnormal and Social Psychology*, 1960, **61**, 365-369.

Hollander, E. P. Emergent leadership and social influence. In L. Petrullo & B. M. Bass (Eds.), *Leadership and interpersonal behavior*. New York: Holt, Rinehart & Winston, 1961. Pp. 30-47. (a)

Hollander, E. P. Some effects of perceived status on responses to innovative behavior. *Journal of Abnormal and Social Psychology*, 1961, **63**, 247-250. (b)

Hollander, E. P. *Leaders, groups, and influence*. New York: Oxford University Press, 1964.

Hollander, E. P., & Julian, J. W. Leadership. In E. Borgatta & W. W. Lambert (Eds.), *Handbook of personality theory and research*. Chicago: Rand McNally, 1968. Pp. 890-899.

Hollander, E. P., & Julian, J. W. Contemporary trends in the analysis of leadership processes. *Psychological Bulletin*, 1969, **71**, 387-397.

Homans, G. C. *Social behavior: its elementary forms*. New York: Harcourt, Brace & World, 1961.

Julian, J. W., & Steiner, I. D. Perceived acceptance as a determinant of conformity behavior. *Journal of Social Psychology*, 1961, 55, 191-198.

Katz, D., & Kahn, R. *The social psychology of organizations*. New York: Wiley, 1966.

Mann, R. D. A review of the relationships between personality and performance in small groups. *Psychological Bulletin*, 1959, 56, 241-270.

Mausner, B. The effect of one partner's success in a relevant task on the interaction of observer pairs. *Journal of Abnormal and Social Psychology*, 1954, 49, 557-560.

Osgood, C. E., Suci, G. J., & Tannenbaum, P. H. *The measurement of meaning*. Urbana, Ill.: University of Illinois Press, 1957.

Raven, B. H., & French, J. R. P. Legitimate power, coercive power, and observability in social influence. *Sociometry*, 1958, 21, 89-97.

Sabath, G. The effect of disruption and individual status on person perception and group attraction. *Journal of Social Psychology*, 1964, 64, 119-130.

Stogdill, R. M. Personal factors associated with leadership: A survey of the literature. *Journal of Psychology*, 1948, 25, 35-71.

Wiggins, J. A., Dill, F., & Schwartz, R. D. On "status-liability." *Sociometry*, 1965, 28, 197-209.

A Further Look at Leader Legitimacy, Influence, and Innovation[1]

Edwin P. Hollander and
James W. Julian

STATE UNIVERSITY OF NEW YORK AT
BUFFALO

In conceiving the line of research presented in our original chapter, we had two main interests. First, we wanted to move toward a more transactional approach to leadership, taking greater account of leader–follower relations over time. Second, we tried to break away from the focus on leader characteristics as such and look more at the bases for the leader's legitimacy as perceived by followers and by the leader.

Legitimacy was a convenient way of summarizing several variables that we saw as part of a process affecting leader–follower relations. These variables were the leader's source of authority, through appointment or election, perceived competence and motivation, and success or failure in producing desired outcomes for the group.

Since that chapter was written, we and our colleagues, many of whom are former students of ours, have done further research in this and related areas. Sometimes those same variables were used, and other times new ones were introduced. In this paper we will be reporting briefly on this research. Its several features include: an extension of

[1] The preparation of this paper was facilitated by Contract N00014-76-C-0754 from the Office of Naval Research, Organizational Effectiveness Research Programs. It is Technical Report No. 4, dated June 1977, under that contract, in which the first author is Principal Investigator. Two of the studies included here were issued as Technical Reports under this contract. They are Hollander, Fallon, and Edwards (1977), which was No. 1, and Gleason, Seaman, and Hollander (1978), which was No. 3.

experimentation on leader appointment or election and success or failure; quantity and quality of participation in situations of emergent leadership; and the interaction of personality and task set in emergent leadership.

I. Further Findings on Appointment versus Election and Success versus Failure

Pursuing the broader line of this research on appointment or election, Hollander, Fallon, and Edwards (1977) reported two experiments in which there was an actual election of leaders, contrasted with those who were appointed, and cross-cut with conditions of success or failure. In the first experiment, twelve groups composed of four male students each were presented with typical urban problems from the city called "Colossus." This is the task that had been used in the experiment by Hollander, Julian, and Sorrentino (1969), reported as the last one in the original chapter. The second of the new experiments also used the Colossus material, with a third phase added to study the effects of having a change in leaders.

Immediately before grouping together for discussion in both experiments, the subjects were provided with a form listing the various urban problems, with four possible action programs for each. They were asked to make their own individual determinations of the order of priority among these programs for each of the problem areas. After the forms were collected, subjects were assigned on a random basis to one of the four adjoining rooms and spent a few minutes discussing which problems they thought were most important. They then either elected a group discussion leader or had one appointed by the experimenter.

Each group's task was to arrive at group rankings of the action programs for each of the problem areas, namely, education, race relations, welfare, transportation, and so on. After the first five problem areas had been completed in about a half hour, one each of the groups with elected and appointed leaders was led to believe that it had done well or not well, that is, had succeeded or failed with respect to the "acceptability" of the first ranking they had assigned in each problem area. All of the groups then continued with the next five of the problem areas, discussing them to reach group consensus on each.

The primary dependent measure from these face-to-face discussions was calculated by an "influence ratio" that took account of the disparity between an individual's original ratings and the group's ratings. The less the disparity, the greater the influence. In general, it was found that the *appointed* leaders were *less* influential, proportional to how much change they could possibly have induced, than were elected ones, thus confirming one aspect of earlier results.

A more compelling result, however, was the difference in the influence of elected leaders as compared to appointed ones in the failure condition. Briefly, the elected leaders showed a substantial gain in influence *after failure*, whereas no such shift was evident for the appointed leaders. Alternatively, the elected leaders, *after* success, were found to be less influential during the second phase than they had been during the first. A further finding, to be noted shortly, shows that the elected leaders' initial gain in influence after failure is short-lived.

The interpretation offered for these findings, which included post-discussion responses, is that failure was seen by the group as a "crisis" (cf. Hamblin, 1958). In dealing with it, the elected leader was able to assert greater influence and the group in turn gave him greater latitude in accepting his position. Under the success condition, however, there was no crisis and, accordingly, group members acted out of a greater security in their own judgments.

As a further probe in the direction of the influence of leaders over time, the second experiment reported by Hollander, Fallon, and Edwards (1977) took a closer look at the followers' tolerance of the leader and their willingness to have him continue. There were thirteen groups, again of four male students each, in this experiment. A major change was that a third phase, of five more trials, was introduced. Before it, group members could choose a new leader or have a new one appointed, depending upon the condition.

The findings of this experiment showed that the heightened influence of the elected leader in the second phase was indeed brief. If the group saw no signs of greater success, then the leader was deposed, despite his rise to greater influence following the crisis. After a point, then, the perpetuation of the crisis did not serve to sustain the leader's position. This finding is in accord with Alvarez's (1968) research that demonstrated that leaders lost esteem more rapidly than nonleaders in groups which experienced failure, since the leaders were held more responsible for the group's outcomes.

These data also indicate that the newly elected leader is more influential in the phase just before he takes over than when he does become the leader. The contrast is sometimes quite dramatic. Furthermore, even a not-too-successful leader is later seen more favorably than his successor, thus suggesting an initial handicap for the successor that somehow must be overcome. Succession therefore offers promising leads as an area for further study.

More than the literal findings of this set of experiments is the dynamic quality it illustrates. The emphasis is on getting at the followers' perceptions in relationship to the perceptions and actions of leaders and, then, the counteractions of followers (cf. Hollander, 1978). More aspects of this relationship will need to be plumbed, beyond the matters of source of authority and success or failure. Among those aspects are the means by which the leader's position is legitimated and the validators to whom the leader must be responsive. Relatedly, there is the matter of task set and the composition of the group.

Another experiment (Edwards, 1975) used the Colossus material to investigate the effect on leadership of the presence or absence of an evaluative set in the instructions given to the group. The guiding hypothesis was that the members' perceptions of an imminent evaluation on the task would increase the leader's latitude to exert influence.

There were 16 groups of male students in the experiment, each of which elected its own discussion leader. The experimental session was conducted in two phases. Before each of the phases, the experimenter informed half the groups that the group choices would be scored for acceptability, but the experimenter said nothing about scoring to the other half, in a cross-over design. Therefore, four treatments were created by combinations of sets: Eval-Eval, Eval-None, None-Eval, and None-None. After the first phase, all groups were given failure feedback indicating that they had not scored well.

The results of the experiment were directly contrary to the hypothesis. An analysis of variance showed a significant main effect for task set ($p < .01$), with the leaders most influential when there was no evaluative set (None-None). This supports the rationale of the research, insofar as the presence or absence of an evaluative set affects leadership. However, the inverse result requires explaining. This takes the form of proposing that the complete lack of an evaluative set is actually a condition of ambiguity. Given the leader's position, the effect of ambiguity is to enhance dependence upon the leader as a unique resource. In short, the leader is valued more as a provider of information and direction. At the very least, this is a suggestive point for further study.

Another area of fruitful research is the effect of sex-composition on leadership in groups. Very little has been done on this as compared with the large number of studies with all-male groups. An experiment by Fallon (1973) addressed this point with groups made up of two male and two female students with either an appointed or an elected leader. The same two-phase Colossus material discussed previously was used. Briefly, the results indicated that, after having been given feedback on group performance, male leaders were more influential than female leaders regardless of their source of authority or the feedback given. Examination of the postinteraction questionnaires supported the conclusion that this patterning of influence resided in sex-typed social expectancies, with leadership considered to be more a male than a female domain.

In a follow-up experiment reported by Fallon and Hollander (1976), 32 groups of two male and two female students each elected leaders. The three-phase Colossus material noted above was used, thereby extending the duration of the interaction. It should also be noted that an equal number of males and females served as leaders, even with a bias toward electing males. This was accomplished by declaring the female the winner when an examination of the closed ballots indicated a tie between a female and a male. In all cases, therefore, the leader was the most or equal to the most chosen person in the group.

After the first and second phases, the groups were given consistent success or failure feedback, cross-cutting the male or female leader distinction. Therefore, there was an equal number of groups in the four conditions of success-male, success-female, failure-male, and failure-female. The results were again determined by an influence measure indicating how much any one person's judgments were paralleled by the decisions made subsequently by the group.

The findings of this experiment were consistent with those of Fallon (1973). Regardless of the type of feedback, the male leaders were significantly more influential than the female leaders in the last two phases. Indeed, female leaders significantly decreased in influence after feedback, whereas male leaders maintained their influence. This effect is shown in Figure 1.

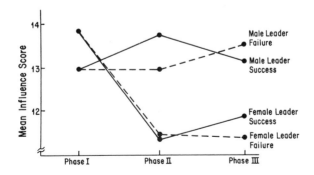

Figure 1. Mean influence scores in three phases for male and female leaders, with success or failure feedback to their groups after Phases I and II. $N = 8$ groups of 4 members for each condition. (From Fallon & Hollander, 1976.)

II. The Importance of Verbal Participation
to Emergent Leadership

The legitimation of the leader in an emergent situation was assumed to rest on the members' perceptions of the "leadership potential" which an individual displays during the group interaction. Thus, a central assumption in this approach is that the interactive behavior of the leader is an important determinant of members' perceptions of his or her abilities and motivations in the context of the group. By implication, the legitimacy of the leader and the members' acceptance of the leader's influence is seen to follow from these perceptions.

The question of the actual behavior which influences attributed leadership ability confronts the consistent finding in the literature that it is the quantity or frequency with which an individual interacts that predicts the ratings of his leadership, rather than the substance of what he or she has to say (e.g., Riecken, 1958). However, what a person says, or the quality of the contributions made, has been confounded in the typical investigation of this relationship. To evaluate this relationship experimentally, Regula and Julian (1973) manipulated both the quantity and the quality of individual verbal contributions to the task, with the aim of investigating their combined influence on perceptions of task ability. In another study, Sorrentino and Boutillier (1975) focused more directly on the issue by manipulating these same variables in an interactive group setting where their impact on ratings of both task and socioemotional leadership could be assessed.

Regula and Julian (1973) manipulated talkativeness, or relative frequency of contributing to the task, by preparing scripts for two confederates. The confederates ostensibly worked together to complete a creativity task before an audience of 94 subject-observers. The two confederates appeared to be chosen at random from the larger group in order to demonstrate a task which all subjects anticipated performing in due course. The creativity task was to suggest as many alternative uses for a wire coat hanger as possible. The actual uses suggested by each confederate and the frequency were completely controlled. The originality of the particular uses was established from previous pilot research in which judges rated the originality of various possible uses.

Three distinct conditions were created: a quantity condition in which all uses were of low originality but one confederate suggested 10 possibilities, whereas the other confederate suggested only 6; a quality–quantity condition in which 6 high quality suggestions were pitted against 10 uses of low originality; and a quality condition in which both confederates suggested 10 uses, one consistently original and the other low in originality. The primary dependent measures were the audience-subjects' ratings of the two confederates, particularly their judged creativity and their ability to influence others.

The results obtained in this experiment provided strong confirmation for the importance of frequency of participation as a factor influencing judgments of task ability. Virtually no differences in observers' perceptions were found in the quality condition, in which only the quality of task contributions differentiated the performances of the two confederates. In each of the other two conditions the confederate who contributed more frequently was seen as more creative and as wielding greater influence. This was the case even in the condition in which his more frequent contributions were of distinctly lesser originality, as calibrated beforehand.

These results, though consistent with earlier investigations, were observed in a setting in which the subjects were observers so that they were not actually participating in a group problem-solving process in which ratings of leadership or influence might have salience. To sharpen the examination of this issue, Sorrentino and Boutillier (1975) manipulated both quantity and quality of participation in an interactive laboratory task. Each of 16 four-man groups was composed of three naive subjects and a confederate. The four experimental conditions were manipulated by having the confederate follow one of four carefully prepared scripts for his interaction during the problem-solving process. Each of the four conditions was a combination of high or low quality and high or low quantity of interaction. To permit the confederate to use a script, the participants were separated upon their arrival at the laboratory for the ostensible reason of more accurate recording of their discussions.

The task was an adaptation of the one used by Hollander (1960) in earlier research. In it, subjects discuss which column of a 7×7 payoff matrix will be next in a supposedly predetermined pattern and then predict the row that maximizes the points gained and/or minimizes losses. All groups began with 200 points and worked at mastering the patterns for a 15-trial game. Each trial included a 1-minute individual choice period in which members were to think about their choices privately, followed by a 3-minute group discussion period during which the group members were to decide collectively on their group's choice. All groups chose to use majority decision for reaching a group choice. The major dependent variables were included in a postsession questionnaire administered at the completion of the game. The questionnaire asked for ratings of task and socioemotional leadership, in addition to judgments of influence on others, contributions to the group goal, competence, and interest.

The scripts which created the manipulations of the quality and quantity of the confederates' participation were developed in extensive pilot work following procedures used by Sorrentino (1971). Quantity was equated with the number of statements the confederate made on a trial, whereas quality of interaction was determined by knowing the number of points the group would accrue if it adopted the particular choice advocated by the confederate. For example, for the high quality–high quantity condition, the confederate was a relatively frequent contributor (as determined by the distribution of comments obtained in the pilot analyses) who advocated consistently the choice that would be successful. The other combinations of quality and quantity were similarly manipulated.

The results showed once again that quantity of interaction was a strong determinant of the perceptions of group members. Quantity of verbal participation was a significant determinant of ratings of the confederate's competence, confidence, interest, influence, and contributions to the group goal, in addition to its strong effects on ratings of both task and socioemotional leadership. Differences in the quality of participation influenced only judgments of competence, influence, and contributing to the goal. Quality of verbal interaction did, however, have a significant interactive effect on ratings of socioemotional leadership.

This intriguing pattern of results provided some initial insights into the dynamics of the ways in which members' contributions influence attributed leader characteristics. Without a doubt, quantity of verbal interaction is overwhelming in its influence on the perception of leadership ability and, as such, emerges as at least a relatively more salient or obvious feature of the interactive process. People who interact more frequently are more apt to receive the initial endorsement of group members for leadership roles. This seems to be the case consistently. The Sorrentino and Boutillier study also suggests that competence is judged to be a function of the quality of interaction as well. The confederate whose contributions were of high quality was seen as more competent. Indeed, competence or ability, in combination with a high quantity of participation, was especially valued for socioemotional leadership. Thus, it would appear that, at least in the early stages of group development, member motivation to perform, as indexed by their quantity of participation, is a more potent determinant of leader emergence than is perceived ability. This plausible inference certainly warrants further investigation.

III. Personality, Task Set, and Emergent
Leadership Processes

As we have previously noted (cf. Hollander & Julian, 1968, 1969, 1970; Hollander, 1974), leadership has been viewed either as a function of the personality characteristics of the individual or of the characteristics of the situation. The newer conception of emergent leadership phenomena that we favor requires consideration of leadership as an outcome of the interaction of personality and situational demands.

An experiment by Sorrentino (1973) examined emergent leadership as a function of group members' achievement orientation and affiliation motives. Twenty-one groups of four male students each were composed by using the scores they had obtained on measures of need for achievement, test-anxiety, and need for affiliation. Four motive categories were thereby created, with each member of the group differing from the others in his combination of achievement-related and affiliation motives.

The groups performed a task adapted from Hollander (1960) that was referred to earlier as a feature of the experiment by Sorrentino and Boutillier (1975). Subjects are presented with a 7 × 7 payoff matrix and are asked to predict the row that will come up next in a supposedly predetermined pattern. The object is to gain or minimize losses. In this experiment, the groups were told either that the ability to engage in future activity was contingent, or, alternatively, that it was not contingent, upon success at the immediate activity. It was found that members who were high on need for achievement, low in test anxiety, and high on affiliation received the highest scores on emergent leadership measures in the noncontingent condition, but not in the contingent condition. This finding suggests that when a group is working for the attainment of future as well as immediate goals, achievement and affiliation motives are important for leadership in task-oriented groups. More broadly, this experiment indicates the strong interaction between personality and situational characteristics.

In an experiment by Gleason, Seaman, and Hollander (1978), another situational factor, task structure, was varied in combination with the personality characteristic known as Machiavellianism. As an interpersonal style, Machiavellianism has shown some success in the prediction of social behavior. It has been found to be associated with "emotional detachment in interpersonal relations, a tendency to exploit situations and others for self-gain, and a tendency to take over *control* in small groups" (Geis, 1968). Past research has shown that those scoring high on the Machiavellianism ("Mach") Scale developed by Christie and Geis (1970) tend to initiate and control structure in a situation when taking initiative is a viable position.

The specific hypotheses investigated in this experiment were that High Machs would show more leadership-related behaviors and would be perceived to be leaders more than would Low Machs when there is a task of Low Structure, and that Low Machs would show more leadership behaviors, and the like, under a task of High Structure. Emergent leadership was measured by both behavioral observations and questionnaire ratings.

Sixteen groups of four male students were used in this experiment. Each group was composed of one High, two Medium, and one Low Mach, based on scores on the Mach V Scale (Christie & Geis, 1970). The groups were randomly assigned to either a High or a Low Structure condition. The procedure consisted of having subjects come together in the experimental room and receive identification tags from Experimenter 1. Experimenter 2, who was blind to the subjects' Mach levels, and two expert observers who were blind both to hypotheses and to conditions, observed through a well-concealed one-way mirror. Experimenter 1 left, and then Experimenter 2 entered and administered the appropriate treatment.

In both structure conditions, subjects were given instructions for a model-building task, using a "Supertransit" Tinkertoy Model Kit. The models which could be made were a jet airplane, a steamboat, and a suspension bridge. Although Low Structure groups were free to make their own decisions about how to organize their construction team and to proceed with model construction, the High Structure groups were given strict instructions in these matters. Group members then had a 10-minute session for planning, followed by a 15-minute task period. For all conditions, there was a plan sheet from which members could choose models and follow their form.

During the planning and task periods, the two expert observers scored group members' interactions as "ascendent" or "accepting," using a modification of the Bales system (1950). Inter-rater reliability was .94. After the interaction periods, the subjects completed a questionnaire dealing with their group experiences.

The two major behavioral measures available in the experiment were: (a) the time of possession of the plan sheet (control of a key resource); and (b) "accepting" and "ascendent" behaviors. Analysis of variance of the total "time of possession" measure for the task and discussion periods together produced a main effect for Structure ($p <$ .02), with subjects holding the plan sheet longer in the Low Structure condition than in the High Structure condition.

Although analysis of "ascendent" statements showed no significant effects, a main effect for Structure occurred for "accepting" statements ($p < .05$). More accepting statements were made under the High Structure condition than under the Low Structure condition.

With regard to the expected interaction between Mach level and Structure, it *was* found on the questionnaire item asking "How well do you imagine your group did relative to other groups at the same task?" The highest mean was for High Mach and Low Structure, and an ANOVA yielded a significant interaction effect at the .05 level. On the other postinteraction measures found to yield statistical significance, only a main effect for Structure was obtained.

The major main effect for Mach level was found for the question: "Other than yourself, which member of the group would you be most willing to have as the leader in a similar group in the future?" It was analyzed by chi-square. Though not consistent with the hypothesis, the results were nevertheless revealing. Medium Machs were chosen as leaders significantly more often than either the High or Low Machs in *both* the High and the Low Structure conditions ($p < .02$).

Machiavellianism did, therefore, make a difference in leader choice. There was at least the suggestion of a preference for the moderate person in this aspect of leader personality. However, the absence of an interaction effect with Structure on any of the behavioral measures suggests that other contingencies were also involved. Perhaps the most important finding regarding such contingencies in emergent leadership was the main effect for Structure. The Low Structure situation evidently provided more maneuverability for emergent leadership, as was shown most clearly in the measure of "time of possession" of the plan sheet.

In sum, therefore, this experiment reveals some of the powerful effects of Structure on emergent leadership and suggests that the importance of personality may have been masked by these effects. It would be incorrect to conclude that a personality variable, such as Machiavellianism, is only of limited importance for emergent leadership, especially in light of the experiment reported previously. Indeed, the leader choice data indicate that persons of moderate Mach level were preferred for the leader role. This finding therefore supports the necessity for an interactive conception of the emergence of leadership.

IV. Conclusions

We have reviewed some of the leadership research in which we have been involved in some way over the years since our chapter was written. In doing so, we discovered a good deal of continuity in this work, linking back to our earlier conceptions about the transactional qualities of leadership. Hollander's (1978) book deals further with these.

Clearly, the source of authority distinction, with respect to appointment or election, shows even more vitality now. It also reveals a persisting interaction with the success–failure variable. Perhaps of even more interest is the finding that this interaction is bounded by the mixed-sex composition of the group. The evidence that male leaders retain an advantage in influence as compared to female leaders strongly suggests a sex-role stereotype effect. There is also an effect on leader influence from the task set given to the group.

The most striking thing about the work on quantity and quality of participation in affecting emergent leadership is the continuity of the basic finding. That is, quantity matters more than quality, at least initially, although longer term studies of leader–follower relations are needed.

The results on personality and task set are also pertinent to emergent leadership, but they are less definitive with respect to their implications for an interactive conception of the basis for leader emergence. This is obviously an important area for additional research.

Finally, we believe that this newer work more nearly approximates the complexity of the transactional features of leadership phenomena. This is consistent with the recent efforts in this direction by Graen (1975), with his Vertical Dyadic Linkages model of leader–follower relationships. It also is seen in Mintzberg's (1973) conclusion from a study of top executives that their situations are not so homogeneous in quality across a day's activity as has been suggested in the past. These are only some of the suggestive new lines of thinking that need to be followed up more thoroughly in further work in this area.

REFERENCES

Alvarez, R. Informal reactions to deviance in simulated work organizations: A laboratory experiment. *American Sociological Review*, 1968, **33**, 895–912.

Bales, R. F. *Interaction process analysis: A method for the study of small groups.* Reading, Mass.: Addison-Wesley, 1950.

Christie, R., & Geis, F. *Studies in Machiavellianism.* New York: Academic Press, 1970.

Edwards, M. T. Leader influence and task set. Unpublished M.A. thesis, Department of Psychology, State University of New York at Buffalo, 1973. Presented as a paper entitled "Effect of group task set on leader's influence" at the Eastern Psychological Association Convention, 1975.

Fallon, B. J. Leader legitimacy and influence in mixed-sex groups with male or female leaders. Unpublished M.A. thesis, Department of Psychology, State University of New York at Buffalo, 1973.

Fallon, B. J., & Hollander, E. P. Sex-role stereotyping in leadership: A study of undergraduate discussion groups. Paper presented at the American Psychological Association Convention, 1976.

Geis, F. Machiavellianism in a semireal world. *Proceedings of the 76th Annual Convention of the American Psychological Association,* 1968, 3, 407–408.

Gleason, J. M., Seaman, F. J., & Hollander, E. P. Emergent leadership processes as a function of task structure and Machiavellianism. *Technical Report No. 3,* ONR Contract N00014-76-C-0754. State University of New York at Buffalo, Department of Psychology, November, 1976. *Social Behavior and Personality,* 6(1), 1978 (in press).

Graen, G. Role-making processes within complex organizations. In M. D. Dunnette (Ed.), *Handbook of industrial and organizational psychology.* Chicago: Rand McNally, 1975. Pp. 1201–1245.

Hamblin, R. L. Leadership and crises. *Sociometry,* 1958, 21, 322–335.

Hollander, E. P. Competence and conformity in the acceptance of influence. *Journal of Abnormal and Social Psychology,* 1960, 61, 361–365.

Hollander, E. P. Processes of leadership emergence. *Journal of Contemporary Business,* 1974, 3(4), 19–33.

Hollander, E. P. *Leadership dynamics: A practical guide to effective relationships.* New York: Free Press/Macmillan, 1978.

Hollander, E. P., & Julian, J. W. Leadership. In E. F. Borgatta & W. W. Lambert (Eds.), *Handbook of personality theory and research.* Chicago: Rand McNally, 1968. Pp. 890–899.

Hollander, E. P., & Julian, J. W. Contemporary trends in the analysis of leadership processes. *Psychological Bulletin,* 1969, 71, 387–397.

Hollander, E. P., & Julian, J. W. Studies in leader legitimacy, influence, and innovation. In L. Berkowitz (Ed.), *Advances in experimental social psychology,* Vol. 5. New York: Academic Press, 1970. Pp. 33–69.

Hollander, E. P., Fallon, B. J., & Edwards, M. T. Some aspects of influence and acceptability for appointed and elected group leaders. *Journal of Psychology,* 1977, 95, 289–296.

Hollander, E. P., Julian, J. W., & Sorrentino, R. M. The leader's sense of legitimacy as a source of his constructive deviation. *Technical Report No. 12,* ONR Contract 4679. State University of New York at Buffalo, Department of Psychology, July, 1969. Also reported in Hollander & Julian (1970).

Mintzberg, H. *The nature of managerial work.* New York: Harper & Row, 1973.

Regula, R. C., & Julian, J. W. The impact of quality and frequency of task contributions on perceived ability. *Journal of Social Psychology,* 1973, 89, 115–122.

Riecken, H. W. The effect of talkativeness on ability to influence group solutions to problems. *Sociometry,* 1958, 21, 309–321.

Sorrentino, R. M. A manual for scoring task-oriented group discussion. *Research Bulletin No. 196,* Ontario, Canada: University of Western Ontario, Department of Psychology, 1971.

Sorrentino, R. M. An extension of theory of achievement motivation to the study of emergent leadership. *Journal of Personality and Social Psychology,* 1973, 26, 356–368.

Sorrentino, R. M., & Boutillier, R. G. The effect of quantity and quality of verbal interaction on ratings of leadership ability. *Journal of Experimental Social Psychology,* 1975, 11, 403–411.

A CONTINGENCY MODEL OF LEADERSHIP EFFECTIVENESS[1]

Fred E. Fiedler

DEPARTMENT OF PSYCHOLOGY
UNIVERSITY OF ILLINOIS
URBANA, ILLINOIS

[1] The present paper is based on Technical Report No. 10, ONR Project "Group and Organizational Factors Influencing Creativity" (NR 177-472, Nonr 1834(36), Fred E. Fiedler, C. E. Osgood, L. M. Stolurow, and H. C. Triandis, Principal Investigators. The writer is especially indebted to his colleagues A. R. Bass, L. J. Cronbach, M. Fishbein, J. E. McGrath, W. A. T. Meuwese, C. E. Osgood, I. D. Steiner, H. C. Triandis, and L. R. Tucker, who offered suggestions and criticisms at various stages of this paper.

Reprinted from *Advances in Experimental Social Psychology*, Volume 1, 149–190.
Copyright © 1964 by Academic Press, Inc.

I. Introduction

Leadership research has primarily been concerned with two major questions: What personality factors determine whether a particular individual will become a leader? and What personality traits or attributes determine whether a leader will become effective? The first question was treated extensively in the excellent reviews by Stogdill (1948) and Mann (1959), as well as in discussions by Gibb (1954), Bass (1960), and Hare (1962). It will not be dealt with further at this time.

The second question concerns the identification of personality attributes which characterize the *effective* leader. This problem did not receive much attention until the 1950's, when substantial support of programmatic research became available. Empirical research in this area is notoriously difficult and expensive, especially if the leader's effectiveness is defined in terms of the group's performance. It requires access to a large number of teams having comparable tasks, and it is frequently if not always difficult to develop reliable and meaningful criteria of group output, especially under "real-life" conditions. Most reviewers (e.g., Browne and Cohn, 1958; Hare, 1962; McGrath, 1962; Steiner, 1964) have also taken pains to point out that we have yet to produce a generally acceptable theory of leadership. No major development in recent years requires a basic change in this statement and our understanding of effective leadership is, therefore, at a fairly primitive stage.

It is hoped that this chapter will contribute to the theoretical integration of the area by proposing a framework for the understanding of factors which determine how a leader's personality attributes affect group performance. While a number of programs have demonstrated the influence of various leader attributes on team effectiveness, the results have not been generalizable from one group situation to another. This is true despite the fact that a large number of studies have converged on two major clusters of leadership, beginning with the classical experiments by Lewin and Lippitt, 1938; Lippitt and White, 1952).

These clusters have been variously labeled autocratic vs. democratic; authoritarian vs. equalitarian; task-oriented vs. human-relations-oriented; self-oriented vs. group-oriented; psychologically close, permissive vs. distant, controlling, managing; or "initiation of structure" vs. "consideration." While these concepts are by no means synonymous, they do have certain features in common. There is little doubt at this time that the attitudes and behaviors involved in these clusters play a crucial part in affecting group performance.

In a summary of relevant research, Hare (1962) noted that autocratic leadership seems to promote greater quantitative productivity, while democratic leadership tends to result in higher morale and qualitative productivity. Autocratic leaders appear, therefore, to be most effective in industrial work situations or the armed forces, in which the task requires strong, centralized control.

However, the data are not as consistent as one might wish. The apparent convergence in selecting relevant leader attitudes or behaviors has not been matched by a corresponding convergence in the results of various investigations. On the one hand, a number of important studies, for example, Katz and Kahn (1953) showed that human-relations-oriented, supervisory attitudes tend to promote increased morale and productivity in a work situation; Argyle *et al.* (1957) found that foremen of highly productive work groups were more democratic and nonpunitive in their supervisory behavior than foremen of low-producing groups (although democratic supervision itself was not related to productivity); Comrey *et al.* (1954) showed that the personal interest of supervisors in their subordinates was positively related to effectiveness of forest management; and McGregor (1960) reported that democratically led organizations function more effectively and creatively.

On the other hand, Spector and Suttell (1956) found no relationship between authoritarianism of the leader and effectiveness of groups in problem-solving situations, and neither did Haythorn (1956) in studies of a leader's authoritarian attitudes and discussion behavior. Hawkins (1962) reported that task-oriented, punitive leaders tend to be more effective in sales display teams and gasoline service station management. Maier and Solem (1952) showed that groups with an active discussion leader were more effective on a problem-solving task than they were with permissive, passive, observing leaders. Halpin (1955) found a positive correlation between rated efficiency of the air crew and the aircraft commander's tendency to manifest friendly, mutually trusting relations defined by the term "consideration." Shaw's (1955) study showed that authoritarian leadership attitudes produced better performance and required less time for correct solution than nonauthoritarian leadership attitudes in communication nets.

While these examples do not represent an exhaustive review of the literature, they indicate clearly that the problem is not simple. Factors other than the strong, centralized control, or the quantitative vs. qualitative nature of task performance are likely to play an important role in determining the type of leader attitude or behavior which is most appropriate in a particular situation.

There are several noteworthy reasons for the low degree of systematization in this field. First, as Mann (1959), McGrath (1962), Steiner (1964), and others have pointed out, most investigators used measures tailored to fit their own research interests and predilections, but which, at the same, time precluded direct comparisons with other studies. Second, definitions of leadership, groups, and even of effectiveness, vary widely from study to study (Janda, 1960). Third, there has been little attempt to systematize the information necessary for identifying the social context within which the groups operate. As a result, it is frequently impossible to tell whether the findings of one investigation do or do not confirm those of a purportedly similar study. In his *Summary of small group research studies,* McGrath (1962) remarked that

> there has been relatively little research which could properly be called systematic, in the sense that it was explicitly designed to explore a wide range of variables drawn from different aspects of the small group field. . . . Nor has much effort been devoted to development and application of a standard battery of measures which sampled the broad aspects of group research and which could provide a direct basis for comparing results from one study to another. Such a procedure seems to represent a first step in the development of something equivalent to standard conditions as utilized in the physical and biological sciences. In the absence of such systematic programming of research, the rate of progress in understanding small group phenomena is likely to remain relatively slow, even though the rate of production of small group research seems to be increasing without abatement.

In light of the existing reviews by Bass (1960), George (1962), McGrath (1962), and others, another attempt to review the literature would not appear profitable. An intensive study of one, unified, programmatic effort is more likely to lead to the development of principles from which to build a general theory. The program directed by the writer and his associates during the last twelve years has involved over 25 major studies with reasonably large numbers of comparable groups and reliable criteria of performance. These studies were based on a unified, if partial, rationale, and are linked together in a cumulative fashion. An intensive examination of this programmatic effort may provide new insights into the problem and it may enable us to integrate other major programs of research on leadership effectiveness.

Before going further it would help to define a few key terms which will be used in this discussion. For the purpose of this analysis, a *group* is defined as a set of individuals who, in Campbell's (1958) terms, have proximity, similarity, and share a "common fate" on task-relevant events. The specific concern is with groups in which the members are, and also perceive each other to be, interdependent in achieving a common goal. It should be noted that this definition explicitly excludes "coacting" groups in which members work individually on a task, even though their performance might later be summed to yield a "group score," and even though coacting group members may indirectly affect each other's performance. According to this definition, a basketball team is a good example of an interacting group, and a track team exemplifies a coacting group.

The *leader* is defined as the individual in the group who directs and coordinates task-relevant group activities, or who, in the absence of a designated leader, automatically performs these functions in the group. It is recognized that leadership functions are frequently shared (Berkowitz, 1953; Cattell, 1951), but only the single individual who meets one of the following criteria is dealt with here: (a) he was appointed by a representative of the larger organization of which the group is a part; (b) he was elected by the group; (c) if there is neither an elected nor an appointed leader, he is the individual who can be identified as most influential on task-relevant questions of a sociometric preference questionnaire.

The leader's *effectiveness* will be defined in terms of his group's performance on its assigned task. This assumes that task-relevant skills and abilities of group members are reasonably similar, or that they were experimentally or statistically controlled.

II. History of the Program

The research program which is the focus of this chapter was initiated in 1951. It hypothesized that the leader's perceptions of his co-workers reflect important, task-relevant attitudes, and that these would materially influence group interaction and performance. This hypothesis was substantiated in a general sense. However, correlations between the leader's interpersonal perception score and his group's performance which were statistically significant and positive in one set of studies were often statistically significant and negative in another. The present theoretical position grew out of the attempt to make psychological sense from the conflicting but highly significant findings obtained in our own research, as well as in related investigations by others.

The interpersonal perception measures which served as predictor variables were first developed in research on psychotherapeutic relations (Fiedler, 1951). The data showed that reputedly effective psychotherapists tended to see their patients as more similar to themselves than did reputedly poor therapists. These results suggested that the individual who perceives another person as similar tends to feel psychologically close, accepting, and permissive toward him. It was, therefore, anticipated that liked persons would be seen as more similar to oneself, and more favorable, than would disliked persons. This hypothesis was clearly supported in an investigation of a college fraternity in which each man was asked to describe his most and his least liked fellow group member (Fiedler *et al.*, 1952), and since then has been substantiated by other studies (e.g., Davitz, 1955; Bieri, 1953). The question remained whether these interpersonal attitudes were also related to team performance.

Early research by Likert (1961) and his associates indicated that human-relations-oriented, supervisory attitudes increased productivity. It was expected, therefore, that our measures of close, interpersonal relations, permissive, and accepting attitudes, would be related to good teamwork. Subsequent research on this problem showed, however, that the prediction of group performance on the basis of these leader attributes is contingent upon the specific, situational context in which the leader operates. The major purpose of this paper is to integrate and explain the currently available findings by developing a generalized model of the relationship between these leader attributes and group performance.

A. PREDICTORS OF LEADERSHIP EFFECTIVENESS

Two main personality measures were used for predicting leadership effectiveness: the "assumed similarity between opposites" (ASo), and the esteem for the least preferred coworker (LPC) scores. ASo and LPC are highly correlated (.70 to .93), and will be used here interchangeably. Because of their importance to the theoretical model, these measures will be discussed in some detail.

The ASo score is obtained by asking a person to think of all the individuals with whom he has ever worked. He then describes: (a) the person whom he considers his most preferred co-worker (MPC), and (b) the person whom he considers his least preferred co-worker (LPC). The descriptions are made on eight-point, bi-polar adjective checklists similar in form to Osgood's (Osgood *et al.*, 1957) Semantic Differential, using items descriptive of personality attributes. For example,

```
Pleasant :—8—:—7—:—6—:—5—:—4—:—3—:—2—:—1—:Unpleasant
Friendly :—8—:—7—:—6—:—5—:—4—:—3—:—2—:—1—:Unfriendly
Rejecting:—1—:—2—:—3—:—4—:—5—:—6—:—7—:—8—:Accepting[2]
```

ASo scores are derived by scoring each of the items from most to least favorable (8 to 1) and computing a measure of profile similarity, D (Cronbach and Gleser, 1953). A person who perceives his most and least preferred co-workers as very similar will, therefore, have a high assumed similarity score (or, in operational terms, a small discrepancy score), while a person who strongly differentiates between these two "opposites" on his co-worker continuum, will have a low ASo (and thus a large discrepancy)

---

[2] Other items in a recent scale were: helpful–frustrating; unenthusiastic–enthusiastic; lots of fun–serious; tense–relaxed; supportive–hostile; distant–close; cold–warm; cooperative–uncooperative; boring–interesting; quarrelsome–harmonious; self-assured–hesitant; efficient–inefficient; gloomy–cheerful; open–guarded.

score. The LPC score is one component of ASo and is obtained by simply summing the item scores on the least preferred co-worker scale sheet.

A person with a high LPC score tends to see even a poor co-worker in a relatively favorable manner (i.e., "Even if I can't work with him, he may still be a very nice and valuable person"). A low LPC person perceives his least preferred co-worker in a highly unfavorable, rejecting manner ("If I cannot work with him, he is probably just no good"). LPC scores have a high internal consistency, with split-half co-efficients of over .90. The scores are reasonably stable over time, although changes do take place depending upon intervening training and experience. In this respect, the scores resemble other attitude measures.

Since LPC and ASo do not correlate consistently with commonly used personality tests, there has been considerable difficulty in deriving an adequate interpretation of these scores. Earlier investigations treated ASo as a measure of psychological distance. This interpretation now appears to be an oversimplification. A recent investigation by Hawkins (1962) showed that individuals who differentiate sharply between their most and least preferred co-workers (low ASo) tend to be more oriented toward the task than toward relationships with others. They also tend to be more punitive, although not necessarily more distant. Studies of group interactions, based on tape-recorded transcripts, support this interpretation. High LPC (or ASo) leaders behave in a manner which promotes member satisfaction and lowers member anxiety; they are more compliant, more nondirective, and generally more relaxed, especially under pleasant and nonthreatening conditions. They are described by their groups as being higher on the Ohio State "Consideration" dimension (Meuwese, 1964).

Low LPC leaders, on the other hand, give and ask for more suggestions, are less inclined to tolerate or to make irrelevant comments, demand and get more participation from members, and are more controlling and managing in their conduct of the group interaction (Fiedler *et. al.,* 1961b). Low LPC leaders also interrupt group members more often contribute more statements to the discussion, and make and receive more negatively toned statements, again indicating less concern with having pleasant relationships with others in their group. As with all attitude measures, the corresponding behavioral manifestations of high and low LPC leaders appear only under conditions which permit the individuals to behave in either of these ways with equal propriety.

B. Previous Findings

The problems posed by the prediction of leadership effectiveness might be most readily understood in terms of the specific difficulties which arose in the course of research. The first group effectiveness study investigated 14 high school basketball teams which competed in leagues of matched schools. Team performance was defined as the percentage of games won by mid-season. The leader of the team was identified by means of sociometric preference questions. As mentioned above, it was anticipated that psychologically close teams would be more effective than teams characterized by task-oriented, psychologically distant, and less accepting relations. Contrary to expectation, team performance correlated negatively (—.69) with the leader's ASo score. Thus, better teams apparently had active, controlling, psychologically distant leaders. A similar result was obtained in a validation study at the end of the season by comparing seven teams from the upper and five from the lower third of the Illinois high school team standings ($r_{p.b.}$ —.58).

A study of 22 student surveying parties cross-validated these findings. These three-to-four-man teams were engaged in measuring predetermined parcels of land. The criterion was the accuracy of surveying, as judged by the course instructors. Again it was found that the effective teams had low ASo leaders (—.51) (Fiedler, 1954). Thus, the ASo score—and the personality attribute which it reflected—was clearly an important variable in the prediction of group performance.

However, did the effective teams choose low ASo leaders, or did low ASo leaders make their teams effective? This question could be partly answered in studies of groups in which the leaders were appointed by higher authority.

The first two investigations in this series dealt with B-29 bomber crews and Army tank crews. The criteria consisted of two uncorrelated bomber-crew tasks (radar and visual bombing accuracy) and two uncorrelated tank-crew tasks, time to hit a target and time to travel to a new target). In these studies, significant relations between the leader's ASo score and crew performance measures occurred only if the leader was sociometrically the most chosen member of the crew. Under these conditions, the correlation was *negative* (—.81, —.52, —.60, —.33) for crews in which the leader sociometrically endorsed his keyman (e.g., the gunner on a tank gunnery task), and *positive* (.42, .27, .60, .43) for crews in which the leader sociometrically rejected his keyman (Fiedler, 1955).

The relationship between ASo and crew effectiveness thus seemed to be contingent upon the sociometric choice pattern within the crew. This interpretation was later supported in studies of anti-aircraft artillery crews (Hutchins and Fiedler, 1960) and infantry squads (Havron et al., 1954).

The importance of sociometric acceptance as a moderator variable became even more apparent in a study of 32 farm-supply service companies (Godfrey et al., 1959). The criterion of success was the percentage of a company's net income compared to total sales over a three-year period.

The formal leader of the executive group was the general manager, and the chairman or most influential member of the board of directors was the leader of the policy- and decision-making body.

This investigation, from a theoretical point of view, demonstrated: (a) that ASo or LPC scores predicted leadership effectiveness to the degree to which the leader had good interpersonal relations in the group; and (b) that the direction of the relationship was contingent upon the leader's relations with key group members, as well as upon the nature of the task. In other words, the effective executive group performed better under an accepted, low ASo leader, while the policy- and decision-making groups operated more effectively under permissive, nondirective, and considerate, high ASo leaders (Table I).

TABLE I

CORRELATIONS BETWEEN COMPANY EFFECTIVENESS CRITERIA AND
ASo SCORES OF BOARD OF DIRECTORS' INFORMAL LEADER
AND COMPANY'S GENERAL MANAGER

| | | Correlations between net income and ASo of | |
| | | | |
Sample	N	Informal leader of board	Company's general manager
All companies	32	−.06	−.14
Informal leader endorses general manager	23	−.08	−.39
Informal leader endorses sociometrically accepted general manager	13	.20	−. 0[a]
Informal leader endorses accepted general manager, who endorses his keyman	8	.62	−.74[a]

[a] Significant at the .05 level of confidence for one-tailed tests.

The results obtained in the study of the board of directors led to an extension of the research to groups having creative tasks. A series of four studies was conducted which showed that the permissive, accepting, high LPC leaders had better group performance on creative tasks under relatively stress-free conditions; the managing, controlling, low LPC leaders had better performance under relatively less pleasant, more tension-arousing group climates (Fiedler, 1962). Thus, the direction of the relationship between the leader's attitude toward co-workers and group performance was again contingent upon his relations with the members of his group.

The nature of the task and the leader's relationship with group members are, however, not the only factors which determine the nature of the optimal leader attitudes. Recent studies have also indicated that the power of the leadership position also plays an important role in determining the type of leadership attitude and behavior which will contribute to group effectiveness. Gerard (1957), and Anderson and Fiedler (1962) have shown that the leader who has a powerful position will behave differently from one who holds a very tenuous position.

In brief, the data, taken as a whole, presented an extremely complex picture. The main problem became one of developing an integrative model which would account for the variations in the relationships which had been obtained.

III. Development of an Integrative Model

A. A System for Classifying Group Situations

The findings in this field made it abundantly clear that different group situations require different leadership styles. Progress in predicting group effectiveness thus seemed to require a meaningful system for classifying groups in terms of relevant situational factors.

Leadership is generally thought of as an interpersonal situation in which one individual in the group wields influence over others for the purpose of performing an assigned task. It will, therefore, be very important to know whether the group environment will make it relatively easy or difficult for the leader to influence the members of his group. A "good" system of classification would then be based on the crucial factors which determine whether a given situation is favorable or unfavorable for the leader.[3]

1. Situational Components

Three, critical, situational components which are likely to affect the leader's influence are postulated: (a) his personal relations with members of his group, (b) the power and authority which his position provides (the legitimate power, in French's term, 1956), and (c) the degree of structure in the task which the group has been assigned to perform.

It is recognized that a variety of other factors may affect the leader's ability to exert his influence. Among these are the relative abilities of the leader and his members, the members' motivation, and the extent to which the group is operating under conditions of external stress. To a large degree, these situational aspects are intercorrelated. A very able

[3] A considerable number of excellent papers (e.g., French, 1956; French and Raven, 1958) are available which deal with various influence processes. However, as Janda (1960) pointed out, there has been little effort to relate these influence processes to leadership effectiveness.

leader tends to be highly respected, and a highly respected and liked task leader is able to motivate his group members.

Only the three situational components listed above will be dealt with here. The elaboration and necessary corrections of the model are left to future research. The three specific measures which define the three, major, situational components are presented below.

a. Affective leader-group relations. The personal relationship between the leader and key members of his group is probably the most important, single determinant of group processes which affect team performance. The liked and respected leader does not need formal power, and he can obtain compliance from his group under circumstances which, in the case of a disliked or distrusted leader, would result in open revolt. As has been shown (e.g., Godfrey *et al.,* 1959; Fiedler, 1961), the liked and accepted leader's interpersonal attitudes influence group performance to a significantly greater degree than similar attitudes of a leader who is sociometrically not accepted by his group.

A number of indices have been used to tap this particular dimension. Although the various measures are by no means identical, they seem to reflect relatively similar relations. The writer's early studies employed an index which reflected the sociometric acceptance of the leader by his co-workers. Later investigations obtained the leader's rating of the group's atmosphere (GA) on simple scales similar or identical to those used for measuring LPC. The leader was asked to describe the group on 10- to 20-item scales consisting of eight-point continua such as:

Pleasant:—8—:—7—:—6—:—5—:—4—:—3—:—2—:—1—:Unpleasant
Bad   :—1—:—2—:—3—:—4—:—5—:—6—:—7—:—8—:Good[4]

The leader who feels (and is) accepted by his group members is obviously able to act more decisively and with more confidence than the leader who feels rejected or distrusted by the members of his group. For purposes of the present analyses, groups in the upper and lower thirds of the group atmosphere or sociometric preference distribution have been chosen. This seems necessary here since only sociometric scores were available for some groups, and only group atmosphere scores for others. Moreover, the dimension is highly subject to shifting frames of reference. What appears relatively pleasant in a stressful situation may seem very unpleasant in a normally relaxed group discussion.

It should also be noted that the group climate and leader-group relations in laboratory studies tend to range from very pleasant to, at worst, moderately unpleasant situations. The most stressful laboratory con-

---

[4] Other items were: worthless–valuable; distant–close; cold–warm; quarrelsome–harmonious; self-assured–hesitant; efficient–inefficient; gloomy–cheerful.

dition we could devise produced a leader-group atmosphere score of only 5.0 (expressed as an item average) on a scale which ranged from a low score of 1.0 to a high of 8.0, and hence fell above the midpoint of the group atmosphere scale.

On the other hand, real-life groups occasionally express strong rejection of the leader, which places him in a very unfavorable position that might well outweigh other factors ordinarily compensating for poor leader-member relations.

   b. Task structure. The second important dimension describes the nature of the task in terms of its clarity or ambiguity. Although it is generally not thought of in this manner, the assigned task in effect constitutes an order "from above." This order might be highly programmed, such as drilling "by the numbers," assembling a rifle, or operating a simple machine; or, it may be a very unstructured, vague order, such as to develop a policy which will maximize the profits of a company. The leader's job will be considerably easier if the job is highly structured than if it is vague and unspecific. This can be readily seen by noting, for example, that enlisted men frequently serve as instructors in officer training courses in which the material can be programmed, viz., in assembling and handling of weapons, in map reading, or in close order drill. The authority of the higher command is implicit in such highly structured tasks and the leader serves primarily to supervise the implementation of the task order.

In contrast, when a committee is given an unprogrammed task such as planning an annual picnic, the leader knows no more than do his members, and he cannot readily order anyone to execute such a task in a specific manner. This holds even in situations in which the leader has considerable formal power, e.g., a professor working with his assistants on a research plan, or an army officer working with enlisted specialists who are experts in their fields.

The task structure dimension is operationally defined here by four of the scales developed by Shaw (1962). It was reliably assessed by four, independent judges who rated 35 tasks on eight-point scales with a resulting interrater agreement of .80 to .88 The four dimensions are: *decision verifiability*—the degree to which the "correctness" of the solution or decision can be demonstrated, either by appeal to authority (e.g., the census of 1960), by logical procedures (e.g., mathematical demonstration), or by feedback (e.g., examination of consequences of decision, as in action tasks); *goal clarity*—the degree to which the requirements of the task are clearly stated or known to the group members; *goal path multiplicity*—the degree to which the task can be solved by a variety of procedures (number of different paths to the goal, number of alternative for solution, number of different ways that the task can be completed) (reversed scoring); *solution specificity*—the degree to which there is more than one "correct" solution. (Some tasks, e.g., arithmetic problems, have only one solution that is acceptable; others have two or more, e.g., a sorting task where items to be sorted have several dimensions; and still others have almost an infinite number of possible solutions, e.g., human-relations problems or matters of opinion.).

   c. Position power. A third major dimension is defined by the power inherent in the leadership position. This includes the rewards and sanctions which are officially or traditionally at the leader's disposal, his authority over his men, and the degree to which this authority is supported by the organization within which the group operates. The leader's power is, generally speaking, inversely related to the power of his members.

   The man who occupies a powerful leadership position may be able to obtain compliance even though he is personally resented by his group members. The chairman of a volunteer committee generally has to influence the group by persuasion or other indirect means suggested by Hemphill's term "consideration."

   The dimension of "leader position power" is defined by a checklist in which all items are given one point, except for 4a, b, c, which are weighted $+5$, $+3$, and -5, respectively.

1a. Compliments from the leader are appreciated more than compliments from other group members.
   b. Compliments are highly valued, criticisms are considered damaging.
   c. Leader can recommend punishments and rewards.
   d. Leader can punish or reward members on his own accord.
   e. Leader can effect (or can recommend) promotion or demotion.

2a. Leader chairs or coordinates group but may or may not have other advantages, i.e., is appointed or acknowledged chairman or leader.
   b. Leader's opinion is accorded considerable respect and attention.
   c. Leader's special knowledge or information (and members' lack of it) permits leader to decide how task is to be done, or how group is to proceed.
   d. Leader cues members or instructs them on what to do.
   e. Leader tells or directs members on what to do or what to say.

3a. Leader is expected to motivate group.
   b. Leader is expected to suggest and evaluate the members' work.
   c. Leader has superior, or special, knowledge about the job, or has special instructions, but requires members to do job.

 d. Leader can supervise member's job and evaluate it or correct it.

 e. Leader knows own as well as members' job and could finish the work himself if necessary (e.g., writing a report for which all information is available).

4a. Leader enjoys special or official rank and status in real life which sets him apart from (or above) group members, e.g., military rank, or elected office in a company or organization. (+5 points)

 b. Leader is given special or official rank by experimenter to simulate for role playing purposes, e.g., "you are a general," or, "the manager." This simulated rank must be clearly superior to members' rank, and must not be just that of "chairman" or "group leader" of the group during its work period. (+3 points.)

 c. Leader's position is dependent on members. Members can replace or depose leader. (−5 points.)

The position power in 35 group situations was rated by four independent judges, who reached interrater agreement of .95.

2. Interrelations of the Task-situation Dimensions

In the sample of 35 tasks (which contains most of the tasks described in Table II), leader-group relations correlated with task structure .03 and with position power −.09; task structure and position power correlated .75. The last two appeared to measure very closely related aspects of the group-task situation. This reflected the commonly observed fact that leaders of groups which perform highly structured or programmed tasks are generally given relatively higher power than their members. Most industrial and military situations are of this nature. Unstructured tasks, such as research and development, or planning and policy-making, tend to be performed under a chairmanship system in which the leader has relatively low position power.

Although this cannot be inferred from the available data, it also seems likely that certain tasks are naturally more conducive to good leader-member relations than are others. Good interpersonal relations tend to develop in group games or party planning, while work under highly competitive conditions or in a stressful environment frequently leads to unpleasant or hostile relations (Sherif and Sherif, 1953; Deutsch, 1949; Myers, 1962; Sells, 1962).

The three dimensions which are postulated here have previously been described in somewhat similar terms by others in the small group area Cartwright and Zander (1960) spoke of task structure, labor division, power structure (closely related to position power), and sociometric or friendship structure. The Ohio State studies discussed structure-in-inter-action and consideration (Halpin and Winer, 1957), and Schutz (1953) suggested the dimensions of control (related to position power) and affec-tion and inclusion (related to leader-member relations).

B. DIMENSIONALIZING GROUP SITUATIONS

A rough categorization leads to an eight-celled cube (Fig. 1). Accord-ing to hypothesis, a group located in one cell or "octant" of this three-dimensional space may require a different leadership style than a group located in an adjacent cell. It is possible to develop a partial order of the various group situations in terms of their favorableness or the degree to which the leader's job of influencing his group will be easy or difficult.

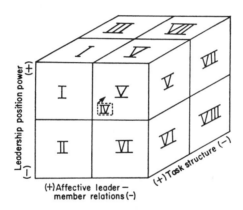

FIG. 1. A model for the classification of group task situations.

As mentioned before, the leader-member relations dimension de-picted in this figure extends only from "good" to "moderately good," since very poor leader-member relations are almost never found in laboratory groups. An Octant V-A, which deals with real-life groups having presum-ably very poor leader-member relations, was later added. This will be described in the section on validation evidence.

There is no difficulty in identifying the extreme points on this dimen-sion of ease of influencing the group (favorableness-to-leader dimension). A leader who is liked by his group, who knows exactly what to do and how to proceed, and who holds a relatively powerful position will find it easier to do his job than one who is disliked by his group, has a vague, un-structured task, and has no power. Difficulty arises because any ordering of intermediate points is to some degree arbitrary. To order these octants, the leader's relationship with his members was postulated to be the most important of the three dimensions. A highly trusted and liked leader usually does not need to rely on the power of his position; he can define the job in the way he sees fit.

The task structure is considered to be next in importance, since it embodies the demands of higher authority, such as a standard operating procedure. Position power is considered third in importance in this model.

The resulting dimensionalization is thus obtained by first ordering the group-task situation on the basis of the leader's relations with his group, then on the basis of task structure, and finally on the basis of position power. An additional "octant" (V-A) includes real-life groups in which the leader's relationship with group members is very poor. This octant was classified as least favorable for the leader, since it seems reasonable to assume that the strongly disliked leader will have a very difficult job. The basic hypothesis underlying the model is, then, that *the type of leader attitude required for effective group performance depends upon the degree to which the group situation is favorable or unfavorable to the leader.*

IV. Empirical Support for the Model

A. LEADER PERCEPTION SCORES (LPC AND ASo) AND GROUP PERFORMANCE

Groups included in previous studies were categorized by octants (Table II), as described above. Table II also includes "validation evidence," consisting of results from new studies, as well as re-analyses of old data that tested hypotheses derived from the model. The validation evidence is described in greater detail toward the end of this section.

Figure 2 graphically summarizes the information contained in Table II. The correlations between leader LPC/ASo scores and group performance were plotted by octants to indicate the range and dispersion of these correlation coefficients. The validation evidence is indicated by triangles. As can be seen, the correlations between leader scores and group performance measures within each octant are quite similar in size and direction. Even granting the post hoc nature of the classification, the consistency of the relations within octants is highly nonrandom in distribution.

The median correlations between LPC/ASo and group effectiveness can now be plotted against a hypothesized "advantage-for-the-leader" continuum which is obtained by collapsing the three dimensions. This leads to a bow-shaped curve showing that controlling, managing, and directive (low LPC) leaders perform best in group conditions which are either very favorable or very unfavorable to the leader, and that permissive, considerate, passive (high LPC) leaders perform best under moderately favorable or unfavorable conditions (i.e., in Octants IV, V, and VI). This is shown on Fig. 3 and Table III.

In very favorable conditions, where the leader has power, informal backing, and a relatively well-structured task, the group is ready to be directed on how to go about its task. Under a very unfavorable condition, however, the group will fall apart unless the leader's active intervention and control can keep the members on the job. In moderately unfavorable conditions, the accepted leader faces an ambiguous task, or his relations with group members are tenuous. Under these circumstances, a relationship-oriented, nondirective, permissive attitude may reduce member anxiety or intra-group conflict, and this enables the group to operate more effectively (i.e., the members would not feel threatened by the leader, and considerate, diplomatic leader behavior under these conditions may induce group members to cooperate).

The pattern of relations which determine the leader attitudes a group requires is also shown in Fig. 4. This graph presents the correlations between leader LPC/ASo and group performance, plotted against the two dimensions of the leader's position power and task structure for good and for moderately poor leader-member relations. The most important result indicated by this figure is again the remarkable reversal in the direction of the correlation coefficients in Octants IV and VIII, as well as in Octants I and V. While many of the interpretations are still highly speculative, it seems quite obvious that the dimensions of leader-member relations and task structure act as very powerful moderator variables in determining the type of leader attitude which maximizes group performance.

B. QUANTITATIVE INDICATIONS OF RELATIONS WITH COMPONENTS

In order to obtain a quantitative indication of the extent and nature of these relationships, a configural multiple regression technique was employed (Horst, 1954; Lubin and Osborn, 1957). The predictors in this multiple regression equation were the three group dimensions (PP, TS, and leader-member relations, R), and the four interactions of these dimensions with one another (PP $\times$ TS, PP $\times$ R, TS $\times$ R, and PP $\times$ TS $\times$ R). The latter terms were obtained from standard scores on the original dimensions and are thus nonlinear functions of these dimensions. A multiple regression was then computed for these seven predictors, the criterion being the correlation between the leader's LPC score and his group's performance. The number of cases for this multiple regression was 68, which is the number of individual correlations between LPC and performance reported in Table II. The validation data were not included in these analyses.

Table IV shows the correlation among the predictors and their validities. This table also presents the beta weights for the predictors and the multiple correlation between the seven predictors and the criterion. The three dimensions taken alone are unrelated to the LPC-performance

TABLE II

Summary Table of Relations Obtained in Research with LPC and ASo Scores

Octants I and V

High leader position power—high task structure
Good/moderately poor leader-member relations

Study			Leader-member relations			
			Good Octant I	Mod. poor Octant V	N_I	N_V
B-29 bomber crews. Military commander in Oct. I is sociometrically most preferred crew member and endorses keyman; is sociometrically most preferred crew member in Oct. V, but rejects keyman.	PP[a]	18.5				
Criterion 1: radar bomb score-circular error average (radar observer or navigator are keyman)	TS[b]	8.0	−.81	.42	11	7
Criterion 2: % satisfactory visual bomb runs (bombardier is keyman) (Fiedler, 1955)	TS	8.0	−.52	.27	6	7
Army tank crews. Tank commander in Oct. I is sociometrically most preferred crew member; in Oct. V is sociometrically most preferred crew member, but does not endorse his keyman.	PP	18.5				
Criterion 1: time to hit target (gunner is keyman)	TS	8.0	−.60	.60	6	5
Criterion 2: time to travel to target (driver is keyman) (Fiedler, 1955)	TS	8.0	−.33	.43		
Antiaircraft artillery crews. Commander in Oct. I is sociometrically most preferred crew member; in Oct. V is among 10 sociometrically least preferred crew members.	PP	18.5				
Criterion: location & acquisition of unidentified aircraft. (Hutchins and Fiedler, 1960)	TS	7.3	−.34	.49	10	10

[a] PP-Leader's Position Power.
[b] TS-Task structure.

184

185

Infantry squads. Squad leader is sociometrically most preferred crew member (no data available for sociometrically not chosen men). Criterion: umpire ratings of field tests (Havron *et al.*, 1951)	PP TS	18.8 7.5	$-.36$	26
Open hearth steel shops. Foremen accepted by crew. Criterion: tap-to-tap time (tonnage per unit of time) (no rejected foremen identified) (Cleven and Fiedler, 1956)	PP TS	18.5 7.2	$-.52$	15
Company management. Gen. mgr. in Oct. I is sociometrically accepted by board and staff; in Oct. V is sociometrically accepted either by board or staff. Criterion: % of company net income over 3 yrs. (Godfrey *et al.*, 1959)	PP TS	18.0 5.6	$-.67$	.23 10
Median correlation			$-.52$	.42

Validation evidence

Sales display teams. Teams with detailed instructions for setting up sales displays and preparing merchandise. Only Oct. I data available. Criterion: ratings by higher supervisors on conformity to performance standards. Tested by analysis of variance; low ASo leaders performed better than high ASo leaders. (Hawkins, 1962)	PP TS	17.0 5.8	$-(F < .10)$	76
Service station management. Managers of gas stations in various communities. Company has detailed operating procedure for servicing, stock control, and reporting. (Only Oct. I data available). Criterion: stock control, sales, monthly audit, and checks by inspectors Tested by chi-square. Low ASo managers performed better than high ASo managers. (Hawkins, 1962)	PP TS	17.0 5.8	$-(X^2 < .05)$	60

TABLE II (Continued)

Octants II and VI

Low leader position power—highly structured task
Good/moderately poor leader-member relations

Study			Leader-member relations			
			Good Octant II	Mod. poor Octant VI[c]	N_{II}	N_{VI}
High school basketball teams. "Leader" is sociometrically the most chosen team member, but is not appointed or elected, although wielding considerable influence.	PP	3.8				
Study I criterion: % of games won by mid-season	TS	7.2	−.69		14	
Study II criterion: 7 good, 5 poor teams tested at end of season (Pt.bis.r) (Fiedler, 1954)	TS	7.2	−.58		12	
Student surveying parties. "Leader" is sociometrically most preferred team member.	PP	3.2				
Criterion: accuracy of surveying pre-selected parcels of land as rated by instructors. (Fiedler, 1954)	TS	7.3	−.51		22	
Median correlation			−.58			
Validation evidence						
Team judgments. Two students were paired to judge which answers are best for *How Supervise Test.* Leader was designated by experimenter, but had no special function.	PP	2.0				
Criterion: # of items completed. Tested by Analysis of Variance, low ASo leaders were better than high ASo leaders (Hawkins, 1962)	TS	8.0	−($F < .05$)		67	

[c] No groups were classified as belonging into Octant VI.

186

Octants III and VII

High leader position power—unstructured task
Good/moderately poor leader-member relations

Study			Leader-member relations			
			Good Octant III	Mod. poor Octant VII	N_{III}	N_{VII}
ROTC creativity study. Three-man ROTC groups with leader officially appointed. Study was part of ROTC training course. Highest ranked cadet chosen as leader. Leader-member relations measured by group atmosphere scores from upper and lower third of distribution.	PP	9.0				
Criterion 1: propose new pay scale for all ROTC services which will equalize pay scales (creativity rated by judges)	TS	3.4	−.43	.04	6	6
Criterion 2: tell fable on need for peacetime army (creativity rated by judges)	TS	2.2	−.72	.24	6	6
ROTC creativity study—high stress condition. Same as above, but groups worked under close supervision of senior army officers.	PP	9.0				
Criterion 1: pay scale proposal	TS	3.4	−.14	.11	6	6
Criterion 2: fable	TS	2.2	−.60	.57	6	6
(Meuwese, 1964)						
Navy ROTC Creativity Study. Four-man NROTC groups participated as part of NROTC leadership class problem. Senior midshipmen were appointed leaders, freshmen and sophomores served as members. Leaders chaired and *participated* in session.	PP	9.0				
Criterion 1: tell 2 stories based on TAT card 11 (creativity rated by judges)	TS	2.4	−.26	−.14	6	6
Criterion 2: develop arguments pro & con tough military training (rated by judges)	TS	4.2	−.07	.07	6	6
Criterion 3: suggest how average person can win fame and immortality (rated in terms of originality and uniqueness of solutions)	TS	4.7	−.44	−.07	6	6

TABLE II (Continued)

Octants III and VII

High leader position power—unstructured task
Good/moderately poor leader-member relations

Study			Leader-member relations			
			Good Octant III	Mod. poor Octant VII	N_III	N_VII
Navy ROTC creativity study. Same as above but leaders *supervised*—were not permitted to contribute to task solutions, but could suggest procedures and veto ideas.	PP	11.8				
Criterion 1: TAT stories	TS	2.4	−.39	.47	6	6
Criterion 2: arguments	TS	4.2	−.43	.01	6	6
Criterion 3: fame and immortality	TS	4.7	.84	−.10	6	6
(Anderson and Fiedler, 1962)						
Median correlation			−.41	.05		

Octants IV and VIII

Low leader position power—unstructured task
Good/moderately poor leader-member relations

Study	Leader-member relations			
	Good Octant IV	Mod. poor Octant VIII	N_IV	N_VIII

"Dutch" creativity study. Four-man groups consisting of Dutch university students. Leader-member relations inferred from tension indicators in content analysis and group composition, viz., homogeneity vs. heterogeneity, and formal or informal leadership.

188

Criterion task: tell 3 stories about TAT picture; find alternative uses or invent plot titles (creativity rated by judges)	TS	1.7	.75		7	
Composition: homogeneous religious membership and formal leaders appointed by experimenter.	PP	5.8				
Composition: heterogeneous groups, appointed leaders	PP	5.5		−.72		8
Composition: homogeneous, informal leaders	PP	2.0		−.64		6
Composition: heterogeneous, informal leaders	PP	2.0		−.23		8
(Fiedler et al., 1961b)						
Hypnosis study. Three-person groups, leader selected by experimenter's confederates. Leader-member relations based on group atmosphere scores.	PP	5.0				
Criterion: tell 3 stories about same TAT card (creativity rated by judges)	TS	1.7	.64	−.72	8	8
(Fiedler et al., 1961c)						
Church leadership study. Four-person groups participating in leadership workshop. Leaders selected by experimenter, groups changed each day. Leader-member relations measured by group atmosphere.						
Criterion 1: justify minister's position on mercy killing (creativity in this and other tasks rated by all participants)	PP	4.8		.03		6
	TS	2.7	.28		6	
Criterion 2: tell fable about separation of church and state	PP	4.5		−.03		6
	TS	2.2	.89		6	
Criterion 3: devise campaign to raise funds for young student minister	PP	4.8		−.40		6
	TS	3.2	.49		6	
Criterion 4: plan and perform skit on music for the worship service	PP	4.5		−.60		6
	TS	2.2	.37		6	
(Fiedler et al., 1961a)						
Mental health leadership study. Three-person groups, with chairman selected by experimenter. Leader-member relations measured by group atmosphere score.	PP	4.5		−.76		7
Criterion task: justify use of elementary schools for approved research (creativity rated by judges)	TS	2.8	.44		7	

TABLE II (*Continued*)

Octants IV and VIII

Low leader position power—unstructured task
Good/moderately poor leader-member relations

Study			Leader-member relations			
			Good Octant IV	Mod. poor Octant VIII	N_{IV}	N_{VIII}
ROTC creativity study—internal stress condition. Three-man groups, two army and one navy cadets. Leader was lowest ranking army man.	PP	4.5				
Criterion task 1: develop new pay schedule	TS	3.4	.49	−.04	6	6
Criterion task 2: tell fable about peacetime army (creativity rated by judges) (Meuwese, 1964)	TS	2.2	−.03	−.47	6	6
Chairman, board of directors. Boards of directors of small cooperatively owned corporations. Leader-member relations estimated on basis of board chairman-gen. mgr. relations as indicated by sociometric ratings.	PP	7.0				
Criterion: company net income over 3 years (Godfrey et al., 1959)	TS	4.1	.21	−.60	10	10
Median correlations			.47	−.43		
Validation evidence						
Church leadership study II. Three-person groups assembled ad hoc, with leader designated by experimenter. Leader-group member relations assessed by leader's GA scores and post-meeting questionnaires.	PP	4.5				
Criterion task: justify your position to children on reading prayers in school	TS	2.2	.27	−.04	19	19

Octant V-A

Very poor leader-group member relations
High leader position power, structured task

| Study | | | Leader-member relations | |
			Very poor Octant V-A	N_{V-A}
B-29 bomber crews. As in Octants I and V, except that crew commander is sociometrically rejected and does not sociometrically choose his keygroup members.	PP	18.5		
Validation evidence	TS	8.0	−.67	7
Antiaircraft artillery crews. As in Octants I and V, but crew commanders are sociometrically most rejected (re-analysis)	PP	18.5		
	TS	7.3	−.42	10
Company management. As in Octants I and V, but general manager is	PP	18.0		
sociometrically rejected by *both*, board of directors and staff (re-analysis)	TS	5.6	−.75	7
Median correlation			−.67	

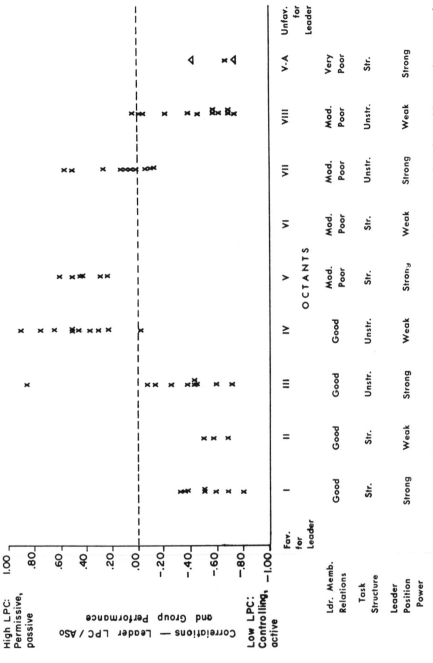

Fig. 2. Correlations of leader LPC and group performance plotted against octants, i.e., favorableness of group task situation for leader.

criterion. However, the two double interaction terms involving leader-member relations (PP $\times$ R and TS $\times$ R) are both significantly related to the criterion. Further, the triple interaction (PP $\times$ TS $\times$ R), while having zero validity for the criterion, has the largest and only significant beta

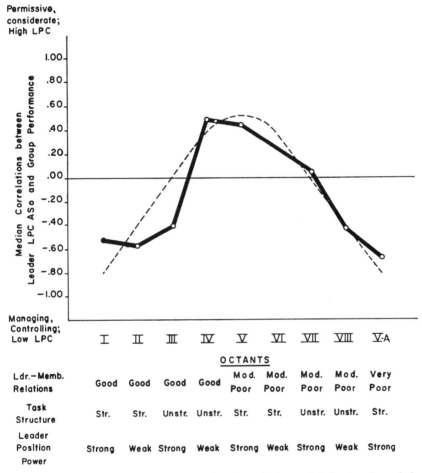

FIG. 3. Curve indicating optimal leader attitudes and behavior (permissive, considerate vs. managing, controlling) required by various group task situations, and median correlations obtained for conditions.

weight. This variable is a suppressor, having relatively high correlations with several other predictors which are components of it. It would appear, however, that this triple interaction is not a particularly important determinant of the LPC-performance correlations. Rather, its relatively large beta weight is a function of its necessarily large relationship with several of the other predictors.

The multiple correlation of all seven predictors and the criterion is .58 ($p < .01$). If we use only the three dimensions as predictors (PP, TS, and R), the multiple R is negligible, namely, .15 (NS). However, if we add to these three predictors the single nonlinear term TS × R, the multiple correlation increases to .45 ($p < .05$). This increase from .15 to .45 is significant beyond the .01 level. Alternatively, if we add to the three original dimensions only the single interaction term PP × R, the multiple R increases from .15 to .42, again a significant increase ($p < .01$) over prediction obtained from the three linear predictors alone.

TABLE III

MEDIAN CORRELATIONS BETWEEN LEADER LPC AND
GROUP PERFORMANCE IN VARIOUS OCTANTS

	Leader-member relations	Task structure	Position power	Median correlation	Number of relations included in median
Octant I	Good	Structured	Strong	− .52	8
Octant II	Good	Structured	Weak	− .58	3
Octant III	Good	Unstructured	Strong	− .41	4
Octant IV	Good	Unstructured	Weak	.47	10
Octant V	Mod. poor	Structured	Strong	.42	6
Octant VI	Mod. poor	Structured	Weak		0
Octant VII	Mod. poor	Unstructured	Strong	.05	10
Octant VIII	Mod. poor	Unstructured	Weak	− .43	12
Octant V-A	Very poor	Structured	Strong	− .67	1

As already mentioned, the task structure and leader position power are highly correlated ($r = .75$). While this may be a peculiarity of the particular sample of tasks used in these studies, it is likely that this relationship is fairly general (i.e., that leader power is higher in those situations in which the task is more highly structured). The above findings may then indicate that the leader who is low in LPC has more effective work groups when either his position power or task structure is high *and* leader-member relationships are favorable. When leader power or task structure is low *and* leader-member relations are poor, then the high LPC leader has a more effective group.

However, the multiple correlation tells only part of the story. As Fig. 4 shows, several areas in this three-dimensional space remain empty, suggesting new hypotheses for further investigations. Thus, the correlations between leader LPC and group performance in the upper left-hand

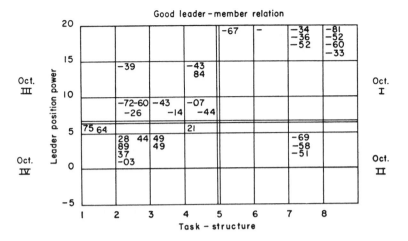

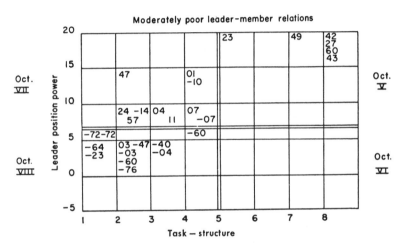

FIG. 4. Correlations between leader LPC/ASo and group performance plotted against leader position power and task structure under good and poor leader-member relations.

corner of Octant III should probably be positive, since an accepted leader with extremely high position power might need to be very permissive and accepting in order to work effectively with a group on a highly ambiguous task. An example might be a general working with two enlisted men on a creative task, or a cardinal with two laymen on a planning committee. The hypothesis is that a large difference in position power would awe low-ranking members and stifle their creativity, unless the leader were extremely permissive and equalitarian in his attitude and behavior. Another

TABLE IV

MULTIPLE CORRELATION ANALYSIS OF GROUP DIMENSION VARIABLES
AS PREDICTORS OF THE LPC/ASo-GROUP EFFECTIVENESS CORRELATION

	(1) PP	(2) TS	(3) R	(4) PP × TS	(5) PP × R	(6) TS × R	(7) PP × TS × R	(Criterion) $r_{LPC\text{-perf.}}$
PP[a]		.75[d]	−.09	.60[d]	−.08	−.17	.02	−.09
TS[b]			.03	.43[d]	−.16	−.01	−.10	−.14
R[c]				−.17	.16	.15	.64[d]	.04
PP × TS					.08	−.15	.01	.08
PP × R						.78[d]	.60[d]	−.34[d]
TS × R							.49[d]	−.41[d]
PP × TS × R								.01

STANDARDIZED REGRESSION COEFFICIENTS

	$r_{LPC\text{-perf.}}$
PP	−.39
TS	.05
R	−.22
PP × TS	.22
PP × R	−.40
TS × R	−.39
PP × TS × R	.59[d]

NOTE: Multiple r = .58.
[a] PP = Position power.
[b] TS = Task structure.
[c] R = Leader-member relationship.
[d] $p < .01$ (N = 68).

hypothesis based on the plot suggests that the leader in Octant VI should be permissive and accepting. This is a situation in which a relatively disliked leader has little formal power. The most appropriate leader behavior in this situation is likely to be one of guidance and suggestion, rather than attempts to manage and control the group.

V. Validation Evidence

In addition to recent investigations which provide partial tests of the model, appropriate data from previously conducted studies were re-analyzed in light of the new hypotheses.

A. RE-ANALYSES OF DATA FROM REAL-LIFE GROUPS

Previous data suggested that managing, controlling leader attitudes appear most effective under group situations which are either *very* favorable or *very* unfavorable to the leader; permissive, accepting leader attitudes are most appropriate under conditions which are only moderately unfavorable. As indicated before, it was hypothesized that the leader-member relations were the most important factor in determining the favorableness of the group situation. It should then be possible to test this hypothesis on real-life groups, which vary widely in leader-member relations while task structure and position power remain constant. This should result in negative correlations between LPC or ASo of the leader and group performance, in which the leader-member relationship is either very good or very poor, and positive correlations when the leader-member relationship is moderately poor.

Such data were obtained earlier in a study of B-29 bomber crews. The leader-member relations were inferred from sociometric preference ratings obtained from all crew members. The groups ranged from highly cohesive crews with a large proportion of mutual choices, to very uncohesive groups in which the leader was strongly rejected and, in turn, expressed low opinions of his ranking crew members. The relationship in these crews was assessed by (a) whether or not the leader was the most chosen member of his crew, and (b) his acceptance or rejection of his key crew members (i.e., the men who performed criterion relevant tasks). Evaluation of crew performance was based on radar bomb scores, which require the aircraft commander to cooperate closely with the radar observer or navigator, who are the keymen in this instance. (Some of these crews were included in Octant I and IV of Table II).

It was assumed that a very favorable situation existed if a crew leader was sociometrically the most chosen crew member and, in turn, endorsed his keymen, while a less favorable condition obtained if a sociometrically endorsed leader did not endorse his keymen (or if an unchosen leader endorsed his keymen), and a very unfavorable condition existed if a rejected leader also rejected his keymen.

Correlations have been made between the leader's ASo score and his crew's radar bomb scores under six conditions of increasingly poorer leader-member relations (Fiedler, 1955). At the time of the original study, these relations could not be properly interpreted, and only recently was there an opportunity to cross-validate the findings.

As Table V shows, the expected curvilinear relationship was obtained between ASo and group performance and the degree to which the leader-member relationship was good. The problem remained to find similar results in other real-life groups in which leader-member relations varied from very good to very poor.

The hypothesis could be tested on data obtained in a study of anti-aircraft artillery crews (Hutchins and Fiedler, 1960). Effectiveness scores were based on rated crew performance in the "location and acquisition" of unidentified aircraft. These data were re-analyzed by separately analyzing the ten crews which fell most nearly in the middle of the distribution and the ten crews which gave their commanders the lowest sociometric ratings. As Table VI shows, the correlations between leader LPC and the crew performance scores again ranged from negative to positive to negative, thus supporting the findings obtained in the study of B-29 crews.

TABLE V

CORRELATIONS BETWEEN AIRCRAFT COMMANDER'S ASo SCORE AND
RADAR BOMB SCORE UNDER DIFFERENT CONDITIONS OF
SOCIOMETRIC CREW CHOICE PATTERNS[a]

Sociometric choice pattern	Rho	N
AC = MPC → VO/N[b]	−.81	10
AC = MPC −−VO/N	−.14	6
AC = MPC ÷ VO/N	.43	6
AC ≠ MPC → VO/N	−.03	18
AC ≠ MPC−−VO/N	−.80	5
AC ≠ MPC ÷ VO/N	−.67	7

[a] Table adapted from Fiedler (1955).

[b] Aircraft commander is (=), or is not (≠) most preferred crew member and sociometrically accepts (→). is neutral to (−−), or rejects (÷) keyman (radar observer and navigator).

TABLE VI

CORRELATIONS BETWEEN LEADER LPC SCORES AND
ANTIAIRCRAFT ARTILLERY CREW PERFORMANCE

	Rho	N
Most highly chosen crew commanders	−.34	10
Middle range in sociometric choices	.49	10
Lowest chosen crew commanders	−.42	10

A second validation analysis was based on a study of 32 farm-supply cooperative organizations (Godfrey et al., 1959). Each of these companies had its own board of directors which, in turn, hired a general manager. The company managers worked closely with the board of directors, as well as with their staff of assistant managers. The criterion of performance was the percentage of net income of the company over the preceding three-year period.

The groups were again divided according to sociometric choice patterns. It was assumed that a general manager who was chosen both by his board of directors and by his staff had a better relationship and hence a more favorable group situation than one chosen only by the board or only by his staff. It was also assumed that his position would be least favorable when both his board of directors and his own staff of assistant managers sociometrically rejected him. As before, the data supported the hypothesis (Table VII).

TABLE VII

CORRELATIONS BETWEEN GENERAL MANAGER'S LPC SCORE
AND COMPANY NET INCOME

	Rho	N
GM most chosen by board and staff	−.67	10
GM chosen by board, but rejected by staff	.20	6
GM rejected by board, but chosen by staff	.26	6
GM rejected by board and staff	−.75	7

It is not yet known what might happen under even less favorable situations in which the leader is not only rejected by his group, but also has an unstructured task and little power. Groups under these conditions might simply disintegrate, or the leader might lose control. It is hoped that studies testing performance under these specific conditions can be conducted in the near future.

B. RESEARCH BY HAWKINS

Three studies were recently conducted by Hawkins (1962) at the University of Minnesota. The first of these involved two-person teams with one individual appointed as leader. These teams were asked to agree on the "right responses" for the *How Supervise?* test. The team's performance was scored on the basis of items completed within a given time. These teams, which obviously had a highly structured task and a leader with low position power, fell into Octant II. Hawkins found that pairs with low ASo leaders completed a larger number of items than did pairs with high ASo leaders. The results thus supported the hypothesis.

A second study utilized sales display teams from a large company. A detailed operating procedure is provided, thus making the task highly structured. The leader, as in most supervisory situations of this type, had fairly high power. He was in charge of the team and could recommend personnel actions. The groups were evaluated by higher supervisors within the district. Although no sociometric measures were obtained in this study, it seems reasonable to assume that most teams fell into Octant I. In line with the model's predictions, Hawkins found a significant difference between teams, with the more controlling, managing, low ASo leaders rated as being more effective.

The third study dealt with a chain of service stations. These stations were also expected to operate in accordance with detailed instructions under the supervision of the station manager. The effectiveness of the stations was evaluated by spot checks of supervisors as well as stock control, accounting, and sales records. Here again, task structure, as well as position power, appeared to be high. We must again assume that personnel in most stations sociometrically accepted their managers. If this assumption is correct, the group situations fall into Octant I, and Hawkins' findings, that low ASo managers were more effective, support the model.

C. Relaxed vs. Stressful Conditions

It will be recalled that groups in which the leader rated the group atmosphere as pleasant performed better on creative tasks under high LPC leaders, while groups rated as having low group atmosphere performed better under low LPC leaders (Fiedler, 1962). These groups were categorized as having low task structure and low position power and as falling into Octants IV or VIII, depending on the leader's GA score.

A study was recently conducted to test this effect experimentally by making the group situation pleasant under one condition and less pleasant under a second condition. This study was conducted within the context of a leadership training conference in which 57 persons participated. The participants were mature men and women who held responsible positions in the Unitarian-Universalist church. The Conference members were assigned to three-person groups which were given two successive tasks. The first required them to write a statement for their children to justify the parents' opposition to the reading of prayers in public schools. This was presented as a relatively easy task, given as a "warm-up" exercise for which the groups had more than sufficient time (35 minutes). The second task required the group to prepare a statement which would explain the Unitarian-Universalist creed to children of another church. This task was presented as a "real test of leadership" which was quite difficult, and which required the groups to operate under considerable time pressure (20 minutes).

LPC and intelligence test scores had been obtained on a previous day. Sociometric preference ratings, group atmosphere scores, and behavior descriptions of fellow group members were collected after each of the tasks. Group task products were evaluated by all conference participants.

This group situation was rated as having low task structure and low position power. As for leader-member relations, they were less strained for the leader in the first than in the second session. Thus the groups were classified as falling into Octant IV during the first task and Octant VIII during the second task.

In line with this expectation, groups of high LPC leaders performed significantly better under the more pleasant, relaxed condition of the first task, while low LPC leaders performed more effectively on the second, less pleasant and more stressful task (Fiedler *et al.*, 1964).

VI. Discussion

This chapter attempts to reconcile the apparent inconsistencies in our work by showing that the prediction of leading effectiveness by means of LPC scores (and perhaps other leader attributes as well) is contingent upon an adequate classification of the group situation. The data raise several questions which merit further discussion. Specifically, how good is the classification system which has been suggested? What does this model say about the effect of leader and member abilities on task performance? What are the implications of this model for leadership strategy? What is the role of quasi-therapeutic leader attitudes in task groups? Finally, how well does this model serve to integrate the findings of other investigators in this field?

A. CLASSIFICATION OF GROUP SITUATIONS

The notion that different types of groups require different types of leaders is, of course, not new. The present model attempts to spell out the specific conditions under which certain leadership attitudes result in effective performance. The first question concerns the adequacy of the proposed classification system.

The underlying basis for categorizing group situations is the degree to which they facilitate the leader's job of influencing his group. This classification system appears to work for the type and range of group situations which have been covered. It remains to be seen whether this classification, based on the three dimensions of leader-member relations, task structure, and leader position power, represents the best solution. Other dimensions, such as group stress or member motivation, may later need to be substituted or added as further data become available.

Several shortcomings of the categorization are readily apparent. One of these is the problem of scaling and ordering the dimensions. Under normal conditions all three dimensions of the group-task situation seem to play an important role. However, some groups dislike the leader enough to sabotage the task. Under these conditions, a leader could not be very effective, even under otherwise favorable circumstances. Such an intense negative feeling by group members toward their leader obviously must be given additional weight. This was done on a rule-of-thumb basis by considering Octant V-A the most unfavorable point on the continuum (cf. Figs. 2 and 3). Similarly, a task might be so highly structured that any permissiveness on the part of the leader would result in degrading the group's performance. Count-downs of space probes or the maintenance of rigid, quality control criteria, are cases in point. These two examples indicate that a better method is needed for weighting the dimensions so as to allow for extreme cases.

B. LEADER AND MEMBER ABILITIES

It is well established that leader and member abilities are among the most important predictors of group performance (Mann, 1959; McGrath, 1962). Teams composed of intellectually inferior and technically unskilled persons will naturally perform worse on almost any task than will groups composed of able and qualified individuals. Fiedler and Meuwese (1963), using commonly accepted intelligence and ability tests, have shown in four different studies that the leader's ability scores correlate highly with performance of military units and laboratory groups *only if the leader is sociometrically accepted.* That is, in groups falling into Octants I and IV, the median correlation between the leader's ability scores and his group's performance was .74 (which is considerably above the comparable median correlation of .25 reported in Mann's 1959 review). In contrast, the corresponding median for uncohesive groups, or those which sociometrically rejected their leader (Octants V and VIII), was only —.22. The difference between these two medians is highly significant.

A high correlation between the leader's ability score and the group's performance presumably reflects the degree of leader influence over the task itself (i.e., either the group followed the leader's suggestions or it permitted the leader to do the job in his own way). The negative correlations found in uncohesive groups suggest that the leader's influence, or his own contribution to the task, is minimal, and that the leader may have to attend primarily to the performance of maintenance functions in these groups.

C. STRATEGIES OF LEADERSHIP

Inasmuch as LPC or ASo are measures of interpersonal attitudes, a leader can, of course, be trained to modify these attitudes. He will probably be able to do so to the same extent to which he can modify other strongly entrenched attitudes. Changes in perceptions of co-workers seem, therefore, possible, but considerable effort might be required on the part of many individuals to make them.

A more practical alternative would involve the training of leaders in diagnosing their group-task situation and in adopting strategies which capitalize on their particular leadership style. For example, a controlling, managing, and directive (low LPC) leader might find himself faced with a relatively unstructured task. According to our contingency model, he should first attempt to structure and clarify the group's problem in order to move his group from Octant IV to Octant I, in which he could operate more effectively. A high LPC leader with poor group-member relations, an unstructured task, and low position power (Octant VIII) should first improve his interpersonal relations to move his group into Octant IV, in which his leadership attitude would be more appropriate.

A different strategy might be desirable for tasks which increase their structure over time. This is well illustrated by research projects which are tentative and unstructured during the planning phases, but become highly structured and programmed once the design is frozen and the experiment is in progress. It is often quite apparent that the group situation changes in other respects as well. While the leader of the project generally plays a very permissive, democratic role during the initial phase of the project, he tends to become managing and autocratic as the experiment gets under way and the need for strict control increases.

D. Quasi-therapeutic Leader Attitudes and Effectiveness

There has been considerable speculation whether the effective leader —or someone in his place—must always perform "quasi-therapeutic" functions which assist the group members to adjust (e.g., Berkowitz, 1953; Clark, 1955; Haernqvist, 1956; Gordon, 1955). The writer's data may throw some light on this question. Studies show that the high LPC leader, who is permissive, non-directive, and considerate in his approach to the group, does tend to promote better group relations and more member satisfaction (Fiedler et al., 1961c; Meuwese, 1964; Hutchins and Fiedler, 1960). It seems likely on the basis of presently available evidence that these quasi-therapeutic attitudes may be especially important in groups which have interpersonal conflict, strained leader-member relations, or an ambiguous, unstructured task. As Neel (1955), Seashore (1955), and others have shown, these conditions tend to arouse anxiety among group members. A quasi-therapeutic leader may then alleviate insecurity and anxiety which might otherwise interfere with efficient performance. Such person-oriented attitudes may be especially desirable in certain co-acting groups in which several individuals independently pursue a common goal with minimal interaction (e.g., a rifle team or a track team). In such coacting groups, the leader has very few directive and controlling functions, and his contribution to effective task performance may consist primarily of providing psychological support and encouragement to the individual team member.

Studies have been conducted on a number of these groups. One investigation dealt with team training of naval aviation cadets. This tends to be an anxiety arousing experience, since it involves not only a demanding ground school program, but also relatively dangerous formation flying. Each "flight" typically consisted of eight trainees who were under the tutelage of a flight instructor team. An informal leader of each flight (who had no power or official status) was identified by means of sociometric indices. The formal leader was the senior instructor in charge of training the flight.

The correlation between the ASo score of the sociometrically preferred member of the flight and the flight's performance was .42 ($N = 36$, $p < .05$). The senior flight instructor's ASo score and the group performance correlated .31 ($N = 14$). These results suggested that the informal leader of the group or the flight's senior instructor promoted effectiveness by his ability to lessen the anxiety of the group's members.

A study by DeZonia (1958) investigated the relationship between the teacher's assumed similarity score and his rated efficiency as a teacher. DeZonia found a correlation of $-.55$ for English teachers, but $+.57$ for teachers of speech. The teaching of grammar and English literature probably requires the teacher to keep the student focused on the task. The ability to teach public speaking may, however, also require that the teacher alleviate the anxiety which is associated with stage fright and pre-speaking jitters. This latter quasi-therapeutic function would be more effectively performed by the permissive, nondirective person.

A coacting group situation was also contained in a laboratory study by Anderson and Fiedler (1962). This investigation utilized naval ROTC midshipmen who were asked to devise unusual uses for two objects, namely, a coathanger and a ruler. The leader of the group under the "supervisory" condition could approve or reject suggestions from his group, but he was not permitted to contribute ideas. Under a "participatory" condition, the leader worked with the group as its chairman.

The task was scored in terms of the uniqueness, i.e., infrequency of the suggested uses by various members. This placed them in a coacting relationship, since the productivity of one member was relatively unaffected by that of his colleagues. Here again, the correlation between the leader's LPC score and his group's performance was positive, i.e., .55 and .31 ($N = 15$) under the supervisory and participatory conditions, respectively.

It can readily be seen why the permissive, accepting leader would promote greater productivity under these conditions. Unique solutions, which are often silly and by definition off-beat, are more likely to occur when the leader is permissive and accepting than when he is critical and threatening.[5]

---

[5] This might also account for the "out-of-place" correlation of .84 between leader LPC and group performance which occurred under the supervisory condition of the Triandis *et al.* (1963) "fame and immortality" problem used in the Navy

We must recognize, of course, that these attempts to relate the leader's task functions and his therapeutic attitudes are highly speculative at this time. Extensive future research will be required to elucidate the role which these therapeutic attitudes play in the group process.

E. GENERALIZABILITY OF THE MODEL

Ideally it should be possible to integrate all other studies of leadership effectiveness traits into the contingency model which has been described here. This is impossible for a number of compelling reasons. Above all, the literature is difficult to interpret, because the profusion of variables does not permit one to infer with any assurance that a particular study fits into one octant rather than another.

However, the model summarized here is consistent with other research. Halpin's finding of negative correlations between the aircraft commander's "consideration" score, and a positive correlation between his "initiation-of-structure" score and crew performance, fit well within the framework outlined here. Likewise, the conclusion is supported that "in an authoritarian setting the most productive group will tend to be authoritarian, whereas in a more democratic situation the equalitarian group will be more productive" (Hare, 1962). Shaw's (1955) finding that authoritarian leadership attitudes led to better performance than did nonauthoritarian leader attitudes is also applicable. Shaw used highly structured tasks and, provided the leaders in his groups were accepted, the groups would then normally fall into Octants I or II.

Permissive, democratic, equalitarian leadership was more effective in Comrey's (1954) study of forest management, and McGregor's (1960) and Likert's (1961) experience with management groups. These, if interpreted correctly, represented situations in which the task was considerably less structured and the leaders tended to be more equal in power to their group members. Given accepted leaders, situations of this nature should fall into Octant IV, which favors the equalitarian, permissive attitudes which McGregor and Likert advocate. Other studies cannot be readily considered as fitting the model (e.g., Argyle *et al.*, 1957; Kahn and Katz, 1960). It is, however, obvious that any attempt to integrate these and similar studies in the absence of specific information about leader-member relations, task structure, and position power becomes speculative to the point of uselessness. If the model is to have any meaning at all, it must be

---

ROTC study (Octant III). In this task, the group members were to suggest ways in which a person of average means and talent could achieve fame and immortality. Each individual contribution was relatively independent of other members' ideas, even though the group later had to agree on the one best solution. The most unusual solution received the highest score.

able to predict a reasonably high proportion of appropriate cases. Where such appropriate data are available, as in the Havron *et al.* (1954) investigation of infantry squads, or the studies by Hawkins (1962), the evidence clearly supports the model's requirements. But we do not know in most other studies how well the leader was accepted, how highly structured his task was, or how much position power was at his disposal. It is hoped that such information will be available in the future, as we learn to identify the specific information about group situations which is crucial for the understanding and prediction of leadership effectiveness.

REFERENCES

Anderson, L. R., and Fiedler, F. E. (1962). The Effect of Participatory and Supervisory Leadership on Group Creativity. Group Effectiveness Research Laboratory, Univ. of Illinois, Urbana, Illinois.

Argyle, M., Gardner, G., and Cioffi, R. (1957). *Human Relat.* **10**, 295–313.

Bass, B. M. (1960). "Leadership, Psychology, and Organizational Behavior." Harper, New York.

Berkowitz, L. (1953). *J. Abnorm. Soc. Psychol.* **48**, 231–238.

Bieri, J. (1953). *J. Abnorm. Soc. Psychol.* **48**, 61–66.

Browne, C. G., and Cohn, T. S., eds. (1958). "The Study of Leadership." Interstate Printers and Publishers, Danville, Illinois.

Campbell, D. T. (1958). *Behav. Sci.* **3**, 14–25.

Cartwright, D., and Zander, A., 1960. *Group Dynamics,* (2nd ed.) Row Peterson, Evanston, Ill.

Cattell, R. B. (1951). *Human Relat.* **4**, 161–184.

Clark, R. A. (1955). Leadership in Rifle Squads on the Korean Front Line. Human Resources Research Unit No. 2, CONARC, Fort Ord, California.

Cleven, W. A., and Fiedler, F. E. (1956). *J. Appl. Psychol.* **40**, 312–314.

Comrey, A. L. (1954). *Personnel Psychol.* **7**, 533–547.

Comrey, A. L., Pfiffner, J. M., and High, W. S. (1954). Factors Influencing Organizational Effectiveness. Univ. of Southern California, Los Angeles, California.

Cronbach, L. J., and Gleser, G. C. (1953). *Psychol. Bull.* **50**, 456–473.

Davitz, J. (1955). *J. Abnorm. Soc. Psychol.* **50**, 173–176.

Deutsch, M. (1949). *Human Relat.* **2**, 199–232.

DeZonia, R. H. (1958). Unpublished Ph.D. Dissertation, University of Illinois, Urbana, Illinois.

Fiedler, F. E. (1951). *J. Clin. Psychol.* **7**, 101–107.

Fiedler, F. E. (1954). *J. Abnorm. Soc. Psychol.* **49**, 381–388.

Fiedler, F. E. (1955). *J. Abnorm. Soc. Psychol.* **51**, 277–235.

Fiedler, F. E. (1961). *In* "Leadership and Interpersonal Behavior" (L. Petrullo and B. M. Bass, eds.). Holt, New York.

Fiedler, F. E. (1962). *J. Abnorm. Soc. Psychol.* **65**, 308–318.

Fiedler, F. E., and Meuwese, W. A. T. (1963). *J. Abnorm. Soc. Psychol.* **67**, 83–87.

Fiedler, F. E., Warrington, W. G., and Blaisdell, F. J. (1952). *J. Abnorm. Soc. Psychol.* **47**, 790–796.

Fiedler, F. E., Bass, A. R., and Fiedler, J. M. (1961a). Tech. Rep. No. 1, Group Effectiveness Research Laboratory, Univ. of Illinois, Urbana, Illinois.

Fiedler, F. E., Meuwese, W. A. T., and Oonk, S. (1961b). *Acta Psychol.* **18**, 100–119.

Fiedler, F. E., London, P., and Nemo, R. S. (1961c). Hypnotically Induced Leader Attitudes and Group Creativity. Group Effectiveness Research Laboratory, University of Illinois, Urbana, Illinois.

Fiedler, F. E., Hackman, R. R., and Meuwese, W. A. T. (1964). Group Effectiveness Research Laboratory, University of Illinois, Urbana, Illinois (in preparation).

French, J. R. P., Jr. (1956). *Psychol. Rev.* **63**, 181–194.

French, J. R. P., Jr., and Raven, B. H. (1958). *Sociometry* **21**, 83–97.

George, C. E. (1962). Research Memorandum #26, U.S. Army Infantry Human Research Unit, Fort Benning, Georgia.

Gerard, H. B. (1957). *J. Pers.* **25**, 475–488.

Gibb, C. A. (1954). *In* "Handbook of Social Psychology" (G. Lindzey, ed.), Vol. II. Addison-Wesley, Cambridge, Massachusetts.

Godfrey, E. P., Fiedler, F. E., and Hall, D. M. (1959). "Boards, Managers, and Company Success." Interstate Press and Publishers, Danville, Illinois.

Gordon, T. (1955). "Group Centered Leadership: A Way of Releasing the Creative Power of the Group. Houghton-Mifflin, Boston.

Haernqvist, K. (1956). "Adjustment: Leadership and Group Relations." Almqvist and Wiksell, Stockholm, Sweden.

Halpin, A. W. (1955). *Harvard Educ. Rev.* **25**, 18–22.

Halpin, A. W., and Winer, B. J. (1957). *In* "Leader Behavior: Its Description and Measurement" (R. M. Stogdill and A. E. Coons, eds.), Bur. Bus. Res Monograph 88. Ohio State Univ., Columbus, Ohio.

Hare, A. P. (1962). "Handbook of Small Group Research." Free Press, Glencoe, New York.

Havron, M. D., Fay, R. J., and Goodacre, D. M., III. (1951). Research on the Effectiveness of Small Military Units. Adjutant General's Department PRS Report 885. Washington, D. C.

Havron, M. D., Lybrand, W. A., Cohen, E., Kassebaum, R. G., and McGrath, J. E. (1954). Tech. Research Note 31, The Adjutant General's Office, Personnel Research Branch, Washington, D. C.

Hawkins, C. (1962). Ph.D. Dissertation, Univ. of Minnesota.

Haythorn, W. W. (1956). *J. Soc. Abnorm. Psychol.* **53**, 210–219.

Horst, P. (1954). *J. Clin. Psychol.* **10**, 3–11.

Hutchins, E. B., and Fiedler, F. E. (1960). *Sociometry* **23**, 393–406.

Janda, K. F. (1960). *Human Relat.* **13**, 345–363.

Kahn, R. L., and Katz, D. (1960). *In* "Group Dynamics" (D. Cartwright and A. Zander, eds.), 2nd ed. Row Peterson, Evanston, Illinois.

Katz, E., and Kahn, R. L. (1953). *In* "Readings in Social Psychology," (G. E. Swanson, T. M. Newcomb, and E. L. Hartley, eds.) Holt, New York.

Lewin, K., and Lippitt, R. (1938). *Sociometry* **1**, 292–300.

Likert, R. (1961). "New Patterns of Management." McGraw-Hill, New York.

Lippitt, R., and White, R. K. (1952). *In* "Readings in Social Psychology," (G. E. Swanson, T. M. Newcomb, and E. L. Hartley, eds.) Holt, New York.

Lubin, A., and Osborn, H. G. (1957). *Psychometrica* **22**, 63–73.

McGrath, J. E. (1962). HSR-TN-62/3-Gn, Human Sciences Research, Inc., Arlington, Virginia.

McGregor, D. M. (1960). "The Human Side of Enterprise." McGraw-Hill, New York.

Maier, N. R. F., and Solem, A. R. (1952). *Human Relat.* **5**, 277–288.

Mann, R. D. (1959). *Psychol. Bull.* **56**, 241–270.

Meuwese, W. A. T. (1964). The Effect of the Leader's Ability and Interpersonal Attitudes on group creativity under varying conditions of stress.

Myers, A. E. (1962). *J. Abnorm. Soc. Psychol.* **65**, 325–332.

Neel, R. (1955). *Personnel Psychol.* **8**, 409–416.

Osgood, C. E., Suci, G. J., and Tannenbaum, P. H. (1957). "The Measurement of Meaning." Univ. of Illinois Press, Urbana, Illinois.

Schutz, W. C. (1953). Studies in Group Behavior. I. Construction of High Productivity Groups. Department of Systems Analysis, Tufts College, Medford, Massachusetts.

Schutz, W. C. (1958). "FIRO: A Three Dimensional Theory of Interpersonal Behavior." Holt, Rinehart and Winston, New York.

Seashore, S. E. (1955). Institute for Social Research, Ann Arbor, Michigan.

Sells, S. B. (1962). "Military Small Group Performance under Isolation and Stress— a Critical Review." Office of Technical Services, U.S. Dept. of Commerce, Washington, D. C.

Shaw, M. E. (1955). *J. Abnorm. Soc. Psychol.* **50**, 127–134.

Shaw, M. E. (1962). Ann. Tech. Rept. University of Florida, Gainesville, Florida.

Sherif, M., and Sherif, C. W. (1953). "Groups in Harmony and Tension." Harper, New York.

Spector, P., and Suttell, B. J. (1956). American Institute for Research, Washington, D. C.

Steiner, I. D. (1964). *Ann. Rev. psychol.*

Stogdill, R. (1948). *J. Psychol.* **25**, 35–71.

Triandis, H. C., Bass, A. R., Ewen, R. B., and Mikesell, E. H. (1963). *J. Appl. Psychol.* **47**, 104–110.

Recent Developments in Research on the Contingency Model

Fred E. Fiedler

UNIVERSITY OF WASHINGTON,
SEATTLE

Since its publication in Volume 1 of this series (Fiedler, 1964), the Contingency Model has stimulated well over 300 theoretical and empirical studies. In briefly reviewing the major developments, I shall here consider four main topics. These are (*a*) methodological questions concerning the stability and interpretation of the Least Preferred Coworker (LPC) score and the Situational Favorableness (or Situational Control) dimension; (*b*) the validity of the model; (*c*) the formulation of a dynamic theory of leadership; and (*d*) the development and validity of a new leadership training program. A more extensive discussion of the current status of the Contingency Model will be found in Volume 11 of this series (Berkowitz, 1978).

I. Methodological Issues

A. Least Preferred Coworker (LPC)

Although the internal consistency of LPC has been quite satisfactory ($\overline{X}$ = .88; Rice, 1978), the retest reliability, or stability, of the score has been the subject of some controversy, since this determines whether LPC can be used as a predictor of leadership performance and whether it can be interpreted as a transsituational personality attribute.

209

The best stability estimate comes from an extensive review of the literature by Rice (1978) that uncovered 23 reported retest correlations. These had a median of .67 and a mean of .64 (SD = .36). The median and mean are well within the range of many widely used personality tests and in the upper range of social psychological measures for which stability coefficients are reported (see Robinson and Shaver, 1973). Though the standard deviation for these retest correlations seems high, it is difficult to say whether this is abnormal for personality or social psychological tests, since there are very few such measures for which more than one or two different retest correlations are available.

There is evidence from some studies that the stability of the score may be influenced by intensive human relations training programs (Arbuthnot, cited in Rice, 1978) or by an unsuccessful or unpleasant group experience (Drucker, cited in Fiedler, 1967). Rice also points out that such experiences as management workshops, T-groups, military training, etc. may provide subjects with a set of implicit demands for "proper" evaluation of coworkers to change their LPC scores. Interestingly enough, Rice's analysis shows only a weak correlation between the stability coefficient and the length of the test–retest interval ($r = .30$, $n = 23$, ns), indicating that time, by itself, does not materially affect test-retest reliability. This is particularly true of mature adults in relatively stable situations. Thus, Bons (1974) reported a retest correlation of .71 over a 5-month period for 45 higher level leaders of an artillery battalion, and Prothero and Fiedler (1974) obtained a retest correlation of .67 over 16–24 months for 18 members of a school of nursing. Low retest correlations are typical of pre- and postexperimental test administrations as well as of individuals who are unmotivated in the job situation (Taylor, 1975). It is clear that we need further research to determine the specific conditions that affect the stability of LPC.

The LPC score has been difficult to interpret, since it does not correlate consistently with other personality tests and behavior observations. It has been viewed variously as, for example, a measure of task- versus person-orientation (Rice, 1978), an assessment of attitude toward coworkers (Fishbein, Landy, and Hatch, 1969), a cognitive complexity measure (Mitchell, 1970; Foa, Mitchell, and Fiedler, 1971), a measure of social distance (Fiedler, 1953), and an index of a motivational hierarchy (Fiedler, 1972b). On the whole, however, the research shows quite clearly that a high LPC score characterizes a person who is primarily oriented toward, and concerned with, relations with others, and that a low LPC score characterizes a person who is oriented toward, and is primarily concerned with, the accomplishment of the task (Rice, 1978).

The LPC score clearly is not correlated with consideration or structuring behavior or any other leader behaviors. Rather, it taps an orientation or a set of priorities and goals. In order to accomplish a task, one may need to be quite considerate and concerned with interpersonal relations under one set of conditions and fairly ruthless under others. To gain the support or loyalty of one's group, it may first be necessary to succeed.

Fairly consistent evidence from a wide variety of studies has shown that high LPC leaders tend to be considerate and concerned with interpersonal relations in situations which are relatively stressful or in which control is relatively low. In these same situations, low LPC leaders tend to become preoccupied with the task to the neglect of their relations with subordinates. However, in situations in which the leader enjoys a high degree of control over the group and the group outcomes, high LPC leaders tend to behave, as one perceptive observer remarked, like "underemployed mothers": They look for things to do and become managing and bossy and no longer concerned with the opinions, feelings, and support of their subordinates. As a result, they are seen as inconsiderate and more concerned with the task, whereas low LPC leaders relax and become considerate and congenial with their coworkers, since they know that the task will get done (e.g., Fiedler, Meuwese, and Oonk, 1961; Meuwese and Fiedler, 1965, cited in Fiedler, 1967; Chemers, 1969; Larson and Rowland, 1973).

The interpretation that best seems to fit these data considers LPC as an index of a motivational hierarchy. It holds that each individual has one basic goal or priority in a leadership situation, as well as one (or more) goals that are of lesser importance. The more basic goal for the high LPC leader is the relationship with his subordinates, the secondary goal is the accomplishment of the task. For the low LPC leader, the basic goal is task performance, and the relationship with subordinates is of lower priority. Only when the basic goal is assured will the individual devote himself primarily to the attainment of his secondary goals.

A number of studies illustrate this point. For example, Sample and Wilson (1965) had small groups of psychology students perform an experiment with a rat in a Skinner box. The leader of each team was surreptitiously observed during each of the three phases of the experiment, that is, planning, running, and reporting the study. In the planning phase, when the task was least structured and the leader had relatively low situational control, the high LPC leaders made a larger proportion of social–emotional remarks, whereas low LPC leaders made a larger proportion of task-relevant remarks. These same leaders behaved in exactly the opposite manner during the running phase (the administration of the experiment), when the leaders had high task structure. Then, the high LPC leaders made relatively more task-relevant comments, and the low LPC leaders made relatively more social–emotional remarks (see Figure 1).

Likewise, a study by Larson and Rowland (1973) showed that low LPC leaders under experimentally induced stress behaved in a task-relevant manner, whereas high LPC leaders behaved in a manner indicating concern for subordinates. Under relaxed, stress-free conditions, however, low LPC leaders expressed more concern and consideration for subordinates, and high LPC leaders expressed concern for the task. Similar results were obtained in laboratory experiments by Green, Nebeker, and Boni (1976); Fiedler, Meuwese, and Oonk (1961); and Meuwese and Fiedler (1965) and in field experiments by Ayer (1968) and Fiedler (1966).

The effect of situational control or favorableness becomes even clearer in a study (Chemers, 1969) in which high and low LPC leaders were given human relations training in how to get along with group members from another culture, whereas a randomly chosen control group was given training irrelevant to the task (i.e., training in the physical geography of the country). If we assume that an effective training program generally increases the leader's control and influence, we would then also predict that the training will increase the low LPC leaders' considerate and relationship-oriented behavior while decreasing the considerate behavior and concern for interpersonal relations of high LPC leaders. As Figure 2 shows, this prediction was

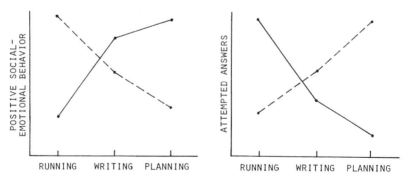

Figure 1. Comparison of behavior by relationship-motivated (high LPC—●) and task-motivated (low LPC—×) leaders.

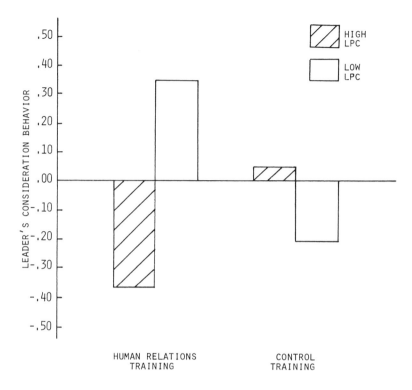

Figure 2. Effects of human relations training on considerate behavior of high- and low-LPC leaders.

supported (no significant differences were found in the behaviors of untrained leaders). A similar study using American students as leaders and Arab students as group members yielded almost identical results (Chemers, Fiedler, Lekhyanada, and Stolurow, 1966). Although other interpretations cannot be ruled out, the results of the various studies on the behavior of high and low LPC persons under high and low control conditions seems most compatible with a motivational hierarchy interpretation of LPC.

As is discussed in more detail elsewhere (Fiedler, 1972b), the behaviors of low LPC leaders meet the requirements of the high control situation by giving the already motivated and willing group encouragement and support, whereas the high LPC leader's unnecessary concern with the task is seen as interfering. In the stressful situation or in one in which the leader has very low control, the task-motivated leader's concern for the task permits him to ignore his poor relations with subordinates, whereas the high LPC leader focuses on the relationship to the neglect of the task.

Moderate-control (or moderately favorable) situations typically involve committees and decision-making bodies and creative teams that accept the leader, or else task groups with high structure and position power that do not support the leader. Both of these situations require the ability to reconcile differences, to engender harmonious relations, and to show concern for the feelings and opinions of group members. The high LPC leader's reaction to a moderate control situation meets these needs. The low LPC leader, however, becomes increasingly concerned with the task to the neglect of his interpersonal relations with group members, which are here crucial to successful performance.

B. SITUATIONAL CONTROL

The measure of situational favorableness or control in the Contingency Model is based on leader–member relations, task structure, and position power scales. Leader–member relations are given four times as much weight as position power, and task structure is given twice the weight of position power. These weights were derived intuitively on the basis of data that predate the 1964 chapter in this series. Recent evidence shows that this system of weighting the three scales is basically correct. A study by Nebeker (1975) on supervisors of two different organizations first showed that the situational favorableness dimension correlated −.56 and −.40 with Lawrence and Lorsch's (1967) measure of environmental uncertainty, that is, lack of clarity of job requirements, difficulty in accomplishing the job, and delay in feedback about job success. (This has also been reported by Mai-Dalton (1975) in a subsequent laboratory study.) Second, Nebeker's study indicated that the theoretical weights (4 × leader–member relations + 2 × task structure + position power) predict almost exactly as well as does an empirically derived multiple regression that provides optimal weighting of these three variables. And a more recent experiment by Beach, Mitchell, and Beach (1978) further confirmed the relative weights of the three components of the situational control dimension and also reported a correlation of .89 between the judged probability of a leader's success and situational control. The relationship between situational control and uncertainty about the outcomes of decisions was also shown by Csoka's (1975) research on army dining halls, which indicated that task-motivated leaders performed more effectively in organizational climates that are mechanistic

(Burns and Stalker, 1961) and, therefore, highly predictable, whereas relationship-motivated leaders performed more effectively in the less predictable and more uncertain organic climates. These findings suggest a strong link between the Contingency Model and approaches based on decision theory.

II. Validity of the Contingency Model

A large number of tests have been conducted by various investigators to assess the validity of the contingency Model. Most competent studies have supported the theory. A review of the literature (Fiedler and Chemers, 1974) listed 47 relationships that tested specific hypotheses of the Contingency Model in each of the seven octants for which predictions had been made. Of these, 38, or 81%, were in the predicted direction. Figure 3 shows the median correlations of the validation studies as connected by a solid line and the original median correlations as connected by a broken line (Table I).

The methodologically most stringent tests were conducted by Chemers and Skrzypek (1972) and by Hardy and his associates (Hardy, 1971, 1975; Hardy, Sack, and Harpine, 1973). In the Chemers and Skrzypek study, 128 West Point cadets served as subjects. Four-man groups were assembled on the basis of LPC scores, and sociometric preference ratings were obtained 3 weeks prior to the experiment. These four-man groups performed one structured and one unstructured task in counterbalanced order. Half the groups had leaders with high, and half with low, position power, and in half the groups the leaders and members had chosen one another as preferred coworkers, whereas the men in the other half of the groups had expressed a desire not to work with others who were assigned to their group. The correlations between the LPC scores and the performance of each of the eight groups is shown on Figure 3 as the dash–dot line. The rank–order correlation between the medians of the original studies and these correlations was .86 ($N = 7$, $p < .01$). Thus, the Chemers and Skrzypek study almost exactly replicated the Model, and, in a reanalysis by Shiflett (1973), accounted for 28% of the variation in performance.

Three studies were conducted by Hardy (1971, 1975) and Hardy, Sack, and Harpine (1973) that utilized college and elementary school students, as well as school administrators. In each of these experiments LPC scores were obtained prior to the group sessions, as also were sociometric scores, so as to preclude the possibility that good or

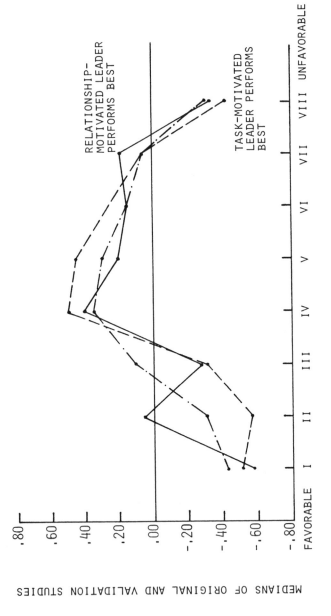

Figure 3. Median correlations between leader performance and group performance for the original studies (———), validation studies (————), and the West Point study (—··—).

TABLE 1

SUMMARY OF FIELD AND LABORATORY STUDIES TESTING THE CONTINGENCY MODEL

	Octants							
	I	II	III	IV	V	VI	VII	VIII
Field studies								
Hunt (1967)	−.64		−.80		.21		.30	
	−.51		.60				−.30	
Hill (1969)			−.10	−.29		−.24	.62	
Fiedler *et al.* (1969)		−.21		.00			.67*	−.51
O'Brien and Fiedler (unpublished)			−.46	.47			−.45	−.14
Tumes (1972)	−.47**			.62**				
Laboratory experiments								
Belgian Navy (Fiedler, 1966)	−.72	.37	−.16	.08	.16	.07	.26	−.37
	−.77	.50	−.54	.13	.03	.14	−.27	.60
Shima (1968)		−.26		.71*				
Mitchell (1969)		.24		.43				
		.17		.38				
Fiedler Exec.		.34		.51				
Chemers and Skrzypek (1972)	−.43	−.32	.10	.35	.28	.13	.08	−.33
Rice and Chemers (1973)						.30		−.40
Sashkin (1972)			−.29*					
Median, all studies	−.59	.04	−.29	.40	.19	.13	.17	−.35
Median, field studies	−.51	−.21	−.29	.47	.21	−.24	.30	−.33
Median, laboratory experiments	−.72	.24	−.23	.38	.16	.14	.08	−.35
Median in original studies	−.52	−.58	−.33	.47	.42		.05	−.43

Number of correlations in the expected direction: 38[a]

Number of correlations opposite to expected direction: 9

p by binomial test: .01

[a] Exclusive of Octant VI, for which no prediction has been made.

*$p < .05$.

**$p < .01$.

poor group performance would influence the leader–member relations ratings.

In each of these studies, all results but those in Octant II were in the predicted direction and were significant. (It should be noted that the results are from laboratory experiments in Octant II. Results from field studies follow the predictions of the model, suggesting that the conditions of Octant II are difficult to develop in a laboratory.)

III. A Dynamic Formulation of the Leadership Process

A major contribution of the Contingency Model is the better understanding that it provides of the dynamic interaction of leader and organization. Leadership is an integral part of the continually evolving organizational process. The Contingency Model enables us to interpret and predict changes in leader behavior and performance in terms of concomitant changes in the leader's situational control.

A schematic representation of the Model is shown in Figure 3. Leadership performance is here indicated on the vertical axis and situational control on the horizontal axis of this figure. The performance of the relationship-motivated leader is shown as a solid line, and that of the task-motivated leader is shown as a broken line. We now see that the performance of the relationship-motivated leader will increase as situational control changes from very low to moderate and will then decrease as it changes further from moderate to high (moving toward the left side of the graph). The performance of the task-motivated leader will first decrease as situational control changes from low to moderate and will then increase as it becomes high.

This is well illustrated by the typical effects of leader experience, which generally increases the leader's ability to control and influence the outcomes of his decisions and actions. That is, in time the job becomes more routine, and the leader no longer needs to develop new methods of handling various problems that arise periodically (see, for example, Bons and Fiedler, 1976). The effects of experience on leadership performance can then be predicted from Figure 3.

For example, one investigation of 28 army squad leaders was conducted in a new infantry division. At the beginning of the study, the leaders were new to the organization and to the men, and their situational control was moderate. Their performance was evaluated by two or more superiors. Five to six months later these same superiors again evaluated each of these sergeants. The results, shown on Figure 4, exactly follow the prediction of the model. This was also shown in a study of general managers of a consumer cooperative organization, in which the criteria were percentage of operating expenses to total sales and net income to total sales (Fiedler, 1975), as well as in a study of elementary and secondary school principals (McNamara, 1968). These findings provide strong support for the theory.

In addition, a well controlled laboratory experiment by Chemers, Rice, Sundstrom, and Butler (1975) yielded similar findings for leadership training. In this case, ROTC students at the University of Utah

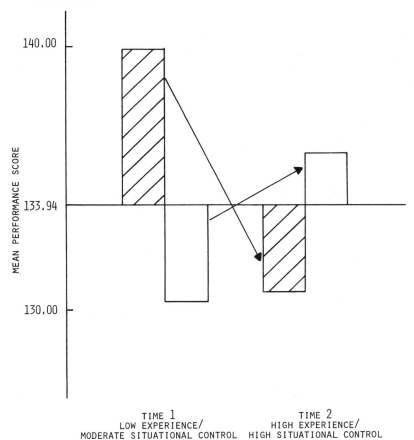

Figure 4.   Change in performance of high and low LPC leaders as a function of increased experience.

were divided into high and low LPC persons and the middle third served as group members along with students from a psychology class. The high and low LPC leaders were randomly assigned to a training and a nontraining group prior to being given the task of deciphering coded messages (i.e., cryptograms). The task training consisted of teaching the leaders the rules for decoding simple messages. For example, they were instructed to count the various letters and assume that the most frequent letter would be an "e," that a three-letter word ending in "e" would be "the," that the only two English words with one letter were "a" and "I," etc.

Because of conflict between ROTC and psychology students and the unexpected presence of officers during the administration of the experiment, the group climate scores in this study turned out to be substantially and significantly below the mean of similar groups. Thus, the untrained leaders had poor leader–member relations, low task-structure, and low position power (Octant VIII). The trained leaders had poor leader–member relations, high task-structure, and low position power (Octant VI). We would expect, therefore, that the task-motivated leaders would perform better than relationship-motivated leaders in the trained condition and poorer in the untrained condition. Figure 5 shows that this was the case. Even more importantly, however, the task-motivated leaders with training also performed significantly less well than did task-motivated leaders who had not received task training (many of these low LPC leaders got into arguments and irrelevant discussions with members of their group). This set of results again supports the Contingency Model.

Predictions from the Model for situations in which the leader's control and influence are decreased as a result of organizational turbulence (e.g., getting a new boss, a new set of subordinates, or a new job) have also been supported (Bons and Fiedler, 1976).

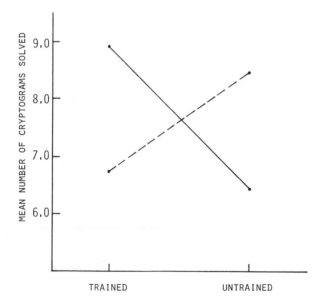

Figure 5. Comparison of task performance by high and low LPC leaders with and without task training.

IV. Leadership Training

The acid test of a leadership theory is its ability to improve leadership performance, that is, to control as well as to describe and predict leadership performance. The Contingency Model's ability to do so is shown by a recently developed leadership training program called "LEADER MATCH." This is a self-paced programmed manual (Fiedler, Chemers, and Mahar, 1976) that lets the leader identify and interpret his LPC score and then provides specific instruction on how to diagnose various types of leadership situations in terms of the control and influence that they provide. In addition, the program gives detailed guidance on how to modify situational control and influence as well as the control and influence of subordinate leaders or managers. The leader is told, for example, that he can match his situation to his personality by such means as volunteering either for highly structured tasks or for relatively unstructured and ambiguous assignments. He can increase or decrease the closeness of his relationship with subordinates by making himself more or less accessible to them, for example, by socializing with them or by being more aloof, and he can modify his position power, for example, by using a participative or a nonparticipative management approach. The total training program requires from 4 to 8 hours of self-instruction.

Eight validation studies have been successfully completed. In each of these, a trained group and a control group were chosen at random from an available pool of leaders or managers. The criterion of performance in all these studies consisted of performance evaluations by two or more superiors.

Two of these studies were conducted independently by other investigators under highly controlled conditions. Csoka and Bons (1978, in press) obtained a sample of 154 army cadets. Before they were assigned as acting platoon leaders to various army field units, one third of this group was randomly selected for LEADER MATCH training and the other two-thirds were used as controls, with one control group informed that members of another group had received training. At the end of the assignment, these men were rated by their superiors in the field, who were unaware that some men had received special training. As shown on Figure 6, the trained group performed significantly better than either of the two untrained control groups. In a second experiment, one of three platoon leaders from each of 27 training companies was chosen at random for leadership training with LEADER MATCH. At the end of approximately 4 months, all platoon leaders were evaluated by their superior officers. Those with

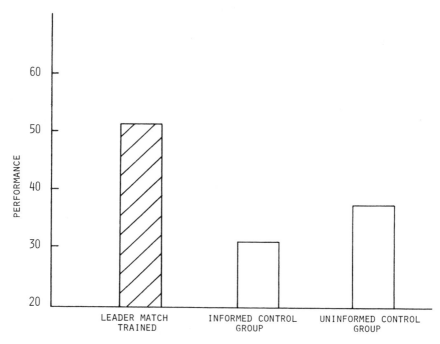

Figure 6.   Acting army platoon leaders evaluated after summer field study.

training were significantly more often considered to be the best of the three platoon leaders in their company (54%) than were those without training (26%). Interviews indicated that trained leaders had been able to modify their situational control as recommended in the training manual.

Two studies of naval personnel compared a randomly selected group of trained officers and petty officers with a similarly chosen control group (Leister, Borden, and Fiedler, 1977). The men were rated prior to training and again 6 months later. The group with training improved to a substantial and significant degree in task as well as in interpersonal performance, whereas the leaders in the untrained group either did not improve or were rated as performing less well than before.

Four other studies (Fiedler, Mahar, and Schmidt, 1975) were conducted in civilian organizations, namely, with middle managers, supervisors and managers of a public works agency, police sergeants, and second-level leaders of a public health volunteer organization in Latin America. In each of these studies, the trained leaders received significantly higher performance evaluations on task-relevant items than did the control group leaders. The effectiveness of this training program thus provides further evidence for the validity of the Contingency Model as well as for its capability to contribute insights and practical applications in the area of leadership.

REFERENCES

Ayer, J. G. Effects of success and failure of interpersonal and task performance upon leader perception and behavior. Unpublished masters thesis, University of Illinois, Urbana, 1968.

Beach, B. H., Mitchell, T. R., & Beach, L. R. Components of situational favorableness and probability of success. *Organizational Behavior and Human Performance*, 1978 (in press).

Berkowitz, Leonard. *Advances in experimental social psychology*, Vol. 11. New York: Academic Press, 1978.

Bons, P. M. The effect of changes in leadership environment on the behavior of relationship- and task-motivated leaders. Unpublished doctoral dissertation, University of Washington, 1974.

Bons, P. M., & Fiedler, F. E. Changes in organizational leadership and the behavior of relationship- and task-motivated leaders. *Administrative Science Quarterly*, 1976, **21**, 433–472.

Bons, P. M., & Fiedler, F. E. Leadership. In L. Bittel (Ed.), *Encyclopedia of professional management*. New York: McGraw-Hill, 1977 (in press).

Burns, T., & Stalker, G. M. *The management of innovation*. London: Tavistock Publication, 1961.

Chemers, M. M. Cross-cultural training as a means for improving situational favorableness. *Human Relations*, 1969, **22**, 531–546.

Chemers, M. M., Fiedler, F. E., Lekhyanada, D., & Stolurow, L. M. Some effects of cultural training on leadership in heterocultural task groups. *International Journal of Psychology*, 1966, **1**, 301–314.

Chemers, M. M., Rice, R. W., Sundstrom, E., & Butler, W. Leader esteem for the least preferred coworker score, training, and effectiveness: An experimental examination. *Journal of Personality and Social Psychology*, 1975, **31**, 401–409.

Chemers, M. M., & Skrzypek, G. J. An experimental test of the Contingency Model of leadership effectiveness. *Journal of Peronality and Social Psychology*, 1972, **24**, 172–177.

Csoka, L. S. Organic and mechanistic organizational climates and the Contingency Model. *Journal of Applied Psychology*, 1975, **60**, 273–277.

Csoka, L. S., & Bons, P. M. Manipulating the situation to fit the leader's style—Two validation studies of Leader Match. *Journal of Applied Psychology*, 1978 (in press).

Fiedler, F. E. The psychological distance dimension in interpersonal relations. *Journal of Personality*, 1953, **22**, 142–150.

Fiedler, F. E. A Contingency Model of leadership effectiveness. In L. Berkowitz (Ed.), *Advances in experimental social psychology*. Vol I. New York: Academic Press, 1964. Pp. 149–190.

Fiedler, F. E. The effect of leadership and cultural heterogeneity on group performance. A test of the Contingency Model. *Journal of Experimental Social Psychology*, 1966, **2**, 237–264.

Fiedler, F. E. *A theory of leadership effectiveness.* New York: McGraw-Hill, 1967.

Fiedler, F. E. Validation and extension of the Contingency Model of leadership effectiveness: A review of empirical findings. *Psychological Bulletin,* 1971, **76**(2), 128–148.

Fiedler, F. E. Leadership experience and leadership training—Some new answers to an old problem. *Administrative Science Quarterly,* 1972, **17**, 453–470. (a)

Fiedler, F. E. Personality, motivational systems, and behavior of high and low LPC persons. *Human Relations,* 1972, **25**, 391–412. (b)

Fiedler, F. E., & Chemers, M. M. *Leadership and effective management.* Glenview, Ill.: Scott, Foresman, and Company, 1974.

Fiedler, F. E., Chemers, M. M., & Mahar, L. *Improving leadership effectiveness: The leader match concept.* New York: John Wiley and Sons, 1976.

Fiedler, F. E., Mahar, L., & Schmidt, D. Four validation studies of Contingency Model training. University of Washington, Organizational Research Group Technical Report 75–70, Seattle, 1975.

Fiedler, F. E., Meuwese, W., & Oonk, S. An exploratory study of group creativity in laboratory tasks. *Acta Psychologica,* 1961, **18**, 100–119..

Fiedler, F. E., O'Brien, G. E., & Ilgen, D. R. The effect of leadership style upon the performance and adjustment of volunteer teams operating in a stressful foreign environment. *Human Relations,* 1969, **22**, 503–514.

Fishbein, M., Landy, E., & Hatch, G. A consideration of two assumptions underlying Fiedler's Contingency Model for prediction of leadership effectiveness. *American Journal of Psychology,* 1969, **82**, 457–473.

Foa, U. G., Mitchell, T. R., & Fiedler, F. E. Differentiation matching. *Behavioral Science,* 1971, **16**, 130–142.

Green, S., Nebeker, D., & Boni, A. Personality and situational effects on leader behavior. *Academy of Management Journal,* 1976, **19**, 184–194.

Hardy, R. C. Effect of leadership style on the performance of small classroom groups: A test of the Contingency Model. *Journal of Personality and Social Psychology,* 1971, **19**, 367–374.

Hardy, R. C. A test of poor leader–member relations cells of the Contingency Model on elementary school children. *Child Development,* 1975, **45**, 958–964.

Hardy, R. C., Sack, S., & Harpine, F. An experimental test of the Contingency Model on small classroom groups. *Journal of Psychology,* 1973, **85**, 3–16.

Hill, W. A. The LPC leader: A cognitive twist. 29th Annual Meeting, Academy of Management Proceedings, August 1969, 125–130.

Hunt, J. G. Fiedler's leadership Contingency Model: An empirical test in three organizations. *Organizational Behavior and Human Performance,* 1967, **2**, 290–308.

Larson, L. L., & Rowland, K. Leadership style, stress, and behavior in task performance. *Organizational Behavior and Human Performance,* 1973, **9**, 407–421.

Lawrence, P. R., & Lorsch, J. W. Differentiation and integration in complex organizations. *Administrative Science Quarterly,* 1967, **12**, 1–47.

Leister, A. F., Borden, D., & Fiedler, F. E. The effect of Contingency Model leadership training on the performance of Navy leaders. *Academy of Management Journal,* 1977 (in press).

Mai-Dalton, R. The influence of training and position power on leader behavior. University of Washington Organizational Research Group, Technical Report 75-72, Seattle, 1975.

McNamara, V. D. Leadership, staff, and school effectiveness. Ph.D. thesis, University of Alberta, Edmonton, Alberta, Canada, 1968.

Meuwese, W., & Fiedler, F. E. Leadership and group creativity under varying conditions of stress. Urbana, Illinois: Group Effectiveness Research Laboratory, University of Illinois, 1965.

Mitchell, T. R. Leader complexity and leadership style. *Journal of Personality and Social Psychology*, 1970, **16**, 166–174.

Nebeker, D. M. Situational favorability and environmental uncertainty: An integrative study. *Administrative Science Quarterly*, 1975, **20**, 281–294.

Prothero, J., & Fiedler, F. E. The effect of situational change on individual behavior and performance: An extension of the Contingency Model. University of Washington Organizational Research Group, Technical Report 74-59, Seattle, 1974.

Rice, R. W. Psychometric properties of the esteem for least preferred coworker (LPC) scale. *Academy of Management Review*, 1978, 3(1), 106–118.

Rice, R. W., & Chemers, M. M. Predicting the emergence of leadership effectiveness. *Journal of Applied Psychology*, 1973, **57**, 281–287.

Robinson, J. P., & Shaver, P. R. Measures of social psychological attitudes. Ann Arbor, Michigan: Institute for Social Research, University of Michigan, 1973.

Sample, J. A., & Wilson, T. R. Leader behavior group productivity and rating of least preferred coworker. *Journal of Personality and Social Psychology*, 1965, **1**, 266–270.

Sashkin, M. Leadership style and group decision effectiveness: Correlational and behavioral tests of Fiedler's Contingency Model. *Organizational Behavior and Human Performance*, 1972, **8**, 347–362.

Shiflett, S. C. The Contingency Model of leadership effectiveness: Some implications of its statistical and methodological properties. *Behavioral Science*, 1973, 18(6), 429–440.

Shima. H. The relationship between the leader's modes of interpersonal cognition and the performance of the group. *Japanese Psychological Research*, 1968, **10**, 13–30.

Taylor, D. An investigation of the relative stabilities of some prominent measures of leadership. Thesis, University of Washington, 1975.

Tumes, J. The Contingency Theory of leadership: A behavioral investigation. Paper presented at the Eastern Academy of Management Meetings, Boston, Mass., May 5, 1972.

WHY DO GROUPS MAKE RISKIER DECISIONS THAN INDIVIDUALS?[1]

Kenneth L. Dion,[2] Robert S. Baron,[3] and Norman Miller[4]

UNIVERSITY OF MINNESOTA

[1]Preparation of this paper was supported in part by National Institutes of Health Predoctoral Research Fellowships to Kenneth L. Dion and Robert S. Baron and both a Sabbatical Leave Fellowship from the University of Minnesota and a National Institutes of Health Special Research Fellowship to Norman Miller. We would like to thank our many colleagues who provided us with prepublication drafts of their recent research and/or comments on an early draft of this paper. In particular we wish to thank Penny Baron, Tom Bouchard, Roger Brown, Eugene Burnstein, Russell Clark III, Karen K. Dion, Calvin Hoyt, James P. Flanders, Jerald Jellison, Nathan Kogan, Helmut Lamm, George Levinger, David Myers, Serge Moscovici, P. Pilkonis, Dean Pruitt, Ivan Steiner, Allan I. Teger, Donald L. Thistlethwaite, Neil Vidmar, Michael A. Wallach, Robert Wolosin, Robert Zajonc, and Mark Zanna. We feel that this review has benefited substantially from their help. Of course, responsibility for its present content is our own and it should not be assumed that any completely agree with our own evaluations.

[2]Now at University of Toronto.
[3]Now at University of Iowa.
[4]Now at University of Southern California.

Reprinted from *Advances in Experimental Social Psychology,*
Volume 5, 305–377.

I. Introduction

In recent years the biggest surge of research on group processes focuses on group decision-making and risk-taking. The typical finding is that a person's willingness to take risks increases after participating in a group discussion of the problem. This change is usually referred to as the "risky-shift" and was first demonstrated by Stoner (1961).

The research impetus that followed Stoner's initial experiment sprang largely from the counterintuitive nature of the findings. At that time, most of the relevant evidence suggested that group interaction and group decision-making would reduce risk-taking. For example, the research on conformity indicated that pressure to reach a compromise or a consensus induces groups to direct influence attempts at "deviates" (e.g., Schachter, 1951). This suggests that group interaction should reduce the variability among individuals' tendencies to take risks and produce an averaging effect among group members. On the face of it, these effects seem incompatible with an overall increase in riskiness. In addition, Whyte (1956) contended that group discussion produces more conservative decisions. Supporting this view, Barnlund (1959) demonstrated that in a logic task, groups are more careful than isolated individuals. Why then did Stoner find a shift to greater risk-taking following group interaction? This chapter reviews the research that has sought to analyze why groups make riskier decisions than individuals. In doing so, it builds on an earlier and more general chapter on risk-taking by Kogan and Wallach (1967c).

A. THE BASIC RESEARCH PARADIGM

Paralleling the Stoner experiment, investigations of group risk-taking typically use a repeated measures design. Private, pretest measures of risk-taking are obtained first. Subjects are then exposed to an experimental situation. A typical treatment, for example, may assemble individuals into a discussion group and have them arrive at consensus decisions on each of the items (hypothetical situations) comprising the initial risk-taking measure. These decisions constitute a second measure of risk-taking (the treatment test). A difference between the mean pretest score and mean group (treatment) score in the direction of increased risk-taking typically defines the magnitude of the risky-shift.[5] After the group is dissolved, the posttest measures of individual risk-taking are often again obtained. While the comparison of this third measure to the other previous measures is interesting, differences between it and other measures generally are not what researchers are speaking of when they use the phrase "risky-shift."

When the treatment requires participation in group interaction, the several "shift" measures derived from the three-staged, repeated measures design may reflect different compliance processes. Wallach, Kogan, and Bem (1962) recognized that differences between the pretest and treatment may be due to the operation of external social pressures. That is, it is possible that such differences might at least in part reflect overt compliance processes. Maintenance of group-induced changes on private, posttest measures of risk-taking (taken after the group is dissolved) is interpreted as covert compliance; it reflects the effect of group exposure on the realignment of individual risk-taking preferences. In a more specific classification scheme, Mackenzie and Bernhardt (1968) proposed that the amount of recidivism observed in treatment-posttest comparisons defines the degree of overt compliance, and that pretest-posttest comparisons provide a measure of covert compliance. The important difference between these two classification systems is that the Mackenzie and Bernhardt scheme specifies the means for assessing the degree of overt compliance in the decision process. While Wallach *et al.* emphasize the role of covert changes, Mackenzie and Bernhardt regard the risky-shift as a mixture of both overt and covert compliance. It is important to note that the repeated-measures format typically used in the risky-shift research is minimally effective in detecting "overt-covert" differences. The Solomon Four-Group design (Campbell & Stanley, 1963)

[5]There are a few isolated exceptions to this rule. For example, Flanders and Thistlethwaite (1967) used the difference between pretest and posttest measures of private risk-taking as their index of the risky-shift.

offers a more effective procedure for determining overt and covert compliance effects.

B. Tasks

The paper-and-pencil Choice-Dilemmas questionnaire (Kogan & Wallach, 1964, Appendix E) has served as the primary instrument for assessing risk-taking preferences in risky-shift investigations. It consists of twelve hypothetical, "real life" situations in which a fictitious person must choose between a risky or a conservative course of action. Of these two alternatives, the risky option was constructed to be more desirable, but, of course, also had less chance of succeeding. For example, a sample item describes an electrical engineer who has a choice between: (a) remaining at his present, secure job—one with a modest salary but little hope of improvement or (b) joining a new firm which has a highly uncertain future but offers the possibility of sharing in the ownership. The respondent's task is to indicate what the odds for success would have to be before he would advise the fictitious person to attempt the risky alternative. On this task the respondent can choose among 1, 3, 5, 7, or 9 chances in 10 that the risky alternative will be successful. There is an additional category for each item which allows respondents to indicate whether they feel that the risky course of action should not be attempted no matter what the odds.

Authors of the Choice-Dilemma instrument have reported corrected split-half reliabilities ranging from .53 to .80 for various samples (Kogan & Wallach, 1964; Wallach & Kogan, 1961) and test-retest reliabilities (1 week) of .78 and .82 (Wallach et al., 1962). In terms of validity, relationships with other risk-taking behaviors (Kogan and Wallach, 1964) and other types of risk-related phenomena (Kogan & Wallach, 1961; Wallach & Kogan, 1961) have been demonstrated. Pruitt and Teger (1967) pooled data from group risk-taking studies conducted in various locations throughout the country. They found that 10 of the 12 items exhibit consistent risky-shift effects across investigations.

A second type of group risk-taking task requires subjects to choose the level of difficulty they wish to attempt on various categories (e.g., antonyms, analogies, and mathematics) of old College-Board examination items (Wallach, Kogan, & Bem, 1964). These questions reflect nine levels of difficulty ranging from 10 to 90% failure in a national sample. The higher the difficulty level chosen, the greater the risk-taking.

Experimental gambling situations constitute the third type of risk-taking task. Zajonc and his associates used a simple light guessing situation in which one light has a high probability of occurrence with a small reward, while the other has a lower frequency but a somewhat larger reward. On a given trial, the aim is to guess which of these two lights will be illuminated. One of the important features of this task is that subjects immediately experience the outcomes of their decisions from trial to trial. A second gambling task (Pruitt & Teger, 1969) differs from the one used by Zajonc and his associates in that it manipulates stake as well as probability preferences.[6] In addition, it uses a wager format that provides no feedback from trial to trial.

In terms of frequency, the Choice-Dilemmas task accounts for approximately 80% of group risk-taking studies. The problem-solving and gambling tasks constitute a much smaller share — approximately 14% and 6%, respectively.

C. Generality of the Risky-Shift[7]

Risky-shifts are not characteristic of any one age or occupational category. They have been found with established professional groups (Siegel & Zajonc, 1967), industrial supervisors (Rim, 1965a), senior executives (D. G. Marquis, 1962), management trainees (Stoner, 1961), male and female undergraduates (Wallach et al., 1962) as well as grade-school boys and girls (Kogan & Carlson, 1969). However, it is apparent that all of these groups (with the exception of the last) probably fall in the upper portion of the intelligence distribution. Although the bulk of research used American subjects, risky-shifts have also been found with other nationalities, including English (Bateson, 1966), Israeli (e.g., Rim, 1963), Canadian (Vidmar, 1970), French (Kogan & Doise, 1969), and German subjects (Lamm & Kogan, 1970).

Most of the aforementioned studies demonstrated risky-shifts under conditions in which group discussion continued until consensus was reached. However, increased risk-taking has also been obtained under

[6] A stake preference refers to the *amount* one will risk (for any given likelihood of a successful outcome). A probability preference refers to the *likelihood of success* (irrespective of the amount waged).

[7] A semantic problem arises in that careful consideration of the problems or tasks suggests that they tell us little about actual risk-taking (cf. Flanders, 1970). Rather, they provide a means of studying group processes — largely within the setting of a single type of experimental design. These issues are discussed more fully in the concluding section. But meanwhile, for the bulk of this paper we use the term used by those performing the research, *viz.*, "risky-shift." In the interests of a more proper conceptualization, the reader might profitably substitute a term like "polarization" or simply think of the term "risky-shift" as a nominal label for the process evoked by the various experimental designs. In other words, we postpone our discussion of the extent to which the results of the various experiments bear on the question of how groups behave when confronted with truly high risk alternatives that portend direct consequences for themselves.

discussion without consensus (Wallach & Kogan, 1965) and under conditions that minimize the impact of group discussion by preventing visual contact (Kogan & Wallach, 1967a) or face-to-face communication (Lamm, 1967; Kogan & Wallach, 1967d). In addition, risky-shifts have also been produced in two situations that completely eliminate any exposure to group discussion: (1) when subjects do not interact verbally but are aware of each others' risk preferences (Blank, 1968; Teger & Pruitt, 1967) and (2) when individual subjects are given a chance to restudy the Choice-Dilemma items and respond to them a second time (Bateson, 1966; Flanders & Thistlethwaite, 1967).

While most studies have used the Choice-Dilemmas task, risky-shifts have also been demonstrated with more realistic tasks involving monetary rewards for successful problem-solving performance (Wallach *et al.*, 1964), and monetary rewards under threat of aversive consequences for failure on a problem-solving task (Bem, Wallach, & Kogan, 1965). On the other hand, the group risk-taking experiments that use gambling or betting tasks yield an equivocal pattern of outcomes. Several studies (Atthowe, 1960; Hinds, 1962; Hunt & Rowe, 1960; Lonergran & McClintock, 1961; Zajonc, Wolosin, Wolosin, & Sherman, 1968) report either no shift or a shift toward caution in group situations; others (Blank, 1968; Pruitt & Teger, 1969; Zajonc, Wolosin, Wolosin, & Sherman, 1969; Zajonc, Wolosin, Wolosin, & Loh, 1970) show shifts toward risk. Nevertheless, it is obvious that all told, the effect generalizes across subjects, situations, and tasks. The one reservation that must be reiterated is the extent to which the tasks lack any (much less severe) consequences for the group members.

D. PLAN OF THE PAPER

This chapter will review the literature manifestly concerned with group risk-taking comprehensively, and examine in detail the major alternative interpretations offered to explain the risky-shift. Four explanations receive particular consideration: (1) diffusion of responsibility, (2) persuasion, (3) familiarization, and (4) cultural values. Later sections spell out these explanations. They were chosen because they are the most frequently cited explanations in the literature and have generated the bulk of empirical research on group risk-taking. Other more specific hypotheses concerning the risky-shift effect have been considered elsewhere (Pruitt & Teger, 1967) and are not dealt with in detail here.[8]

[8] Likewise, this review does not consider recent experiments dealing with risk-taking in *intergroup* negotiation situations (Hermann and Kogan, 1968; Lamm and Kogan, 1970) or ethical risk-taking in social situations (e.g., Rettig, 1966).

The sections to follow use a standard format in considering each of the four selected interpretations. In each case, derivative hypotheses are specified and underlying mechanisms presumed responsible for the risky-shift effect are examined in detail. Relevant research is brought to bear upon these elements and serves as the basis for a judgment of the adequacy and validity of each of the four hypothesized explanations.

II. The Diffusion of Responsibility Explanation

According to the responsibility-diffusion explanation, the risky-shift represents a *true* group effect, namely, one that cannot occur with isolated individuals (Pruitt & Teger, 1967; Secord & Backman, 1964). It assumes that fear of failure primarily deters an individual's tendency to take risks. Group decision-making, in contrast to individual decision-making, presumably diffuses responsibility among the group members. This diffusion of responsibility reduces fear of failure and thereby enables people to make riskier decisions.[9]

The relationship between fear of failure and responsibility-diffusion has not been given a great deal of emphasis by those who advocate the responsibility-diffusion interpretation. However, they do imply that diffusion of responsibility reduces or eliminates fear of failure and thereby increases risk-taking. For example, consider the following excerpts from Kogan and Wallach (1967b):

> . . . failure of a risky course is *easier to bear* when others are implicated in a decision; . . . consider a homogeneous group composed of test anxious individuals, that is, individuals uniformly fearful of failure . . . [such people] might be especially willing to diffuse responsibility in an effort to *relieve the burden* of possible fear of failure. (p. 51).

> The burden of *fear* [carried by high test anxious subjects] will show *a sharp decline* when one does not have to bear all alone, the responsibility for failure of a risky course. [p. 56, all italics ours].

The major proponents of the responsibility-diffusion interpretation — Kogan, Wallach, and their associates — have consistently viewed responsibility-diffusion primarily as an "anxiety melioration" process. On the other hand, their view of the necessary and sufficient conditions for the occurrence of responsibility-diffusion has undergone substantial change. Their original formulation (Wallach *et al.,* 1964) stressed that

[9] Fear of failure has been characterized as "a capacity for reacting with shame and embarrassment when the outcome of performance is failure . . ." and is thought to be aroused "whenever it is clear to a person that his performance will be evaluated and failure is a distinct possibility" (Atkinson, 1964, p. 244).

the process of reaching a unanimous group decision diffuses feelings of personal responsibility for failure among group members and, therefore, leads to increased risk-taking. With the subsequent discovery that group consensus and joint decision had no additional impact over and above mere group discussion in producing a risky-shift (Wallach & Kogan, 1965), the emphasis upon group decision was obviously no longer tenable. As a result, they pointed to other consequences of group discussion or membership that might produce a feeling of shared responsibility for a risky decision and thereby lead to a risky-shift. They suggested that stronger affective bonds between group members was one consequence of group discussion that could increase feelings of shared responsibility. This notion will be referred to as the "affective bonds" hypothesis.

In summary, the diffusion of responsibility interpretation of group risk-taking now suggests the following causal chain: (1) group discussion creates affective bonds; (2) affective bonds permit diffusion of responsibility; (3) diffusion of responsibility reduces fear of failure; (4) reduced fear of failure produces the risky-shift. Those who advocate the responsibility-diffusion interpretation focus upon three basic derivations from this causal chain. (a) Group discussion is the necessary and sufficient condition to produce the risky-shift (from statements 1-4). (b) The affective bonds produced by group discussion permit diffusion of responsibility (from statements 2-4). (c) Responsibility-diffusion is the essential reason for the risky-shift (from statements 3 and 4).

These derivations are not necessarily basic or central points for a responsibility-diffusion explanation. For instance, one can easily argue that both hypothesis (a) and (c) are stated too strongly, and that neither (a) nor (b) need be true for (c) to remain valid. Nevertheless, these propositions are listed here because they have been most frequently espoused by those who favor the responsibility-diffusion explanation. Although most of the research on responsibility-diffusion explores or examines these hypotheses, other closely related hypotheses can be derived (and tested). For example, if as Wallach and Kogan (1965) maintain, the affective bonds formed in discussion enable the individual to feel less blame when he entertains the possible failure of a risky decision, then the magnitude of risky-shift should be directly proportional to the strength of these bonds. Although not considered by responsibility-diffusion theorists, another derivation is that the magnitude of risky-shift monotonically increases as a function of group size (e.g., Mackenzie & Bernhardt, 1968). When a large group of people share in a decision, each individual member should feel little responsibility for the outcome and further should feel relatively protected from retaliation or punishment since his own individual behavior is not apt to be a point of focus. The proposition that large groups are riskier than smaller ones suggests still another derivation — that anonymity might contribute to the risky-shift. This notion will be referred to as the anonymity hypothesis. Note that in contrast to the affective bonds hypothesis, the anonymity hypothesis seems to focus attention on fear of punishment as well as on guilt and fear of failure as inhibitors of risk-taking. Anonymity should increase risk-taking by facilitating deindividuation and lowering fear of punishment.

The following sections separately consider hypotheses (*a*), (*b*), and (*c*) (above) in terms of relevant empirical evidence. In addition, they discuss the evidence relevant to fear of failure and the anonymity hypothesis.

A. THE RESPONSIBILITY-DIFFUSION HYPOTHESIS

Wallach *et al.* (1964) tested the responsibility-diffusion mechanism by systematically manipulating features of individually shared and group shared responsibility within the basic discussion-to-consensus paradigm. The experimental task was a problem-solving situation in which rewards for correct solutions were commensurate with the difficulty of the problem. Feelings of responsibility for others were experimentally manipulated by varying whether or not individuals felt that either their own problem-solving performance or choice of difficulty level might determine the group outcome. Thus, the experimental conditions outlined in Table I can be viewed as varying along a continuum of responsibility.

For example, subjects in the "group-group" condition should feel least responsible in that no single individual is liable for the level of difficulty chosen, and, if a subject is chosen as group representative, it is the responsibility of those others in the group who chose him. In the "group-lot" condition, a given subject should feel slightly more responsible, for, although he is not responsible for the level of difficulty chosen, he cannot blame other members of the group if he is chosen as group representative. The results provide several sources of direct support for the responsibility-diffusion interpretation. First, a risky-shift only occured in those conditions in which there was a group decision. Second, when responsibility for others was experimentally created, a cautious shift occurred. Finally, the monotonic cell orderings array themselves in precisely the pattern that would be predicted from the responsibility-diffusion hypothesis.

On the other hand, several studies seem to challenge the responsibility-diffusion interpretation. D. G. Marquis (1962) designated as group leader the person whose initial risk-taking score on a set of hypothetical items was closest to the average for that group. This appointed leader was then given authority for conduct of the group discussion as well as sole responsibility for the final decision. Significant increases in risk-taking were found for both leaders and other group members after group discus-

TABLE I
DESCRIPTIONS AND RESULTS OF WALLACH et al. (1964)[a]

	Low Responsibility (Experimental Condition)			High Responsibility	
	Grp–Grp[b]	Grp–Lot[c]	Grp–Indiv[d]	Indiv–Indiv[e] (control)	Indiv–Grp[f]
Mean Change in Risk-Taking (over all items)	12.5	9.4	5.6	2.4	−1.6

N.S. (between Grp–Grp and Grp–Lot)
N.S. (between Grp–Lot and Grp–Indiv)
Sig. (over Grp–Grp, Grp–Lot, Grp–Indiv)
Sig. (between Grp–Indiv and Indiv–Indiv)
Sig. (between Indiv–Indiv and Indiv–Grp)

[a]Scores are differences between mean initial risk level and mean risk level chosen for posttest stage. Higher scores denote a greater risky-shift.

[b]Group-group: After completing pretest, Ss reach unanimous consensus about the degree of difficulty that a representative of the group will attempt (on College Board items) where that representative is to be subsequently chosen by the group and where his performance determines the group outcome (monetary reward).

[c]Group-lot: Same as group-group condition except that group representative is chosen by lot.

[d]Group-individual: After completing pretest, Ss required to reach unanimous consensus about what level of difficulty they, as individuals, would attempt on College Board items where each individual's performance determines his outcome (monetary reward).

[e]Individual-individual: After completing pretest, Ss required to individually choose the difficulty level that they as individuals would attempt on College Board items.

[f]Individual-group: After completing pretest, Ss required to individually choose difficulty level and were told that they might be chosen to represent a group at whatever level they, as individuals, chose.

sion, with no significant difference between them. Because leaders were charged with sole responsibility for the group decision, Marquis argued that diffusion of responsibility could not operate for them. However, since group discussion preceded the leader's decision, there nevertheless was an opportunity for responsibility to be diffused. In other words, even though the leader had formal responsibility for all decisions, he may not have felt personally or psychologically responsible. He may have still viewed the decision as a group product.

More recently, Pruitt and Teger (1969) failed to find a risky-shift in a situation in which responsibility for the group decision was objectively shared by the entire group. After making individual private decisions, groups (of individuals) then discussed risk-irrelevant filler items. Before making their final decisions on Choice-Dilemma items, subjects were instructed that their preferred risk levels would be averaged with those of other group members to determine the group decision. Since everyone in the group objectively participated in the group decisions as to what risk to take, responsibility should have diffused and allowed riskier decisions. Instead a nonsignificant shift toward caution was observed. These data indicate that an objective sharing of responsibility does not necessarily lead to increased risk-taking. Although the data certainly seem incompatible with the Kogan and Wallach contention that diffusion of responsibility leads to increased risk-taking, the study is vulnerable to a criticism that applies in general to the research investigating the relationship between responsibility and risk-taking. Specifically, an apparently direct manipulation of responsibility may not result in a psychological state that corresponds to the intended manipulation. For example, although averaging individual decisions to form a group discussion may objectively diffuse responsibility, it does not necessarily follow that corresponding feelings of subjectively shared responsibility accompany this operationalization. The use of indirect situational manipulations of responsibility and the virtual absence of manipulation checks undermine our confidence that the specific operations do, in fact, result in actual variations of the desired psychological states. This criticism seems particularly important when the experimental tasks have no implication or consequence for the group.[10] Thus, whether or not a sharing of responsibility leads to increased risk-taking remains an empirical question.

B. THE ROLE OF GROUP DISCUSSION

Stoner's (1961) original experiment was criticized for using managerial trainees as subjects—a sample that may have been more prone than average persons to make risky decisions. As indicated, Wallach and co-workers' (1962) subsequent replication with male and female undergraduates suggested to them that processes of group interaction and interpersonal confrontation (rather than the proclivities of business students) produced the preference for riskier decisions. To demonstrate further the importance of actual group discussion as the antecedent condition for the risky-shift, Bem et al. (1965) compared a discussion-to-consensus condition to several other treatments which led subjects to anticipate a discussion that never actually materialized. Only the actual discussion situation yielded a significant risky-shift. While these results support the idea that some form of group interaction increases risk-taking, they fail to pinpoint the specific aspect of interaction necessary and/or sufficient for the effect.

In a more analytical experiment, Wallach and Kogan (1965) contrasted two discussion situations (with and without consensus) to a consensus-without-discussion situation in which subjects communicated their risk preferences by written messages. Significant risky-shifts occurred in both discussion conditions but not in the consensus-only treat-

---

[10]This is not to argue that manipulation checks are a panacea. They are often particularly susceptible to the image-making or face-saving whim of the subject. When administered before the treatment they may have either a sensitizing effect or a commitment effect; when administered afterward, they may no longer reflect the psychological state that existed at the time of the treatment.

ment. On this basis, Wallach and Kogan concluded that group discussion is the minimal, necessary, and sufficient condition for the risky-shift effect.

However, as we noted earlier, subsequent research has not confirmed their contention that participation in group discussion is the only sufficient condition for producing the risky-shift. For example, Teger and Pruitt (1967) criticized the methodology of the consensus-without-discussion in the Wallach-Kogan study for appearing to ". . . encourage group convergence on the mean of initial decisions. . . ." Teger and Pruitt required groups to reach consensus by successive balloting and found a small but significant risky-shift in a nondiscussion "information exchange" condition in which subjects merely revealed their risk preferences publicly without verbal communication. Blank (1968) also obtained a risky-shift in an information exchange condition. He used a gambling task that made subjects aware of each other's risk choices but prevented them from interacting verbally.

In another related research procedure, subjects observed actual discussion situations but did not participate themselves. Kogan and Wallach (1967a) compared interacting discussion groups with groups that only listened to a tape recording of another interacting group. Although the magnitude of shift was greater for interacting groups, significant risky-shift effects were found for the listening as well as the interacting conditions. In addition, Lamm (1967) compared two vicarious discussion conditions: (1) observation of an interacting group through a one-way mirror, and (2) listening to group discussion over a loudspeaker. Significant increases in risk-taking were found under both conditions. Observation produced a numerically but not significantly greater increment in risk-taking than mere listening.

To explain these various results, Kogan and Wallach (1967c,d) suggested that vicariously experienced emotional interchange as well as the modeling and social facilitation effects that arise from exposure to actual group interactions may increase risk-taking in situations that only permit the observation of a discussion. The fact that a more complete exposure to actual interaction results in a greater risky-shift supports their argument. For example, it has been demonstrated that purely vocal content provides emotional cues as well as objective information (Levy, 1964). Apparently, however, emotional communications are judged more accurately with vocal, visual, facial information than with mere vocal cues (Levitt, 1964). The relative magnitude of risky-shifts in the experimental conditions of the Lamm study supports this interpretation (cf. Mackenzie & Bernhardt, 1968). However, in order to assess the importance of vicariously experienced affective interchanges fully, it would be necessary to control for modeling and information effects. If risky-shifts in vicarious exposure to discussion situations can be shown to be largely attributable to vicariously experienced emotional exchanges, the findings from this research paradigm could be accommodated by the responsibility-diffusion notion. Nevertheless, these results as well as Kogan and Wallach's account of them clearly contradict their earlier emphasis on the necessity and sufficiency of group discussion for producing riskier decisions.

Although Kogan and Wallach (1967d, p. 82) currently admit that information exchange and vicarious experience acquired through exposure to actual interaction may contribute to a risky-shift, they contend that: ". . . group interaction . . . introduce(s) still another determinant . . ., the effect of which is a further enhancement in risk-taking shifts." This generalization has been substantiated in two investigations (Kogan & Wallach, 1967d; Teger & Pruitt, 1967). Lamm, however, has found observation to produce as robust a risky-shift as participation in actual group discussion.

Studies investigating the familiarization interpretation are also relevant to the issue of the importance of group discussion as a necessary condition for the risky-shift. Although the familiarization explanation is discussed later in more detail, a brief mention of the relevant findings is appropriate here.

By having isolated individuals reconsider their initial decisions on Choice-Dilemma items, several investigators (Bateson, 1966; Flanders and Thistlethwaite, 1967) were able to produce significant increases in risk-taking in the complete absence of group discussion. However, a number of subsequent attempts to replicate reliable risky-shifts with familiarization procedures for individuals (Miller & Dion, 1970; Pruitt & Teger, 1967; Teger, Pruitt, St. Jean, & Haaland, 1970; Ferguson & Vidmar, 1970) have failed. Anticipating our conclusion in the familiarization section, we believe that it remains to be demonstrated that familiarization per se produces a risky-shift with *isolated* individuals.[11] Furthermore, it now seems unlikely that familiarization is the critical factor in risky-shifts produced by group discussion.

In sum, the available research indicates that participation in group discussion is not a necessary prerequisite for the risky-shift. Nonverbal information exchange and exposure to discussion by other persons constitute sufficient conditions for producing the effect. However, it appears that group discussion is still capable of producing increments in risk-taking over and above these sufficient conditions. Given the uncertain reliability of familiarization treatments and the lack of research on other variables that might increase risk-taking in isolated individuals, the answer to whether the risky-shift effect can occur in the complete absence of interaction or exposure to it awaits further research. Yet, as already indicated, it should surprise us if it cannot, and the demonstration that it can would not by itself strike us as particularly interesting.

[11]Of course, there are other variables besides familiarization that might increase risk-taking in socially isolated individuals. For instance, it seems quite feasible to apply procedures or instructions to isolated persons that directly reduce responsibility or fear of failure.

C. THE AFFECTIVE BONDS HYPOTHESIS

According to Kogan and Wallach (1967c,d; Wallach and Kogan, 1965), the development of affective bonds between individuals during group discussion forms an essential basis for the operation of the responsibility-diffusion mechanism which, in turn, presumably mediates any risky-shift. Two recent studies of personality factors present evidence for this assumption.

In one study, Kogan and Wallach (1967b) composed discussion groups of subjects who scored either high or low on the personality variables of defensiveness and test anxiety. Defensiveness is the personality variable of primary relevance to the present discussion. According to several theorists (Rogers, 1959; White, 1964), highly defensive persons approach others with a careful, guarded manner because they fear exposing personal weakness. Conceptualized in this manner, defensiveness should interfere with the formation of the affective bonds that are presumably essential for the diffusion of responsibility. Therefore, defensiveness should inhibit the shift toward risk normally produced by group discussion. Consistent with this expectation, groups of persons low in defensiveness showed greater risky-shifts on consensus decisions than groups of persons high in defensiveness.

In the second investigation, Wallach, Kogan, and Burt (1967) used scores from the Embedded Figures Test to compare discussion groups with either field-dependent or field-independent persons. Prior research indicates that the field independence-dependence distinction closely parallels a contrast between cognitive versus affective styles. While field-independent persons supposedly display interpersonal aloofness and a predominantly cognitive orientation, field-dependent individuals emphasize affective features in their self-descriptions and react more responsively to social stimuli (Witkin, Dyk, Faterson, Goodenough & Karp, 1962; Messick & Damarin, 1964). Therefore, the affective bonds hypothesis suggests that groups of field-independent persons will exhibit greater shifts toward risk than groups of field-independent persons. Contrary to predictions, field-dependent and field-independent groups produced risky-shifts of equal magnitude. An internal analysis revealed that considerably different outcomes emerged under conditions of high involvement — as inferred from the length of discussion time. Longer discussion times were associated with stronger risky-shifts for field-dependent groups and conservative-shifts for field-dependent groups. Since differences emerged only in an internal analysis, these data are less compelling; nevertheless, they are consistent with the affective bonds hypothesis.[12]

A further derivation from the affective bonds hypothesis is the prediction of a positive relationship between group cohesiveness and magnitude of the risky-shift. In the context of an experimental gambling situation requiring subjects to discuss actual bets, Pruitt and Teger (1969) included sociometric items on a postquestionnaire to derive an index of cohesiveness. Correlations between cohesiveness and risky-shift were all positive and, in one case, reached an acceptable level of significance. Albeit the evidence is weak, the results suggest that ". . . more cohesive groups shift more toward risk . . .," as would be expected from the affective bonds hypothesis. In a subsequent experiment, Pruitt and Teger (1969) established a similar level of cohesiveness by having groups of individuals discuss risk-irrelevant items. When individuals subsequently made private decisions, nonsignificant shifts toward caution resulted. If nothing else, these results suggest that the mere presence of affective bonds is not a sufficient condition for the risky-shift.

Unfortunately, none of the above studies provides an *experimental* test of the affective bonds hypothesis: *viz.*, that the affective bonds formed or present during discussion of risk-relevant items produce a risky-shift. Specifically, none of the previous investigations directly manipulated the strength of "affective bonds" among members of discussion groups. In the two personality investigations, independent variables were established by selecting subjects on the basis of scores on personality factors. Such a selection procedure precludes the possibility of randomly assigning subjects to experimental conditions. As a consequence of nonrandom assignment, extraneous variables may be systematically confounded with the independent variable. Thus, any observed effects on the dependent measure could be potentially attributable to a third, uncontrolled variable associated with the treatment. The positive correlations between cohesiveness may be either a cause or a consequence of the risky-shift. Finally, Pruitt and Teger's (1969) subsequent study primarily indicates that, in the absence of discussion, the formation of affective ties between persons is insufficient to increase risk-taking. However, it provides no information about the role of "affective bonds" in the paradigmatic situation in which group members discuss risk-relevant problems.

To remedy these difficulties, Dion *et al.* (1970) experimentally induced high and low levels of group cohesiveness. Contrary to the affec-

---

[12]It is important to note that any explanation that stresses the role of influence processes during group interaction would not find these results incompatible. In other words, this type of finding should not be thought of as providing exclusive support for a diffusion of responsibility explanation. They could easily be incorporated into leadership or cultural value explanations.

tive bonds hypothesis, however, the less cohesive groups exhibited greater risky-shifts than high-cohesive groups on both consensus and post consensus shift measures. Although the risky-shift is a robust phenomenon, high cohesiveness actually inhibited the shift toward risk in discussion groups. To account for these results, Dion, Miller, and Magnan suggest the intuitively plausible explanation that, as group members become more attracted to one another, they also become more loathe to minimize *personal* responsibility or displace responsibility for failure onto their fellow group members. Although the results diametrically oppose the predictions from the affective bonds hypothesis, this interpretation has the virtue of being consistent with the responsibility-diffusion notion. Nevertheless, this outcome and interpretation is paradoxical in the sense that the affective bonds hypothesis was initally proposed as a means of salvaging the responsibility-diffusion explanation from empirical disconfirmation. And in this respect, it is important to note that the affective bonds hypothesis was originally proposed without a careful examination or explanation of its relationship to the responsibility-diffusion notion (Pruitt & Teger, 1969).

D. THE ANONYMITY HYPOTHESIS

As previously indicated, several investigators (Burnstein, 1967; Mackenzie & Bernhardt, 1968) suggest that the responsibility-diffusion explanation implies a positive relationship between the number of group members and the magnitude of the risky-shift. A larger group presumably permits greater diffusion of responsibility. Present evidence on this point is scanty. Teger and Pruitt (1967) found a strong positive relationship between group size and magnitude of risky-shift. Triads exhibited a nonsignificant shift toward risk, but a moderately strong significant risky-shift occurred in four-man groups, and the most robust risky-shift occurred in five-person groups.[13] Both four-person and five-person groups exhibited greater risky-shifts than three-person groups. although they did not differ from each other.[14]

Although the Teger and Pruitt results indirectly support the notion that anonymity increases risk-taking, the anonymity hypothesis itself

---

[13]Unfortunately, no differences between larger and smaller groups were found on postdiscussion, self-ratings of perceived responsibility. However, without information on the validity of this responsibility measure and with the possibility of floor or ceiling effects due to social desirability factors, the failure to find differences in responsibility does not clearly contradict the present line of thought.

[14]In an unpublished study, Marquis and Forward (cited in Burnstein, 1967) found no relationship between the risky-shift and group size with two, three, and five-person groups.

raises several conceptual issues. One of the inherent theoretical problems with any structural variable such as group size is that many other functional variables inevitably are associated with it (Kelley & Thibaut, 1954). In other words, it is likely that there is an underlying social psychological process responsible for any obtained group-size effect, and it is difficult to discover exactly what this is. Second, other explanations of the risky-shift (e.g., the leadership and cultural value interpretations) make essentially the same prediction regarding the effect of group size on the risky-shift. For that matter, it is not even clear that the responsibility-diffusion explanation necessarily predicts increased risk-taking as a result of increments in group size. Several investigators (e.g., Carter, Haythorn, Meirowitz, & Lanzetta, 1951; Gibb, 1951) report increased feelings of threat and increased inhibition to participate when groups are enlarged. Since these factors would undermine any sense of shared common fate, they should likewise inhibit the process of responsibility-diffusion and curtail risk-taking.

In sum, though plausible in principle, the anonymity hypothesis appears to be a theoretical blind alley when tied specifically to group size. However, anonymity can be directly manipulated in a variety of other ways that would avoid this theoretical ambiguity. To determine an effective experimental procedure, it is necessary to consider the characteristic qualities of anonymous social situations. Similar to a state of deindividuation (Festinger, Pepitone, & Newcomb, 1952; Singer, Brush, & Lublin, 1965), anonymity is basically an individual's subjective feeling of minimal self-consciousness and lowered identifiability. A feeling of anonymity can be created by allowing persons to communicate by means of written messages or intercoms. In addition, the relatively simple procedure of altering the seating pattern in discussion so that subjects do not directly face one another should lower identifiability. These manipulations of anonymity, however, may impair the interaction process somewhat (i.e., interfere with communication) in addition to reducing the salience of individual identity.

Other alterations of the experimental situation however, might lower individual feelings of self-consciousness while retaining the essential features of unrestricted interaction among group members. (a) For example, with the Choice-Dilemma task, one could heighten the feeling of anonymity by instructing subjects that their decisions are to be anonymous and forbid placing names on booklets. (b) Instead of the standard procedure in which each individual is given another blank copy of the Choice-Dilemma booklet on which to inscribe the consensus decisions in the group discussion situation, the experimenter might provide only a single booklet to the discussion group to increase the feeling of group participation and de-emphasize individuality. (c) In many group risk-taking situations, subjects are made to feel that their discussions are being monitored. A standard procedure, for example, is to have the experimenter remain present, but not participate, during discussion (e.g., Wallach & Kogan, 1965). Although the experimenter was not physically present, Teger and Pruitt (1967) told subjects that their discussions were being monitored by hidden microphones. Removing the experimenter's surveillance and minimizing other situational cues associated with monitoring (microphones, one-way mirrors, etc.) should enhance anonymity. (d) On the other hand, feelings of self-consciousness could be increased by having subjects wear name tags and/or telling them that their discussions are being monitored by a panel of experts who are rating members individually on their contribution to group problem-solving.

E. Fear of Failure

As noted earlier, one of the core ideas in the responsibility-diffusion explanation of the risky-shift asserts that an individual's fear of failure inhibits his willingness to take chances. Related to this proposition is the further assumption that responsibility-diffusion decreases fear of failure and, as a consequence, increases risk-taking. Some evidence substantiates the claims of the responsibility-diffusion interpretation concerning fear of failure. In one study, Kogan and Wallach (1967b reviewed above) systematically composed discussion groups with persons either homogeneously high or low on the traits of test anxiety and defensiveness. According to Atkinson (1964), the test-anxiety variable reflects an individual's predisposition or tendency to avoid failure and high test anxiety should therefore indicate a high fear of failure. Since group discussion presumably allays subjective fears of failure, Kogan and Wallach predicted that groups of individuals with high test anxiety would exhibit greater shifts toward risk than groups of persons low on test anxiety. These expectations were borne out. Furthermore, in an experiment discussed earlier, Bem et al. (1965) found that individuals who expected to suffer the aversive consequences of their decisions with others experiencing the same fate shifted to more conservative positions. When decisions with possible aversive consequences were made by the group, however, risky-shifts were obtained.

Although the above evidence is consistent with the responsibility-diffusion interpretation, there are some notable sources of uncertainty concerning the role of fear of failure in producing the risky-shift. In the first place, no study directly demonstrates that either the hypothetical or more realistic tasks used in the present research actually elicit fear of failure. Second, neither of the studies discussed above provides any *indepen-*

dent evidence that group discussion mitigates this fear. Even if fear of failure were found to be markedly lower in the group discussion setting than in the initial individual phase, this result might be due merely to the fact that group discussion follows individual decision-making in the usual risky-shift experimental paradigm. That is, a decline in fear of failure after (or during) group discussion might simply be attributable to an habituation process; individuals might become less responsive to the fear-evoking potential of the risk-taking situation after additional exposure or familiarity. To discount the possibility of habituation, one would have to examine either (*a*) the fear-reducing potential of group discussion without prior, individual exposure or (*b*) the comparative reduction in fear produced through group discussion versus individual reconsideration of initial decisions.

Further, in some instances, derivations from the responsibility-diffusion interpretation that deal with fear-of-failure are not supported by the available evidence. For example, one might predict that Choice-Dilemma items having the most serious consequences in the event of failure would exhibit the strongest risky-shifts. Yet the exact opposite occurs. Clark and Willems (personal communication) had subjects imagine that the hypothetical person had chosen the risky option and that it had failed. High negative correlations were found between mean ratings of the gravity of consequences and the Pruitt and Teger (1967) national risky-shift norms for Choice-Dilemma items. That is, stronger risky-shifts tend to be associated with items having the least serious repercussions in the event of failure.[15]

Finally, the unidirectional prediction between fear of failure and increased risk-taking is empirically disconfirmed in certain instances. For example, two of the twelve Choice-Dilemma items exhibit relatively consistent shifts toward caution with group discussion (Pruitt & Teger, 1967). With its present formulation, the responsibility-diffusion interpretation is totally inadequate to explain these instances of conservative shifts with group discussion. In this respect, the responsibility-diffusion interpretation is analogous to a one-tailed statistical test with all of its explanatory power focused on increased risk-taking. Consequently, instances of conservative shifts following group discussion strongly disconfirm the responsibility-diffusion thesis. However, since most of the Choice-Dilemma items are "risk-oriented," the responsibility-diffusion interpretation can generally predict main effects with some accuracy across Choice-Dilemma items.

In addition to this empirical problem, the responsibility-diffusion hypothesis concerning fear of failure stands in opposition to two alternative

---

[15] In fairness, one could argue that the success of any given discussion in reducing fear will be inversely related to the amount of fear aroused.

and opposite theoretical hypotheses about fear of failure. One is Atkinson's (1964) theory of risk-taking behavior, and the other can be called the Constriction Hypothesis.

According to predictions from Atkinson's theory, persons who exhibit little fear of failure should prefer intermediate risks. On the other hand, persons highly motivated to avoid failure should prefer either extremely conservative or extremely risky options. The choice of high or low odds presumably represents different strategies for coping with the aversive possibility of failure. By selecting particularly low odds, an anxious individual shields himself from the onus of failure by seemingly attempting the impossible. Blame for the failure on these very difficult tasks can be attributed to the relatively low likelihood of success. On the other hand, extremely favorable odds virtually guarantee the probability of success. Several experiments on individual risk-taking (e.g., Atkinson & Litwin, 1960; Edwards, 1953) confirm the prediction that persons with little fear of failure do, indeed, prefer intermediate risks. According to Atkinson's conception of the relation between fear of failure and risk-taking, reduction of fear — whether a consequence of group discussion or any other variable — should tend to reduce the *variance* among individual level of risk-taking *without* necessarily altering the mean level of risk. In other words, on the initial (individual) measure of risk-taking, fear of failure is presumably high and subjects select more extreme levels of risk or caution. The fear-reducing effect of group discussion should elicit more moderate or intermediate levels of risk among those at formerly extreme levels, thereby reducing the variance among individual decisions.[16] Yet the essential characteristic of the risky-shift is that groups of individuals become more extreme in their risk-taking. Of course, the predictions of achievement theory to risk-taking do not apply unless people view the possible outcomes largely as a function of their own skill. Perhaps they believe that the outcomes of (both hypothetical and real) tasks in risky-shift research are primarily under the control of chance factors.

A second and opposing hypothesis concerning fear of failure emphasizes the constricting functions of anxiety — the extent to which it pushes people to (a) look for social support (Schachter, 1959) and (b) behave in socially approved ways (Crowne & Marlowe, 1964). That is, high fear of failure might inhibit creative, atypical, or complex thinking and learning. This hypothesis argues that high fear of failure should lead people to

---

[16]Atkinson's theory doesn't state the relative frequencies with which persons fearing failure choose conservative versus high risk alternatives. If conservative or low risk choices were more normative, then of course mean shifts following a reduction in fear of failure would support the Kogan and Wallach predictions. At the same time however, the predictions regarding decreased variability following group discussion should still be confirmed. Yet needless to say, there are numerous other theoretical considerations that also predict reductions in variability following group discussion.

choose a level of risk-taking that is normative, intermediate, and not extreme in one direction or the other—one that inhibits or masks individualistic whims or idiosyncracies. When fear of failure is subsequently reduced by group discussion, such constricting tendencies no longer suppress individual, non-normative responses. Subjects should feel free to take risks that deviate more drastically from the level of risk ordinarily perceived as normative. Consequently, while the mean level of risk might show no change, the variability of risk-taking among individuals should be greater after group discussion.[17]

The preceding paragraphs demonstrate that at present there are no compelling theoretical or empirical reasons to support Kogan and Wallach's arguments about the role of fear of failure in group risk-taking. That is, when considered alone, neither Atkinson's theoretical approach nor the one outlined directly above predicts an increase in risk-taking as a consequence of reduced fear of failure.

F. CONCLUSION

The diffusion of responsibility explanation has not fared well. In the light of accumulating, contradictory evidence, its proponents have continually had to modify its central, underlying assumptions concerning the roles of group discussion and affective bonds as well as the necessary conditions for the occurrence of responsibility-diffusion. In another sense, however, this continual modification of the theory merits praise rather than reproval. Persistent theoretical modification is the preeminent symptom of the working scientist. It reflects the proper marriage of theory and experiment.

Nevertheless, many of the theoretical details proposed at one time or another have faltered under experimental scrutiny. For example, it is now clear that an actual group discussion experience is not the sole necessary precondition for the risky-shift—although some evidence suggests that group discussion produces an increment in risk-taking over and above that produced by information exchange or mere exposure to discussion. Similarly the fact that the risky-shift fails to occur when subjects (a) engage in risk-irrelevant discussion (Pruitt & Teger, 1969; Clark & Willems, personal communication) or (b) are part of a highly cohesive group (Dion et al., 1970) seriously detracts from the Wallach and Kogan contention that affective bonds per se underlie diffusion of responsibility. While the positive relationship observed between group size and the magnitude of the risky-shift (Teger & Pruitt, 1967) supports some of the an-

---

[17]When combined with the data indicating that individuals perceive themselves as riskier than they actually are relative to others, the Constriction Hypothesis does predict a risky shift as well as increased variance.

cillary theorizing by proponents of the responsibility-diffusion notion, it can also be derived from several other explanations of the risky-shift (e.g., cultural values). Moreover, methodological and conceptual considerations suggest that manipulation of group size is not in itself a very fruitful or efficient means for investigating the mechanisms that underlie the risky-shift.

We must recognize, however, that most of the reported data contradicting the responsibility-diffusion interpretation focus on the roles of group discussion and affective bonds. Despite the problems with these two variables, nevertheless, the hypothesized relation between responsibility-diffusion and risk-taking may be basically sound. That is, although neither group discussion nor affective bonds may effectively diffuse responsibility, successful diffusion of responsibility might nevertheless increase risk-taking. The recent disaffection of researchers with the responsibility-diffusion interpretation seems to stem in part from confusion about this point. In other words, confidence in the responsibility-diffusion explanation has waned primarily because some of the derivative and supporting hypotheses appear to be invalid. Notably absent, however, is evidence on the terminal process in the causal sequence outlined at the beginning of Section II (*viz.* responsibility-diffusion and fear of failure).

In fairness, certain aspects of this confusion are due to the Wallach and Kogan failure to specify clearly just what this causal sequence is. For example, as previously indicated, they have maintained that group discussion is the necessary and sufficient condition to produce the risky-shift, a statement that appears to be empirically false. Yet a broader view of their interpretation would emphasize instead that the risky-shift is mediated by a reduction in *individuals'* fear of failure. Since it should be possible to decrease fear of failure in the total absence of group interaction (e.g., via a manipulation of confidence in task ability), a reduction in fear of failure—not the occurrence of group discussion—should be considered the necessary and sufficient condition for the risky-shift.

Of course, even this emphasis on fear of failure leaves the responsibility-diffusion notion open to serious criticism for, as indicated, it is not at all clear from theoretical considerations that reducing fear of failure will enhance the mean extremity of group decision-making. It is possible, however, that diffusion of responsibility enhances risk-taking without reducing fear of failure. Perhaps group participation increases one's willingness (or ability) to engage in impulsive or irresponsible behavior and this—not a reduced fear of failure—increases risk-taking. In accord with this conjecture, a risky-shift has been found even when the risky alternatives had a lower expected value than the cautious alternatives

(Bem *et al.* 1965). Thus, in this sense the risky-shift can be viewed as an irrational phenomenon. In addition, some evidence implies that people behave more irresponsibly in groups than as individuals. For example, Darley and Latané (1968) demonstrated that in emergency situations subjects are less likely to offer help to people in need if they (the subjects) are in the presence of others rather than alone. Furthermore, Zimbardo (1969) recently reported that when anonymity is enhanced in a group setting, subjects engage in more impulsive aggression than non-anonymous subjects. This effect, however, appears to obtain only when anonymity is created within a group. At any rate it appears that at least under some circumstances irresponsible, impulsive behavior occurs more readily in groups than in individual settings.

The discussion above emphasizes that despite conceptual problems with the fear of failure concept and despite the contradicting evidence relating to group discussion and the role of affective-bonds, there is still reason to suspect that responsibility-diffusion may create increases in risk-taking behavior. Thus, it is unfortunate that most of the research relevant to the responsibility-diffusion interpretation only indirectly tests the basic premise of the explanation.

III. The Persuasion Explanation

A. THE LEADERSHIP HYPOTHESIS

According to the leadership hypothesis, the risky-shift is basically due to personality differences among the group members. Those people who initially make riskier decisions than others in the group presumably exert more influence in the subsequent discussion (Collins & Guetzkow, 1964; D. G. Marquis, 1962). In other words, the leadership explanation hypothesizes a relation between two dimensions of individual difference — viz., riskiness and persuasiveness. By virtue of their greater persuasiveness and initiative in social situations, high risk-takers lead other group members toward a riskier decision. Thus, a unique feature of the leadership interpretation is its stress upon personality factors as determinants of the risky-shift effect.

In order to establish the leadership hypothesis as viable, we must consider two types of related evidence: (1) that influence processes operate during group risk-taking, and (2) that those individuals favoring risk are the primary initiators of influence attempts and/or are more persuasive when they speak. Obviously, if the second proposition is true, so too is the first.

With respect to the first point, there is evidence that social influence processes do occur during group risk-taking. Several investigators report significant reductions in variability following either group decision (Flanders & Thistlethwaite, 1967; Teger & Pruitt, 1967; Wallach & Kogan, 1965) or group discussion (Wallach & Kogan, 1965). Since decreased variability is typically interpreted as a reflection of social influence, these findings imply that influence processes do operate in risky-shift situations. On the other hand, closer inspection of actual experimental procedures suggest that increased uniformity produced in group decision situations is often merely artifactual; the experimental procedures used in the risky-shift research typically instruct groups to reach consensus. However, Wallach & Kogan (1965) found equal amounts of convergence following a group decision procedure that required consensus and one that did not explicitly require consensus but instead encouraged "diversity of opinion." While this parallel outcome superficially suggests that all variability reduction is not artifact, closer inspection of the Wallach and Kogan procedure reveals further difficulty. The reduced variability produced by group discussion is easily attributable to "demand characteristics"—implicit suggestions to group members that they really should end up agreeing with one another. While there is an implicit demand for either exaggerated change or consistency in any multiple measurement design, the implication for change seems to be particularly evident in the Wallach and Kogan procedure. The instructions were given after group discussion and required subjects to reconsider their initial decisions *in light of the opinion diversity* that had been generated. As is obvious from this succession of arguments and counterarguments, the question of whether or not influence processes (separate from artifact and demand) do operate in group risk-taking is an extremely complex one. However, since the effect of demand characteristics in this particular situation is likely to be an accentuation of influence processes already present, we find it hard to imagine that no true influence occurs in group discussion situations (in spite of the problems mentioned above).

But granting that influence does occur, a more critical question is whether or not group pressures toward uniformity are related to the risky-shift. Teger and Pruitt (1967) reported significant positive correlations between magnitude of risky-shift and the decrease in variability produced by group decision—the more the decrease, the greater the shift toward risk. Although correlations were not computed, Wallach and Kogan (1965) also noted the relation between marked convergence in decisions and large risky-shift for both group decision and group discussion. In light of this evidence, they suggested that influence processes are directed primarily toward conservative members and that these persons exhibit more change during group interaction than initially risky individuals. Obviously, however, a causal interpretation of correlational data always invites suspicion. The demonstration of an intrinsic connection between influence processes and the risky-shift requires experimental manipulation of the strength of influence processes and an independent assessment of the magnitude of risky-shift.

Consistent with the leadership hypothesis, several studies (Flanders & Thistlethwaite, 1967; Wallach *et al.*, 1962; Wallach, Kogan, & Burt, 1965) report significant, positive correlations between individuals' initial riskiness and postsession ratings of perceived influence by group members. This latter evidence, however, is still indirect. A more satisfactory procedure for demonstrating the relation between influence processes and the persuasiveness of the high risk-taker would use external observers who independently rate the amount and source of influence during the process of group discussion.

To test the hypothesis that initially risky persons exert more influence toward a risky decision than other group members, a number of studies attempt to isolate individual difference factors that might contribute to this relationship. Rim found that those high in extraversion (Rim, 1964c), high in need achievement (Rim, 1963), and more tolerant of ambiguity (Rim, 1964b) take higher risks on initial measures of risk-taking.[18] Since group decisions shifted in their direction, Rim *inferred* that the high initial risk-takers exerted influence over the decisions of other group members. The truth of this conjecture obviously remains untested, however. Rim also extended his program of research to interpersonal attitudes. Those who tended to take greater initial risks valued leadership and recognition (Rim, 1964a). In addition, they scored higher on personality tests measuring adeptness at manipulating interpersonal relations (Rim, 1966) and affective expressiveness (Rim, 1965b). On the other hand, Stoner (1961) correlated level of participation—a dimension of leadership related to task ability—with initial riskiness and found no relationship.

In a somewhat different vein, Burnstein (1967) suggested that persons initially prone toward risk may achieve greater influence in group discussion due to their greater commitment to, and confidence in, their prior decisions. Using a problem-solving task that required subjects to bid for problems of varying difficulty, Clausen (1965) reported a few significant, positive correlations between initial riskiness and confidence

---

[18]With respect to extraversion, however, Flanders and Thistlethwaite (1967) reported nonsignificant correlations between extraversion and initial riskiness as well as between extraversion and perceived influence ratings from group members.

in their decisions on the initial measure of risk.[19] In a later phase of the experiment, Clausen systematically varied the structural composition of groups to manipulate confidence and initial risk-taking. All groups were constructed to contain a risky person and two conservative members. In high-confidence risky groups (HCR), the risky member expressed high confidence in his initial decision; in low-confidence risky groups (LCR), the risky person expressed little confidence in his initial choice. Similarly, half the groups contained highly confident conservative members and, the other half contained conservative members with little confidence in their initial decisions. In addition, the design included two other conditions: groups homogeneously confident and groups homogeneously unconfident. Strongest risky-shifts were predicted for HCR groups and conservative shifts for LCR groups. No shift was expected for groups homogeneous in confidence. As predicted, HCR groups did exhibit greater increases in risk-taking than LCR groups. However, confidence level of conservative members did not mediate shifts toward either risk or caution. These latter two instances in which Burnstein's predictions were not borne out may be attributable to a "natural" correlation between risk and confidence. As discussed earlier, there is some evidence for a positive relationship between initial riskiness and reported confidence. If so, it would not be possible to compose equivalent groups. At any rate, the overall results of the Clausen study give some support to a direct relationship between confidence in initial risk-taking and risky-shift.

Further research on the leadership hypothesis might profitably explore other aspects of confidence. For example, the importance of differential confidence in initial decisions might well be more pronounced in larger groups in which participation in discussion is less evenly distributed (Allen, 1965). The amount of confidence exhibited during actual group discussion might be a key variable. In a recent study of persuasion, London, Meldman, and Lanckton (1970) assembled dyads to discuss issues and attempt to reach agreement. They found no relation between persuasion and either initial position or initial conviction; but expression of confidence *during* discussion did predict influence.[20]

Detractors of the leadership hypothesis propose that the relation between initial riskiness and attributed influence is more apparent than

[19]These findings may simply be variations of the oft-reported relation between extremity of position and certainty of belief (Suchman, 1950). However, both Stoner (1961) and Teger and Pruitt (1967) found no evidence between confidence and risk-taking with Choice-Dilemma items.

[20]Instead of self-reports, they derived objective measures based upon judge's ratings of the transcribed discussions in terms of pre-established categories of confidence and doubt.

real. For example, Kelley and Thibaut (1968, p. 81) suggest that: "The correlations between initial riskiness and influence may simply reflect what has happened: subjects observe the shift to occur and infer from it that the initially risky persons must have been more influential." Enhancing the credibility of this interpretation, Wallach et al. (1965) demonstrated that subjects are aware of a shift toward greater risk-taking during discussion but inaccurately estimate its magnitude. Moreover, in studies that report conservative shifts (Nordhøy, 1962; Rabow, Fowler, Bradford, Hofeller & Shibuya, 1966), initially conservative persons are rated as more influential. Although these results suggest that retrospective perceptions of influence may merely reflect the direction of shift, they do not by themselves discredit the notion that high risk-takers exert greater influence over group discussion.

On the other hand, direct experimental tests of the leadership hypothesis generally yield negative results. In order to provide an opportunity for social influence processes to occur, Wallach et al. (1968) systematically formed groups of individuals who represented a wide variability in risk-taking propensities. They used items that did not involve risk assuming that if initially high risk-takers generally exert more influence, they should do so regardless of the specific content of the discussion. A weak significant tendency to perceive high risk-takers as more influential was found for groups of females; however, no such relation obtained for groups of males. A second study (discussed earlier in this chapter) examined groups uniformly composed of either field-dependent or field-independent persons (Wallach et al., 1967). Presumably, the perceptual dependence of field-dependent persons also reflects a greater susceptibility to social influence (Witkin, Lewis, Hertzman, Machover, Meissner, & Wapner, 1954; Witkin et al., 1962). Therefore, the leadership hypothesis predicts a stronger impact of persuasion attempts and influence processes on risk-taking in groups of field-dependent than field-independent persons. However, both types of groups exhibited similar levels of initial and subsequent risk-taking. Contrary to the leadership hypothesis, high initial risk-takers were perceived as more influential only in field-independent groups. Although a similar relationship failed to obtain for field-dependent groups, group members did agree on who was the most influential. Therefore, field-dependent groups apparently made systematic distinctions among themselves but not in the direction predicted by the leadership hypothesis.

In a recent study, Hoyt and Stoner (1968) attempted to devise a crucial test of the leadership hypothesis. To control for any initial leadership differences among group members, they assembled discussion groups composed of individuals with highly similar, initial risk-taking scores. Under these conditions, the leadership hypothesis predicts little or no shift. However, significant risky-shifts occurred on the majority of Choice-Dilemma items. In addition, comparisons with two previous studies (Stoner, 1961; Wallach *et al.*, 1962) that used groups heterogeneous in initial risk-taking revealed no appreciable differences in magnitude of obtained risky-shifts. Although the investigators interpret these results as failing to confirm the leadership hypothesis, inadequacies in their homogeneity manipulation may impugn their conclusion. Individual levels of initial risk-taking were determined on the basis of total scores across items. Consequently, the supposedly homogeneous, groups retained a considerable amount of within-group heterogeneity on each item. As the authors note, although their homogenization procedure reduced the range of *total* individual scores, it decreased the range on individual items by only one-seventh.

B. The Rhetoric-of-Risk Hypothesis

A recent variant of persuasion theory does not consider riskiness as an individual difference factor and avoids the leadership issue entirely. Essentially, it views persuasiveness as intrinsic to a *position* rather than a person. This explanation—the rhetoric of risk hypothesis—is more inclusive than the original leadership hypothesis; it presumably explains *both* the positive correlations between initial riskiness and perceived influence as well as the risky-shift itself (Kelley & Thibaut, 1968). As outlined by its major proponents, the rhetoric of risk position maintains that:

> There are two related aspects of the risky position that may give the proponent of such a position a disproportionate weight in open discussion: (1) the "rhetoric of risk" is more dramatic, and (2) the conflicts and uncertainties entailed in accepting the riskier alternative might lead the proponent of such alternatives to state his arguments with a heightened intensity and amplitude. In short, he may have the advantage of a more potent language, more intensively produced (Kelley & Thibaut, 1968, p. 82).[21]

Evidence for the rhetoric of risk interpretation is meager, although suggestive. For example, Rettig (1966) observed that the group discussions that led toward greater risk-taking also enhanced the reinforc-

---

[21] In some respects the rhetoric of risk hypothesis is similar to the cultural-value explanation which is discussed later. The major point of departure is one of emphasis. It stresses the intensity of the influence attempts by the riskier group members. While the cultural-value explanation also implies a greater influence of risk-oriented statements, it does not specifically attribute the effect to the forcefulness of particular group members. Values are seen as censoring the flow and quantity of information, and not as determining the intensity of forcefulness with which different positions are presented.

ing value of the desirable features of a risky decision. Presumably, "rhetorical advantage" would operate in a similar manner and increase the group's valuation of the gains arising from increased riskiness. Kelley and Thibaut also cite the Lonergran and McClintock (1961) study that failed to find a risky-shift in a betting situation as indirect support. Supposedly, the simplicity of the experimental situation and its sparse potential for discussion allowed little room for the rhetorical advantage of risk to manifest itself.

As argued in the case of the leadership explanation, ratings or evaluations of segments of the group interaction and decision-making process offer the most direct means of assessing the validity of the rhetoric of risk hypothesis. Judges listening to segments of the interaction should judge statements uttered in support of risky positions as more forceful or confident in the absence of knowledge about the level of riskiness being advocated. They should also (or alternatively) judge statements specifically advocating a risky decision as more powerful or persuasive. Alternatively, when subjects are asked to role-play different positions, judges should be able to detect these same differences as a function of the riskiness of the position that is role-played.

C. CONCLUSION

The proposition that high initial risk-takers more persuasively argue their own position during group discussions receives little support from existing research. The demonstrated relationship between initial risk-taking and attributed influence constitutes the major support for the leadership hypothesis and counter-evidence (Nordhøy, 1962; Rabow *et al.*, 1966) suggests that this relationship may be a consequence, rather than a cause, of the risky-shift. Furthermore, since correlations between initial riskiness and perceived influence are generally low, and since risky-shifts occur even when subjects have had no exposure to group discussion (e.g., Teger & Pruitt, 1967), investigators have been prompted to relegate the leadership position to the role of a limited, partial explanation of the risky-shift effect (Jones & Gerard, 1967; Kelley & Thibaut, 1968; Kogan & Wallach, 1967c; Mackenzie & Bernhardt, 1968). On the other hand, low correlations may merely reflect lack of reliability or validity of the relevant measures. The type of evidence that could most directly speak on the leadership hypothesis is some type of interaction process analysis of the group decision. With independent raters evaluating who contributes most persuasively to the group discussion, one could directly assess whether it is indeed those whose prior initial decisions are the riskiest. At present, such evidence is simply not available.

The rhetoric of risk hypothesis, although interesting, awaits empirical support. Evidently, direct tests of this view would be difficult to devise (Jones & Gerard, 1967; Kelley & Thibaut, 1968). Furthermore, even if direct tests were feasible and proved satisfactory, Kelley and Thibaut feel that an explanation based on "rhetorical advantage" would nevertheless require supplementation from additional explanations such as responsibility-diffusion. But, more importantly, in their present form, both the leadership and rhetoric of risk hypotheses suffer the defect of considering only unidirectional tendencies toward risk and not accounting for conservative shifts.[22]

Although social influence processes do seem to occur in the group discussions that induce risky-shifts, little systematic effort has been devoted toward directly assessing these persuasion and/or influence processes. Although greater use could be made of the research strategy of assembling groups that systematically vary in terms of personality factors thought to be related to social influence, the inconsistent outcomes of research relating personality differences to persuasion (McGuire, 1968) as well as the evidence that personality factors play a minor role in determining persuasibility (Cohen, 1964; Hovland & Janis, 1959) argue against this approach. Instead, as suggested above, observational analysis of interactions in group risk-taking situations seems a better alternative. It might reveal the importance of such unexplored factors as talkativeness, expressed confidence, and volume of arguments as persuasive elements that induce greater risk-taking.[23] Other sources of personal influence such as status and authority have similarly been neglected in risky-shift investigations. Norms may also be a potent source of influence in risk-taking situations, either enhancing or substituting for the expression of personal influence. Whereas Rabow *et al.* (1966) suggest that an individual's effectiveness at interpersonal persuasion in a risk-taking situation is contingent upon eliciting normative support for his advocated position, Thibaut and Kelley (1959) propose that norms may serve as surrogates for personal influence. In short, an investigation of normative processes in group risk-taking might well reveal the operation of personal and/or impersonal sources of social influence related to the risky-shift.

[22]Of course, the major premise of persuasion theory can be altered so that it can account for conservative as well as risky-shifts. For instance, if the hypothesis is altered to state that the most extreme person in the group is most influential, regardless of his direction of extremity, both risky and conservative outcomes can be expected with frequencies that depend on the relative natural occurrence of these two types in the population (see Burns, 1967).

[23]Kogan and Doise (1969) attempt to grapple with some of these issues.

IV. The Familiarization Explanation

Contrary to other explanations of the risky-shift, Bateson's (1966) familiarization hypothesis implies that the risky-shift is a pseudo group effect — explainable in terms of individual processes and not restricted to group situations. The familiarization hypothesis argues that enhanced risk-taking results from increased familiarity with risk-related items. In other words, any procedure that increases familiarity with the elements relevant to a decision will increase the riskiness of the decision; it matters little whether increased familiarity occurs with isolated individuals or among members of a discussion group. Presumably when individuals consider novel or ambiguous problems, their initial uncertainty produces caution. With increased familiarization, either through further private study or group discussion of the problem, uncertainty (and caution) subsides, thereby freeing individuals to make riskier decisions (Bateson, 1966). In a similar formulation, Flanders and Thistlethwaite (1967) propose a comprehension interpretation of the risky-shift, in which private study produces greater comprehension of risk related problems. This, in turn, increases risk-taking by reducing uncertainty.

Whereas Bateson's interpretation seems simply to emphasize the effect of increased familiarity or experience without mentioning intellectual comprehension of the problem, Flanders and Thistlethwaite place emphasis primarily on the effect of greater comprehension. Although Flanders and Thistlethwaite consider their formulation primarily as a restatement of Bateson's position, the difference in emphasis suggests two distinct interpretations. One can conceptualize the familiarization hypothesis as essentially focusing on the emotional process of uncertainty reduction as the explicit mediator of risk-taking or, alternatively, one might conceptualize enhancement of risk-taking as the consequence of a cognitive process that has little relation to emotional states. For example, it is possible that familiarization leads to a fuller understanding of the relative merits and drawbacks of the various choice alternatives and that risk-taking increases only when risky choices become more attractive for rational reasons (e.g., the risky choices might have greater expected value than conservative choices).

Nevertheless, the formulations considered above imply that familiarization with the task necessarily precedes risky-shifts; that is, increased familiarity is a critical mechanism underlying instances of heightened risk-taking — whether in group situations or with socially isolated individuals.[24]

---

[24]An alternative and more limited position is that familiarization is a sufficient condition for increased risk-taking only under certain specified conditions such as with isolated

There are two separate issues raised by a familiarization interpretation of the risky-shift. The first is whether the risky-shift is a true or pseudo group-effect. A second separate consideration is whether the explanatory mechanisms postulated by advocates of the familiarization hypothesis adequately account for shifts in risk-taking. The following sections consider each of these issues separately.

A. THE RISKY-SHIFT: TRUE OR PSEUDO GROUP EFFECT?

In terms of empirical support, two studies found an increase in risk-taking after exposing individuals to a familiarization procedure. In both, the increase approximately equalled that produced by group discussion. Bateson (1966) instructed individuals to restudy their initial decisions carefully on Choice-Dilemma items and generate arguments for and against the risky course of action. He compared this familiarization procedure to group discussion and control conditions. Postsession recollection tests were administered to determine how much specific information subjects remembered from the hypothetical problem. As indicated in earlier sections of this chapter, risky-shifts occurred for familiarization and group discussion with no significant differences between them. But, further, these two conditions did not differ on the measure of recollection. Bateson views the similar outcomes on both measures as support for the familiarization hypothesis. Since the two procedures—individual familiarization and group discussion—produced equal familiarity as measured by recall, he expected equal shifts toward risky decisions.

In a more sensitive study, Flanders and Thistlethwaite (1967) systematically varied familiarization and group discussion in a 2×2 factorial design to examine interactions or additive effects. The familiarization manipulation required subjects to reconsider their original decisions on Choice-Dilemma items and improvise "pro" and "con" arguments for the risky alternative as preparation for a subsequent discussion in which they would argue forcefully for their own decisions. The four conditions of the design were: neither familiarization nor discussion; familiarization without discussion; discussion without familiarization; and familiarization followed by discussion. A significant risky-shift occurred in the three conditions containing discussion and/or familiar-

---

individuals. This less inclusive view depicts familiarization as one among several mechanisms necessary to account for the risky-shift under different conditions. These different perspectives should be kept in mind while the relevant data are considered.

ization. However, these three conditions did not differ from each other. Since the joint presence of familiarization and discussion did not enhance the magnitude of risky-shift, Flanders and Thistlethwaite concluded that not only was familiarization sufficient to produce the risky-shift but they also seem to imply that it was the sole mediator of the effect. That is, they interpret the finding of "no difference" between familiarization alone, and either group discussion or familiarization plus group discussion as discrediting the possibility that familiarization and responsibility-diffusion are either independent or complementary processes, both capable of producing a risky-shift.[25]

The "strong interpretation" of the Bateson (1966) and Flanders and Thistlethwaite (1967) studies—that familiarization is the source of risky-shifts—seems unjustified. Both studies use a design in which partial support for the hypothesis requires an outcome of no difference (between cells). Although the observed, within-cell differences lend some support to the familiarization interpretation (i.e., familiarization did produce a risky-shift), the unknown likelihood of a Type II error makes the "no difference" outcome a weak basis for making inferences about the relative effects of the two separate treatments.

A more serious problem for any form of the familiarization interpretation is the fact that some later investigations fail to confirm its initial support. (1) In a series of five replications using both the Bateson and Flanders and Thistlethwaite manipulations of familiarization, Pruitt and Teger (1967) found either weak conservative or weak risky-shifts, none of which reached significance.[26] (2) Miller and Dion (1970) tried to analyze the separate components of the familiarization procedure. They suggested that the typical familiarization procedure might contain three separate processes: improvisation, comprehension, and serious reconsideration. Although the process of comprehension and serious reconsideration have received attention in prior interpretations of familiarization (Flanders and Thistlethwaite, 1967), the possible role of improvisation merits some discussion. In both Bateson and Flanders and Thistlethwaite's studies, subjects actively engaged in listing relevant points for and against the risky alternative. In the latter study, subjects also antici-

---

[25]Although we initially interpreted them as advocating this strong stand of "necessary and sufficient," both Flanders and Thistlethwaite (personal communication) have recently made it clear that they meant to interpret their results as simply indicating that familiarization is a sufficient condition for producing a risky-shift.

[26]However, these studies used few subjects and with one exception were performed with intact classes. As a consequence, involvment in the task might have been reduced.

pated using the list of points in a subsequent discussion. Such activity closely resembles the active improvisation procedures used in studies of self-persuasion (e.g., King & Janis, 1956). Thus, it seemed possible that increased risk-taking following familiarization might simply be an instance of self-persuasion arising from improvisation.[27] None of the comparisons between conditions supported a familiarization effect. Second, the relative effects of familiarization and group discussion upon risk-taking were investigated in the two different sequences in which these two treatments can be combined. Contrary to Flanders and Thistlethwaite (1967), group discussion produced a significant risky-shift following familiarization. However, familiarization failed to increase risk-taking over and above that produced by prior discussion. (3) Using a familiarization procedure designed to parallel that of Flanders and Thistlethwaite (1967), Myers (1967) obtained a within subject shift toward increased risk on risk-oriented items. However, (a) a between group comparison of the familiarization condition and a control group that worked on irrelevant problems before a readministration of the risk-taking measure showed no difference; (b) between group comparisons of both the familiarization condition and control condition to a group discussion yields stronger risky shifts than familiarization; and (c) a familiarization condition requiring overt individual rereading and discussion in front of an observer failed to show even a within-subject shift. This last lack of effect points away from comprehension or intellectual understanding as critical and points instead toward the inhibiting effects of emotional factors such as anxiety or evaluation apprehension. (4) In several attempts at exact replications of Bateson's and Flanders and Thistlethwaite's familiarization procedures, Ferguson and Vidmar (1970) also failed to obtain risky-shifts when subjects familiarized themselves with risk-oriented items. Thus, there is certainly no clear support for the hypothesis that increasing familiarity with the task is a *sufficient* condition to enhance risk-taking. But additionally, a number of studies fail to replicate the original familiarization results.

[27]To isolate the effects of these processes, Miller and Dion created three experimental conditions: (1) a replication of Flanders and Thistlethwaite's familiarization manipulation that presumably engaged all three mechanisms; (2) a private-study condition that required subjects to engage passively in further study of each risk problem without improvising arguments. This treatment presumably elicited the processes of comprehension and serious reconsideration (Flanders & Thistlethwaite, 1967) and (3) a reconsideration procedure that required subjects to reconsider their decisions in considerably less time than that provided in the familiarization and private-study conditions. For purposes of comparison, a group discussion condition was also included. Miller and Dion found that only group discussion produced a significant risky-shift.

B. EXPLANATORY MECHANISMS: UNCERTAINTY-REDUCTION AND COMPREHENSION

Not only is the present evidence currently weighted against the validity of the familiarization interpretation, but there are conceptual and empirical difficulties with the underlying processes posited to explain the risky-shift.

1. Uncertainty-Reduction

Bateson (1966) postulates that decreasing uncertainty produces the risky-shift. Akin to our arguments in the responsibility-diffusion section about reducing fear of failure, there does not appear to be any logical reason why there should be a necessary relationship between uncertainty-reduction and shifts toward risk unless one includes additional assumptions or propositions. One expectation about uncertainty-reduction is that it might encourage individuals to behave non-normatively. That is, a reduction of uncertainty might liberate subjects from choosing a middling or normative response; consequently, in comparison to another condition in which uncertainty is not lowered, they would exhibit greater variability in risk-taking. However, as argued before, such an increase in variability would not necessarily create differences in *average* risk-taking between conditions in which the amount of uncertainty-reduction varies. Thus, predictions of differences in mean risk-taking on the basis of the uncertainty-reduction explanation seem unwarranted.[28]

Setting aside our arguments above for the moment, let us grant for argument's sake that reducing uncertainty does promote risky decisions. If so, the uncertainty-reduction explanation encounters further difficulty in that it fails to account for conservative shifts. That is, if study or comprehension only affects risk-taking by reducing uncertainty, presumably what is studied or what is comprehended is irrelevant. Thus, increasing familiarity with Choice-Dilemma items that typically show a conservative shift (following discussion) should reduce initial uncertainty and produce a risky-shift. D. G. Marquis (1968) provided some suggestive evidence against this latter point in a recent experiment in which individuals familiarized themselves with several of the "caution-oriented" items developed by Stoner (1968). Individuals shifted in a cautious rather than a

[28] If Bateson's conceptualization of uncertainty is similar to "fear of failure" then our previous discussion of the Atkinson model (in the section on responsibility-diffusion) would also apply here. But, as previously indicated, that model does not predict main effects for the direction of shift.

risky direction on caution-oriented items. Myers (1967) and Ferguson and Vidmar (1970) provide additional contrary data.

2. Comprehension

According to the comprehension interpretation, familiarization clarifies the relative merits and drawbacks of the various choice-alternatives. Presumably, this is a cognitive nonemotional process. Thus, we interpret Flanders and Thistlethwaite's emphasis upon comprehension as stressing the notion of rationality. With its underlying assumption of rationality, the comprehension interpretation – in contrast to an explanation such as uncertainty-reduction that seems to include some emotional components – appears better equipped to predict both risky- and conservative-shifts. If the rationality of a decision favors conservatism, one would expect a cautious shift after additional familiarization. On the other hand, if the rationality is weighted toward riskineᴊs, increased familiarity should produce a risky-shift.

Unfortunately, some evidence contradicts the expectation that group discussion leads to a more rational decision. Two experiments examined the "rationality" of risky decisions achieved through group discussion by manipulating the payoffs and probabilities on the problem-solving task so as to discourage riskiness. In one study (Wallach *et al.*, 1964) there were *positive*, monetary incentives for all choices among the decision alternatives but all alternatives had a constant expected value (i.e., probability value multiplied by reward value). Thus, failure on the problem-solving tasks meant forfeiting the prize but did not incur any further loss from prior winnings. However, since the expected values were the same for the various decision alternatives, there was no rational reason, in terms of objective value, to prefer the riskier alternative. Nevertheless, a significant risky-shift occurred following discussion. In another experiment, Bem *et al.* (1965) arranged expected values to favor the conservative decision. That is, greater risk-taking was associated with increased aversive consequences in the event of failure. Yet, here too, group discussion enhanced risk-taking.

On the other hand, Flanders (1970) argues that in the absence of independent empirical information we cannot know whether lower probability (or for that matter, higher probability) outcomes are riskier in equal expected outcome tasks. In other words, without independent evidence on the meaning of "rationality" for these situations, we cannot look at these studies as disconfirming the notion that either group discussion or increased comprehension increases the rationality of a decision. It is also possible that expected value is not considered by subjects when they contemplate the efficacy of a course of action. Indeed, Kogan and Wallach (1967) stress this point themselves. If so, this too argues that the studies presented directly above are not necessarily inconsistent with a familiarization position that emphasizes increased comprehension.

C. CONCLUSION

In sum, neither of the explanations underlying the familiarization interpretation fare very well. The uncertainty-reduction explanation does not appear to have an unambiguous logical or psychological basis, and it is also disconfirmed by instances of conservative shifts. The studies on expected values can be interpreted as contradicting Flanders and Thistlethwaite's hypothesis that increased comprehension mediates risk-taking by emphasizing the rationality of various choice alternatives. But more importantly, there is no powerful direct evidence supporting the comprehension explanation. Recall data on Choice Dilemma items probably cannot persuasively speak on it because their contents are too brief and simple. Lastly, the fact that various investigators fail to replicate the early familiarization results adds to our conclusion that the familiarization hypothesis does not adequately explain the risky-shift.

Of course, it is possible that as an independent mechanism operating out of the context of group discussion, some characteristic of familiarization procedures will increase risk-taking in individuals. If so, this would not, of course, explain the risky-shift literature, but it would have to be integrated into a larger theory of risk-taking. Evidence that familiarization produces a pattern of risky-shift across Choice-Dilemma items that differs markedly from the standard pattern produced by group discussion supports this point. Teger *et al.* (1970) found significant positive correlations for the mean risky-shift across items for the group discussion conditions in several investigations (Teger & Pruitt, 1967; Wallach *et al.*, 1962; Flanders & Thistlethwaite, 1967). However, in the Flanders and Thistlethwaite study, correlations between the familiarization-induced pattern of risky-shift and the group-induced patterns of risky-shifts were low, positive, and nonsignificant. Assuming that the pattern of risky-shift across items reflects the underlying mechanism, Teger *et al.* concluded that different mechanisms must increase risk in the individual familiarization and the group discussion procedures. However, the kind of familiarization mechanism that might increase risk-taking in isolated individuals remains unknown.

While failures to produce risky-shifts with familiarization procedures cast doubt upon familiarization as a sufficient condition for increased risk-taking, they do not at the same time establish the risky-shift as solely a group effect. To a large extent, theoretically oriented research on the risky-shift has been misguided by a simplistic belief that there exists a single unique antecedent condition which when discovered will explain the phenomenon completely (Zajonc et al., 1970). In contrast, the existing literature suggests that numerous social (e.g., group decision, group discussion) and quasisocial (e.g., information exchange, vicarious exposure to discussion) conditions may enhance risk-taking. Likewise, other untested variables may increase the risk-taking of socially isolated individuals.

V. The Culture-Value Explanation

A fourth interpretation of the risky-shift proposed by Brown (1965)[29] is the cultural-value explanation. This explanation invokes two separate propositions to account for the risky-shift. These will be referred to as the *value hypothesis* and the *relevant information hypothesis*.

According to the value hypothesis, when an individual confronts a decision problem that contains an element of risk, the problem elicits one of two cultural values: risk or caution. Although Brown originally maintained that the American value system primarily encourages risk and daring, he later recognized the existence of cultural values for caution in response to Nordhøy's (1962) finding that group discussion of certain hypothetical decision items generally yielded conservative shifts. Accordingly, the value hypothesis assumes that in certain situations the culture condones or values risky behavior while in others it values cautious behavior. Presumably, a risky-shift only occurs in those situations that elicit values favoring risk.

Once a cultural value becomes salient, two mechanisms postulated by the relevant information hypothesis presumably lead group members to shift their final decisions in a direction consistent with the value. First, the value censors the flow of information during discussion so that more verbal statements support the value than oppose it. Since no single group member possesses all the relevant information, the value-oriented discussion provides each individual with additional reasons for agreeing with the underlying cultural value. Second, group discussion informs each group member of the level of risk other group members are willing to take. By revealing the distribution of risk preferences, group discussion establishes a social reality against which individuals can concretely define "riskiness." By providing an opportunity for

[29]Nordhøy (1962) also presented this view at approximately the same time. Our presentations will focus on Brown's analysis.

social comparison, group discussion enables an individual to determine the level of risk-taking he must choose in order to appear or act risky or cautious in respect to others. For example, if a decision situation elicits a cultural value of risk, individuals will initially make decisions they privately believe are risky. In a subsequent discussion, however, an individual may discover that he is not as risky as (or not substantially more risky than) others. Consequently, he will tend to adjust his level of risk-taking in a more risky direction in order to be consistent with the underlying cultural value. Presumably, these adjustments in risk-taking are motivated both by the desire to feel that one is adhering to the elicited value and also because the group enhances the salience of the value.

A. The Value Hypothesis

To confirm the value hypothesis, one must establish: (1) that situations typically found to provoke a risky-shift elicit values that favor risk and (2) that situations which typically provoke conservative shifts elicit values that favor caution. The present section reviews, in sequence, the data relevant to each of these two points. Following this, we discuss two studies that investigate a more general question concerning the importance of value considerations in group risk-taking.

To test either of these two points, one must develop operations for assessing values that are distinct from simply observing the direction of shift. The fact that a given item consistently shifts in either a risky or cautious direction does not by itself constitute particularly compelling evidence that cultural values are at work. One would prefer some independent data that more directly reflect what we commonly conceptualize as a value. Moreover, such data would permit an *a priori* specification of what value a given item (or situation) will elicit. For instance, since values can be conceptualized as referring to ideal behavior, one operation for assessing the value elicited by an item is to ask the subject for the "best" or "ideal" level of risk on a given item (as well as that which he would personally choose). Items for which the "ideal" is riskier than "own" can then be defined as eliciting risk, whereas the opposite direction of difference suggests that caution is the elicited value. This procedure would enable one to predict the direction of the effect of group discussion—a risky-shift or a cautious-shift—on any single Choice-Dilemma item.

Using this type of procedure, several investigators (Levinger & Schneider, 1969; Pilkonis & Zanna, 1969) find that individuals typically select odds riskier than their own when asked to indicate that level of risk-taking which they most admire. Consistent with the value hypothesis, this outcome occurred only on items that typically exhibit a risky-shift following group discussion. By means of a somewhat different procedure, Madaras and Bem (1968) demonstrated greater preference for risk than for caution on Choice-Dilemma items that previously produced risky-shifts (e.g., Wallach *et al.,* 1962). They used a series of semantic differential scales to obtain evaluations of a fictitious person who had supposedly accepted or rejected a risky decision at various levels. Fictitious persons who were "risk-takers" were rated more favorably than those depicted as "risk-rejectors."

These data are in accord with the value hypothesis. The value hypothesis holds that in certain settings, risky behavior is encouraged by the value structure of a culture. That is, supposedly risk-taking in certain settings implies that the risk-taker possesses certain personal characteristics that are valued in the culture. If one assumes that people like to view (and present) themselves favorably, this explains why riskiness relative to others becomes desirable. In short, the cultural-value explanation emphasizes, not that people like to take risks per se but instead that they like to view themselves as daring, adventurous, etc. Further, taking risks allows them to attribute such qualities to themselves. It is important to note that the personal quality of daring or riskiness is not the only valued personal characteristic that risk-taking denotes (Brown, personal communication). Indeed, daring or riskiness probably is just one of the qualities typically implied by risk-taking behavior. Moreover, to the extent that risk-taking implies *any* valued trait (e.g., selflessness, strength, confidence, and intelligence), the *behavior* of risk-taking would be culturally valued. Indeed, this is true even if the behavior does not denote daring at all. Accordingly, one would expect that in situations provoking the risky-shift, risk-takers would generally be evaluated more positively than risk-rejectors. One would expect this effect to be particularly pronounced on those traits that seem directly related to the behavior in question, but quite conceivably a halo effect would cause this difference in rated favorability even on relatively unrelated traits.[30]

[30]The results of Madaras and Bem (1968) have recently been corroborated by Jellison and Riskind (1970). In one experiment the favorability of subjects' ratings of others were positively related to the latters' riskiness on Choice-Dilemma items, although this pattern was reversed for traits related to responsibility. In a second study, others who presumably were high in academic ability were expected to take higher risks than persons low in ability. These results also support the value hypothesis. If risky behavior implies certain valued characteristics (see text above), it is reasonable that valued characteristics in turn should imply risky behavior.

Another means for determining the value elicited by a given item is to gauge how people view themselves relative to others. People generally try to present themselves favorably. Therefore if risk or caution is particularly valued, one would expect people to make decisions which deviate from those of others in the valued direction. In accord with the value hypothesis, individuals perceive themselves as riskier than their peers in initial decisions on those Choice-Dilemma items that consistently produce risky-shifts (Baron, Dion, Baron, & Miller, 1970; Hinds, 1962; Levinger & Schneider, 1969; Pruitt & Teger, 1967; Wallach & Wing, 1968). A similar outcome occurs on risk-oriented life situation items[31] (Stoner, 1968). This evidence on *relative perceived riskiness* supports Brown's notion that for certain Choice-Dilemma situations, our culture values risk. That is, we would hardly expect people to exaggerate the extent to which they are risky relative to others, unless they think that riskiness is a desirable trait.

There is one instance, however, in which individuals did not consider themselves to be riskier or more conservative than others. Zajonc *et al.* (1970) found that in a gambling situation, individuals' estimates of their own and a typical peer's level of risk-taking did not differ. Ordinarily there is little point trying to account for no effect when one is expected. However, one possible explanation for this discrepant finding is that experimental gambling situations may severely limit the opportunity for the elicitation of cultural values. That is, with trivial stakes and payoffs, cultural values may not indicate a particular course of action (Stoner, 1968). On the other hand, cultural values may be more salient in hypothetical situations than under more realistic conditions in which riskier decisions may actually incur negative outcomes.

There have also been efforts to demonstrate that caution is culturally valued in certain situations. As indicated, Nordhøy (1962) found that some of the items used by Stoner (1961) and Wallach *et al.* (1962) elicited cautious shifts following group discussion. Consistent with the notion that caution is positively valued under some circumstances, individuals perceive themselves as more cautious than peers on the two Choice-Dilemma items that typically show conservative shifts (Levinger & Schneider, 1969; Pruitt & Teger, 1967). In addition, Levinger and Schneider (1969) find that on these same two items, subjects are less likely to admire a decision riskier than their own.

---

[31]Life-situation items are similar to Choice-Dilemma items in that they describe a series of hypothetical, risk-taking situations. In addition, however, Stoner specifically attempted to design these items so that they would engage "widely held" values of either risk or caution.

Unfortunately these data on relative perceived caution and admiration constitute the only evidence that conservative shifts are due to the elicitation of values favoring caution. One's confidence in this relationship would be considerably strengthened if compatible data resulted from the application of other independent methods such as the semantic differential technique described above.

Another possible procedure for demonstrating cultural values of risk and caution with hypothetical decision-items would be to factor analyze the item-by-item correlation matrix of risk-scores for large samples of subjects. The value hypothesis predicts that a single bipolar factor (with risk and caution as the opposite poles) accounts for substantial variance among the variables. If Stoner (1968) succeeded in devising items that engage either the value of risk or caution, unifactor solutions (with all positive loadings) should be found for his risk-oriented and caution-oriented items respectively. In addition, factor scores derived from a factor analysis of initial decisions should predict the magnitude of shift with discussion. Moreover, the relative magnitude of factor loadings for initial risk-taking scores on Choice-Dilemma items should correspond closely to Pruitt and Teger's (1967) national risky-shift norms for each item. Finally, factor analysis would provide a reasonably precise determination of the amount of variance in risk-taking accounted for by a factor for cultural values – if such a factor indeed appears in the analysis.

So far, the discussion in this section has focused on whether the Choice-Dilemma items elicit values that coincide with the direction of shift. A broader issue concerns the importance of value considerations in the group risk-taking situation. In several instances, advocates of the responsibility-diffusion explanation tried to discredit experimentally the notion that cultural values influence group risk-taking. These studies approach the problem by varying the salience or importance of value considerations. Presumably, if values influence group risk-taking, one would expect particularly strong shifts in those situations in which value considerations are highly salient. In an "anticipated disclosure" condition, Bem et al. (1965) told subjects that each person's private decisions would be made public. If group discussion enhances risk-taking because subjects desire to adhere to cultural values in the presence of others, one might expect this condition to produce a risky-shift. Bem et al. found a weak nonsignificant risky-shift. Unfortunately, little can be made of negative results, and, furthermore, this experimental condition does not adequately test the cultural-value explanation subsequently proposed by Brown (1965). Since it omits any actual group discussion, subjects remain unaware of other group members' risk levels. As a result, subjects lack a reference point and consequently cannot know what odds to choose in order to maintain a cultural value of risk (relative to other group members). In other words, without discussion or information exchange, the cultural-value interpretation predicts no shift. A further criticism of the Bem *et al.* study concerns the subjective meaning of "anticipated disclosure." One effect might be that of increasing feelings of anxiety or fear of failure – the very state that responsibility-diffusion protagonists proclaim as the factor that inhibits risk-taking.

In another case, Kogan and Carlson (1969) compared group risk-taking under noncompetitive and competitive conditions. They used a problem-solving format that required subjects to choose problems varying in difficulty and payoff. Most investigations of group risk-taking consist of cooperative situations, e.g., generating a unanimous consensus concerning the level of risk-taking on Choice-Dilemma items. In the competitive situation, risk may become a more salient cultural value. Kogan and Carlson told subjects that only one group member would receive the monetary reward for a given trial. That person choosing the most difficult problem (i.e., the riskiest alternative) attempted the task first. If he failed, the individual with the next riskiest choice made his attempt to solve his chosen problem, and so on. In the noncompetitive situation, group consensus determined the difficulty level of the problem attempted by either a member delegated by the group or by the group as a whole. The responsibility-diffusion explanation predicts risky-shifts only in noncompetitive situations in that only in these situations do individuals share a sense of common fate. However, if risk is positively valued, individuals in a competitive situation might also exhibit increased risk-taking. Among undergraduates, the noncompetitive group risk-taking task produced significant risky-shifts; whereas competitive situations did not differ from a control condition. However, the unusual, sequential nature of the competitive situations of the experiment probably encouraged individuals to avoid being too risky. Given the plausible assumption that an individual choosing a very difficult problem would fail, a rational strategy dictates moderate risk-taking. By choosing a moderately difficult problem, an individual would maintain an advantageous position in terms of the sequencing of attempts yet more realistically expect to achieve a correct solution. As Kogan and Carlson noted, greater risk-taking might well be found in a competitive situation in which *only* that person choosing the most difficult problem is given an opportunity to try for the monetary prize. In addition, the average expected value across problems of different difficulty levels was negative, i.e., weighted to favor caution. Since the payoff structure is probably a more salient determinant of risk-taking in a competitive than in a cooperative group condition, this negative expected value might have inhibited risk-taking in the competitive cells. Although the results of the above experiment remain inconclusive with respect to the value hypothesis, we do agree that competitive group situations would be a fruitful site to test implications of the cultural-value explanation.

In short, the accumulated evidence provides impressive support for the basic assumption that the content of Choice-Dilemma items elicits values favoring either risk (primarily) or caution (occasionally). Prior to group discussion, subjects act in a manner consistent with expectations derived from a cultural-value explanation of group risk-taking. Whether one examines semantic differential scale data, subject perceptions concerning their own relative risk or caution, or subject expressions of admiration, the data are consistent. Each of the items typically used in studies of group risk-taking consistently elicits values of either risk or caution that coincide with the direction of shift. Moreover, studies suggesting that cultural values do not influence risk-taking (Bem *et al.*, 1965; Kogan & Carlson, 1969) are vulnerable to either conceptual or methodological criticism.

Unfortunately, the validity of the value hypothesis cannot by itself explain the risky-shift; nor does its validity necessarily threaten alternative explanations of the risky-shift.[32] The cultural-value explanation rests primarily upon the second proposition—the relevant information hypothesis. Do elicited values of risk or caution determine the direction and outcome of the subsequent group discussion?

B. THE RELEVANT-INFORMATION HYPOTHESIS

According to Brown's relevant-information hypothesis, group discussion increases the salience of the values elicited in initial decision-making. Two mechanisms presumably produce this effect: (1) discussion provides an opportunity for group members to exchange information concerning the distribution of risk-preferences among them and (2) elicited values bias and sensitize the ensuing discussion so that the greater proportion of arguments favor the salient value. To date, most of the available research focuses on the information-exchange mechanism and the general proposition that discussion increases the strength of elicited values.

Considerable evidence supports the notion that group discussion

---

[32]Indeed Kogan (personal communication) would anticipate these results and would concur with our conclusions in that he claims that the Choice-Dilemma items were specifically constructed so that for each item the risky (or uncertain) alternative was more valued than the conservative (or certain) alternative.

increases the salience of elicited values. According to Pruitt and Teger (1967), levels of initial risk indicate the extent to which a given item elicits a value of risk. In these terms, a positive correlation between initial risk-taking and risky-shift supports the notion that group discussion enhances the initial directional tendencies arising from the content of the decision items. Stating it another way, items that initially elicit tendencies toward risk produce even riskier decisions following discussion; those that initially produce cautious tendencies produce even more conservative decisions after discussion.

Stoner (1968) provides perhaps the strongest support for the notion that values affect the direction of a group decision. Under the guise of a different experiment, subjects initially ranked the importance of a series of value statements. They then completed a set of risk-oriented and caution-oriented "life-situation" items. As mentioned previously, these life situation items presumably engage "widely held" values so that either risky or cautious considerations predominate. The value statements described the alternative outcomes implicit in each of the life situation items. That is, for each item, the risky and conservative outcome corresponded to a given value statement. On "risk-oriented" items, subjects ranked value statements corresponding to the risky alternative as more important than value statements associated with the cautious alternative. On the other hand, for "caution-oriented" items, subjects ranked value statements associated with the conservative alternative as significantly more important than those relating to the risky alternative. As previously indicated, all items engaging values favoring the risky alternative elicited initial tendencies toward risk. But more importantly, group discussion significantly increased these tendencies. Although trends were not as strong or consistent, opposite patterns were observed for caution-oriented items.

In a recent series of investigations, Pruitt (1969) suggests an important conceptual link for understanding the way in which group discussion enhances values elicited by initial individual decisions. When making their initial decisions on Choice-Dilemma items, subjects in risky-shift investigations must not only decide whether a given hypothetical situation warrants a risky or cautious approach; they must also determine acceptable odds for the risky alternative if chosen. In the initial experiment, Pruitt separated these steps by first requiring subjects to label each of the probability alternatives on every Choice-Dilemma item as either risky or cautious. In a *second* administration of the Choice-Dilemma situations, they advised the fictitious person confronting the decision what level of risk-taking they felt was appropriate. On all items, subjects responded with a level of risk-taking they previously labeled as cautious.

On the face of it, these results contradict the prediction from the cultural-value interpretation which suggests that individuals view their initial decisions as risky on "risk-oriented" Choice-Dilemma items and conservative on "caution-oriented" items. In subsequent studies, Pruitt replicated this initial finding and investigated further conditions in which subjects: (1) *simultaneously* make judgments of risk and caution on alternatives of Choice-Dilemma items and choose an appropriate level of risk-taking or (2) make judgments of riskiness or caution *after* making their initial decisions. Under these two conditions, Pruitt did confirm the prediction of the cultural-value interpretation. That is, subjects generally considered their decisions on risk-oriented items to be subjectively risky and those on caution-oriented items, subjectively conservative. In sum, when the labeling operation preceded decision-making, individuals subsequently selected answers that they had previously considered cautious. However, when the decision occurred simultaneously with (or prior to) labeling, subjects subsequently labeled their answers risky.

Pruitt interpreted these results as depicting a "Walter Mitty" effect. Prior to discussion, the value of risk compels subjects to be risky. On the other hand, a fear of "putting themselves out on a limb" restrains them.[33] As a result, initial risk-taking represents a compromise between these opposing forces. Similar to Walter Mitty, subjects resolve their conflict by imagining and representing themselves as "risky" while making decisions that they themselves judge to be truly cautious. In Pruitt's initial study, however, subjects labeled the various alternatives before making a decision. As a result, these subjects had little need to bias their labeling in any way. That is, they had no decisions to misrepresent at that time. Pruitt's analysis provides insights into the particular manner in which group discussion may strengthen initial tendencies toward risk elicited by values, *viz.*, by demonstrating that others are equally or more risky and thereby providing social support for riskiness.

One of the theoretical notions underlying the relevant-information hypothesis is that group discussion primarily provides an occasion for members to exchange information about their levels of risk-taking. Presumably, information-exchange alone (i.e., the knowledge of the pattern of risk-taking among group members) is sufficient to produce a risky-shift (Brown, 1965). As discussed earlier in the context of responsibility-diffusion theory, several experiments using Choice-Dilemma items

---

[33]Pruitt's speculation that a fear of putting one's self out on a limb inhibits risk-taking prior to discussion seems similar to the diffusion of responsibility notion that fear of failure restricts risk-taking.

(Kogan & Wallach, 1967; Teger & Pruitt, 1967) find that mere information-exchange produces increased risk-taking. Also, recent experiments (Clark & Willems, personal communication; Pruitt & Teger, 1969), which contrast the responsibility-diffusion and cultural-value explanations, demonstrate that group discussion primarily produces a risky-shift by allowing an exchange of information about each member's level of risk. Thus, at least with decisions involving Choice-Dilemma items, information exchange appears to be sufficient to produce the risky-shift. A related question is whether this effect holds for situations involving something other than Choice-Dilemma items.

Blank (1968) reported a significant risky-shift in a gambling situation that made subjects aware of others' risk preferences but did not allow them to interact verbally. In contrast, Zajonc et al. (1970) present evidence that information concerning others' risk preferences in a gambling situation is not sufficient to account for risky-shifts induced by group discussion. Kogan and Carlson (1969) found that in a problem-solving situation, overt competition (subjects made public bids for problems that varied in difficulty) did not produce greater risky-shifts than covert competition (private bids). In both instances, they found only weak nonsignificant trends toward increased risk-taking. Therefore, the information exchange occurring in the overt competition situation was apparently insufficient to produce a risky-shift. However, unusual features of the competitive situations, discussed earlier, were probably responsible for the failure to find any difference between overt and covert competition conditions as well as for the attenuation of the risky-shift in these situations.

In summary, it is an open issue whether information-exchange is sufficient to produce the risky-shift in either gambling situations or under competitive conditions. On the other hand, the results of those studies that use Choice-Dilemma or hypothetical items do substantiate the relevant information hypothesis. There is evidence (Stoner, 1968) that the direction of the shift observed after group discussion is a function of the dominant value in the situation. Moreover, the positive correlation observed between initial decisions and subsequent risky-shifts indicates that the more salient or important the value, the greater the risky-shift. That information exchange per se is sufficient to produce the risky-shift (Kogan & Wallach, 1967d; Teger & Pruitt, 1967) is quite in accord with the relevant information hypothesis — as is the finding that group discussion elevates risk-taking even more than mere information exchange. Finally, Pruitt's recent "Walter Mitty" explanation is consistent with an interpretation emphasizing the role of cultural values in decision-making.

On the other hand, this type of datum sheds little light on the mechanism presumably causing the risky-shift — the unidirectional social comparison process. The following section examines the data relevant to this process in detail.

C. THE SOCIAL COMPARISON MECHANISM

One of the more important, implicit features of the cultural value explanation is the idea that information exchange induces social comparison processes. Essentially, this interpretation assumes that social comparison pressures operate in conjunction with elicited values. Presumably, prior to discussion, individuals privately believe that compared to others they are more consistent with the implicit norms elicited by different types of decision items. For example, with risk-oriented items, individuals initially consider themselves to be riskier than their peers. Upon entering into discussion, an individual may discover that he is, in fact, not as risky as others present. To maintain consistency with his self-perception as a high risk-taker, he must then shift to a riskier position. Thus, the value explanation presumes that social comparison pressures operate selectively in the direction of the value made salient by the type of item or situation; i.e., it hypothesizes a "unidirectional push" toward maximal endorsement of the elicited value. In short, subjects who deviate in the valued direction should not shift. Thus, the cultural value explanation considers only a relatively specific instance of the operation of social comparison factors in risk-taking.[34]

One could ignore the possible effects of elicited values and still apply social comparison theory (Festinger, 1954) to group decision-making. Even though no one attempts to do so, the intellectual exercise highlights some interesting points. First, if decisions about appropriate levels of risk are matters of opinion, and if a value is not normatively elicited by

[34]Recently Jellison and Riskind (1970) apply Festinger's notions about the social comparison of *abilities* to the risky shift literature. In this case the predictions coincide with those of the cultural-value explanation. To support their application of Festinger's theorizing on the social comparison of ability, Jellison and Riskind attempt to show that people infer that those who take higher risks possess greater ability. Their efforts support Brown's position that risk is a positive value. Whether increased risk-taking occurs because risk-taking implies ability remains moot. That is, the unidirectional social comparison effect that Festinger postulates for abilities need not be restricted to ability. Instead, as Brown implies, it may apply to any trait that can normatively be placed or located on the good-bad evaluative dimension. As previously indicated, Brown's position implies that when a group discusses a decision problem that elicits any positive value — e.g., generosity — polarization toward the value will ensue. Need this be true only if generosity implies ability? It seems unlikely. If so, Jellison and Riskind may only add unnecessary baggage to the cultural-value explanation (see the discussion and footnote on page 344).

a given decision problem, social comparison theory cannot adequately explain the risky-shift. If group risk-taking solely reflects a social comparison process that permits assessment of others opinions, individuals would accommodate their level of risk-taking to those most similar to them. Considered over the entire group, an averaging effect with little or no shift would result. Second, if one assumes that the decision problem does not normatively elicit a value, social comparison theory generates predictions that compete with those of cultural value theory. For example, considering the range of initial risk-taking on risk-oriented Choice-Dilemma items, the cultural-value position predicts shifts toward risk for all individuals in a group except for the high initial risk-taker. This person should show no shift as a consequence of discussion since the revealed pattern of risk-taking among group members confirms his initial self-perceptions as a high risk-taker. In contrast, Festinger's theorizing about the social comparison of opinions predicts regression-like shifts toward the mean for both initially high and low risk-takers with little or no change on the part of the medium risk-takers. The purpose of this discussion is to demonstrate that social comparison and value considerations are not necessarily always congruent. It is possible to imagine situations in which they might be made to conflict, and in which the issues raised above are relevant.

The bulk of recent research on the cultural value explanation focuses upon identifying the nature and impact of social-comparison factors as determinants of individual and group shifts in risk-taking. With respect to individual shifts, one factor receiving attention is perceived relative initial riskiness. Given that individuals perceive themselves to be riskier than their typical peers in their initial decisions, Pruitt and Teger (1967) reasoned that the *difference* between one's initial risk preferences and those risk preferences attributed to others should predict the amount of risky-shift. Their logic rests on the unidirectional social comparison process implicit in the cultural-value explanation. They assumed that: "The farther ahead of the pack one initially thinks he is, the more catching up he has to do when he finds that he is performing in an average fashion" (Pruitt & Teger, 1967, p. 16). However, only weak nonsignificant correlations were found between relative perceived riskiness and risky-shift. Partialing out the common factor of initial risk further reduced the weak correlations between these two variables. In sum, these results suggest that elicited values are irrelevant to subsequent individual shifts in risk-taking following discussion.

The independent variable of *true relative initial riskiness* provides more favorable results for Brown's social comparison mechanism. While *perceived relative riskiness* refers to subjects' subjective opinions concerning how risky they are relative to others, *true relative initial risk* is the subjects' actual, relative position in the distribution of initial risk decisions in the group. On this measure, regardless of the within-group range of scores, Vidmar (1970) found an inverse relation between the extent to which an individual shifts toward risk and his true initial riskiness relative to others in the group. In other words, the greater the level of risk-taking, the less the subsequent increase in risk-taking following discussion. Furthermore, persons with the highest initial risk scores also exhibited a risky-shift after discussion. Considered by itself, the social comparison process implied by the cultural value explanation predicts no change for the initially riskiest group member. However, the cultural-value explanation also hypothesizes that group discussion will increase the salience of the elicited value. While increased salience may account for this latter result, it is more probably an artifact of the procedure of summing risk scores across items to obtain an overall index of initial risk for individuals. That is, the risky-shift for high initial risk-takers is attributable to specific items on which they were not the riskiest member of the group. An internal analysis of individual items showed that the riskiest person on any given item exhibited no shift or a cautious shift. These findings on relative initial risk provide some support for the importance of social comparison factors as determinants of individual shifts. On the basis of social comparison theory's predictions about the comparison of opinions, the initially riskiest person should exhibit a cautious shift during discussion. On the other hand, if value considerations interact with social comparison processes as Brown suggests, the high risk-taker should show no shift (or considering the salience-increasing effect of group discussion, a risky-shift). Vidmar's evidence favors the first of these two predictions. On the internal analysis, some of the initially high risk-takers exhibited a cautious-shift. However, since regression effects also predict these cautious-shifts, their meaning remains ambiguous.

In an attempt to assess the validity of Brown's social comparison mechanism, Madaras and Bem (1968) had groups discuss only half of a set of hypothetical risk items after individuals indicated initial risk preferences on the entire set. They assumed that knowledge of members' risk preferences on the discussed items would generalize to undiscussed items (thus allowing social comparison on all items). Brown (1965) originally stated that, beyond providing information concerning the level of risk-taking of other group members, the content of discussion is "irrelevant." That is, he placed greater emphasis upon the (unidirectional) social comparison function of group interaction than upon persuasion resulting from the biased group discussion favoring the elicited value. If, however, the content of discussion transmits item-specific information in addition to each group member's typical tendency to be risky, and, if such item-specific information facilitates the risky-shift (say through persuasion), risk tendencies on undiscussed items should be weaker.

Madaras and Bem found risky-shifts only on discussed items. They interpreted this outcome as disconfirming Brown's unidirectional, social comparison mechanism. To account for their results, they suggested instead that the risky-shift occurs because ". . . individuals are culturally disposed to generate and favor risk arguments when considering risky dilemmas in detail" (Madaras & Bem, 1968, p. 360). Unfortunately, this conclusion assumes that subjects use discussed items to infer the distribution of risk preferences on undiscussed items. If they do not, the obtained outcome may simply reflect subjects' inability to engage in social comparison on the undiscussed items. To test the two hypotheses in question critically, subjects who discuss only half of the items should receive information about the pattern of risk-taking among group members on the undiscussed items. Comparison of shifts on discussed and undiscussed items would reveal whether the biased flow of arguments during group discussion provides additional impetus for increased risk-taking over and above the risky-shift produced by mere information exchange. In sum, it is difficult to conclude that the Madaras and Bem study invalidates Brown's social comparison mechanism.

Other attempts to assess the effects of social comparison upon group shifts focus upon the extent to which the discussion group contains members with homogeneous or heterogeneous initial risk scores. The greater the initial dispersion in levels of risk-taking, the more likely it is that most group members will note that *someone* is more risky. As a consequence, since individuals initially consider themselves riskier than others, group discussion should provide greater disconfirmation of initial self-percep-tions and a greater risky-shift when group members' initial risk scores are heterogeneous (i.e., when one group member is particularly risky). In other words, the cultural-value explanation predicts a positive correla-tion between initial range of risk scores and magnitude of risky-shift. Unfortunately, this prediction is not unique to the cultural value inter-pretation. Alternative explanations of the risky-shift, such as the leader-ship hypothesis, clearly suggest the same outcome (i.e., members homo-geneous in riskiness should be equally persuasive and no shift should result). However, in the case of perfect homogeneity, differential predic-tion is possible. The leadership explanation predicts no shift in this situation, but the cultural-value explanation can be interpreted as imply-ing increased risk-taking. In a homogeneous group, all individuals find that they are not riskier than anyone else. Since it is better to have more of a desired trait than others, each individual member must shift toward risk in order to fulfill the desire to be riskier than other group members.[35] Furthermore, as previously indicated, since the group discussion should enhance the salience of the elicited value, this too should promote greater risk-taking even in the case of homogeneous group composition.

There is considerable evidence for a positive relationship between the range of initial risk scores among group members and the magnitude or risky-shift produced by group discussion. Hermann and Kogan (1968) note that shifts toward risk are generally stronger in groups with large divergences among individuals. More concretely, Hoyt and Stoner (1968) report a significant, positive correlation between within-group range of risk scores and the magnitude of shift induced by the group. In a recent experiment, Vidmar (1970) compared three types of homogeneous groups — persons who were high, medium, or low on the initial measure of risk-taking — to heterogeneous groups. The homogeneous groups were not precisely identical, but their range of risk scores was small relative to heterogeneous groups. Except for the homogeneous groups of medium risk-takers (which inadvertently had the smallest range of initial scores), all other types of groups produced a risky-shift following discussion. In addition, heterogeneous groups shifted more toward risk than any of the homogeneous groups (who did not differ from one another). Brown's cultural-value interpretation cannot easily account for the absence of a risky-shift in the most homogeneous groups. As previously argued, not only does it imply that people who find that they are only "average" in riskiness should increase their risk-taking so as to conform to the elicited value of risk, but also the group discussion should increase the salience of risk as a value and thereby promote riskiness. Yet, this does not occur. Moreover, these results have recently been replicated. Using a more stringent criterion of homogeneity than Vidmar (1970), Clark and Willems (personal communication) compared homogeneous and heterogeneous groups under two different experimental conditions: information exchange and group discussion. Using a set of three Choice-Dilemma items, individuals in homogeneous groups could not differ by more than two scale points. Those in heterogeneous

---

[35]This interpretation states that it is a cultural value to be more risky than other members. Brown appears to take a more modest position on this particular point: "each individual answering the Stoner problems means to be at least as risky as people like himself. . . . Each man, on his own, guesses the norm to be at or below his own selection. . . . Those who find themselves below the mean of the six members of the group discover they are failing to realize the ideal of riskiness that they thought they were realizing. Consequently, they feel impelled to move in a risky direction both in accepting the decision of the group and in changing their private opinions. Subjects at or above the group mean feel no impulsion; they are relatively risky just as they meant to be."

groups exhibited risky-shifts under both experimental conditions. In contrast, homogeneous groups did not differ from a control condition.

Thus the data contradict the predictions made by Brown's social comparison mechanism for groups highly homogeneous in their initial risk preferences.[36] Furthermore, recent evidence suggests that social comparison pressures in group risk-taking situations need not always be biased toward the implicit values of the decision items. In a pilot investigation, Steiner (personal communication) initially required subjects to complete risk-oriented and caution-oriented items and also to indicate what they consider to be the average response of a self-selected, reference group (e.g., "20 to 30 persons like yourself"). In a subsequent phase of the experiment, he presented subjects false norms about the actual distribution of others' responses. This false feedback disconfirmed the subject's initial expectations concerning the responses of others. For one risky item, the norm was one scale position riskier; whereas for another, the norm was one scale position in a more conservative position. A similar procedure was used with caution-oriented items. The manipulated norms produced marked conformity with virtually no trace of an interaction between type of item and direction of deviation from the subject.

Baron *et al.* (1970) also conducted a study in which social-comparison pressures were pitted against value considerations. Subjects discussed either risk-oriented or caution-oriented items. Confederates created a majority consensus which was either two scale positions riskier or more conservative than the naive subject's initial decision. Similar to Steiner's findings, subjects exhibited marked conformity to the artificially established group consensus—regardless of whether the group norm coincided or conflicted with the underlying value elicited by the type of the item. Further, these shifts toward majority consensus largely remained intact on private, posttest measures of risk-taking.[37]

These preliminary results do not, on the face of it, argue well for the cultural value explanation of risk-taking as the only relevant theory. If risk and caution are indeed cultural values and if cultural-value theory

---

[36]Vidmar (personal communication) points out that although the above results do fall short of significance (i.e., show no effect) for homogeneous groups, these groups do exhibit shifts in the expected direction. He argues that the magnitude of effect will simply be smaller if group members are similar to begin with, and, consequently, it is more difficult to obtain significance. By implication, when these studies are replicated with large samples, this conclusion will require modification.

[37]Unfortunately, the possibility of regression effects might also account for these observed outcomes; analyses assessing this possibility have not yet been completed and, therefore, these results are not conclusive.

is the only theory relevant to this situation, individuals should *selectively* adjust their levels of risk-taking to coincide with the dominant values underlying risk-oriented and caution-oriented items, respectively. That is, on risky items, the riskiest subject should not conform to a more conservative group norm; conversely, on conservative items, the most cautious subjects should not conform to a riskier group norm. In both cases these subjects presumably deviate in a valued direction. Yet experimental results show that responsiveness to disconfirmation does not depend solely upon direction of deviation from the value orientation of the different types of decision items. Subjects conform dramatically to the majority consensus provided them. These results suggest that, when the effects of value are pitted against those of normative influence, the latter primarily determine risk-taking.

Another means of evaluating the importance of social comparison processes in group risk-taking is to examine the parameter of "certainty-uncertainty." Without any objective information for an appropriate course of action in a given situation, social comparison pressures should strongly influence risk-taking because "social reality" provides the only guidelines for behavior. The hypothetical situations depicted in Choice-Dilemma items are "uncertain" risk problems in that one cannot assign definite values to outcomes or provide objective specification of consequences for alternative outcomes. Under these circumstances, the pressures of normative influence should prevail. Stooges primed to exert influence (or merely display opinions) in a particular direction should produce dramatic influence effects. On the other hand, situations such as gambling tasks provide more concrete and objective cues for deciding an appropriate course of action. In these settings, the effect of others' opinions about the optimal level of risk should be diminished. Using a gambling task with an element of uncertainty (i.e., probabilities of events were unspecified), Zajonc *et al.* (1970) investigated the effect of information about other's risk preferences upon individuals choices in isolation and coaction situations. They conclude that information about others' level of risk-taking "had a numerically negligible" effect even on this gambling task—one that did have some element of uncertainty (Zajonc, *et al.*, 1970, p. 44). Extrapolating from these results, knowledge of others opinions should have little or no effect upon risk-taking in a "pure-risk" gambling situation (i.e., one in which outcomes, event probabilities, and stakes are all clearly specified). That is, the more minimal the uncertainty in a risk-taking situation, the less an individual must depend upon social reality as a means of validating a course of action.[38]

[38]These remarks suggesting that ambiguity or uncertainty increases the impact of group discussion—increases social influence effects, should not be interpreted as con-

D. SUMMARY

The research findings pertaining to cultural-value theory reveal a reasonably clear-cut pattern. The existing evidence strongly confirms the value hypothesis — more strongly for the value of risk than for the value of caution — and basically supports the general predictions from the relevant-information hypothesis. On the other hand, its implicit social-comparison mechanism does not fare so well. For example, the variable "relative perceived riskiness," a cornerstone of the cultural-value explanation, is sadly deficient as a predictor of individual shifts (Pruitt and Teger, 1967). Similarly, the absence of a risky-shift with groups composed of members homogeneous in their initial risk scores (Clark & Willems, personal communication; Vidmar, 1970) embarrasses the theory. Finally, some evidence (Baron et al., 1970; Steiner, personal communication) suggests that persons who initially deviate from others in a culturally valued direction will nevertheless subsequently conform to manipulated group norms that conflict with the initially elicited value. In fairness, however, many of these "contradictory" data are tentative; they stem from preliminary studies that investigate the role of social-comparison processes in group risk-taking. Also, there is no reason to suspect that the operation of other social psychological principles (e.g., normative influence) should be completely precluded by the group processes postulated by the cultural-value explanation. Instead, these studies in conjunction with other social comparison models (e.g., Levinger & Schneider, 1969) may well define the boundaries of the cultural-value explanation. Finally, a variety of other questions about the generality of the cultural-value explanation also await exploration. The possibility that the cultural-value explanation applies to decision problems that elicit any one of a variety of other values besides risk and caution awaits testing. Another consideration is the extent to which the explanation applies when one alters structural properties of the group (other than homogeneity of group members in respect to original propensity toward risk or caution). For instance, would status differences preclude the postulated effects concerning the flow of relevant information? Likewise, would groups comprised of individuals who know one another well and presumably have little to learn about one another's values show smaller risky shifts? Clearly, these are problems for future research.

---

tradicting our earlier comments about uncertainty and fear of failure in the sections on responsibility diffusion and familiarity. Here we make the point that to the extent that tasks differ in the extent to which they offer an unambiguous "right answer," individual decisions are more susceptible to normative influence.

VI. Recent Theoretical Trends

Most recent theoretical efforts conveniently fall into two categories: enhancement and social comparison models. Respectively, they attend to the polarization and the social comparison functions of group discussion. Enhancement models assert that group discussions enhances (or polarizes) whatever tendency was present on initial (individual) decisions. This position superficially resembles the cultural-value interpretation, i.e., that group discussion increases the salience of whatever cultural value a decision problem happens to elicit. Enhancement models, however, explain polarization by invoking explanatory mechanisms suggested by leadership, familiarization, and involvement interpretations of the risky-shift. In contrast to enhancement models, social comparison models more directly extend cultural-value theory. More specifically, they precisely examine, emphasize, and extend the social comparison processes implicit in the cultural-value explanation. Despite their differences, both enhancement and social comparison models exhibit a similar attention to variables only tacitly assumed important by previous explanations.

A. ENHANCEMENT MODELS

Burns (1967) proposes an extremity-variance model that assumes groups behave more decisively (i.e., more extremely) than individuals. This notion resembles the leadership explanation; it assumes that those who hold a more extreme view or position exert a greater influence in group decision-making. This broad assumption enables Burns to also take into account the possibility of cautious-shifts. Using this model, he accurately predicted the outcome of over half of 361 group decisions while using only two parameters: initial extremity of the most extreme group member and the within-group variance among group members' initial decisions.

Marquis and Reitz (1969) also stress the enhancing function of group discussion; however, their position has closer conceptual ties to the familiarization explanation. Specifically, they suggest that group discussion has two effects. First it enhances the prior expected value of various possible outcomes. This characterization of group discussion is similar to heightening the reinforcement value of the decision alternatives; a process Rettig (1966) postulates to be a function of group discussion. Second, they hypothesize that when there is initial uncertainty, subsequent discussion increases risk-taking. The two assumptions that underlie this second hypothesis also underlie the familiarization and responsibility diffusion interpretations: (1) uncertainty inhibits risk-taking and (2) group discussion reduces uncertainty. Finally, they assume that enhancement of expected value and uncertainty-reduction operate independently. For example, group discussion presumably increases the salience of the expected value regardless of whether the risk-taking situation is uncertain or not. Conversely, reducing uncertainty presumably heightens risk-taking regardless of whether the expected value is positive, zero, or negative. In sum, Marquis and Reitz's model explicitly specifies two independent parameters of group risk-taking, both of which require consideration before generating a prediction.

Marquis and Reitz tested various predictions and assumptions of their model by using gambling problems varying in expected value and uncertainty. They manipulated uncertainty in the following manner. For "pure risk" (i.e., certain) problems, they precisely specified for each problem the stake, the probability of winning, and the prize. For uncertain problems, they only indicated the value of those elements of the risk-taking situation with a range of possible values. In support of their model, they cited previous research (Hubbard, 1963; Reitz, no year indicated) in which individuals made larger bets on certain than on uncertain gambling items. These data, then, support the assumption that uncertainty reduces the willingness of individuals to take risks. Nevertheless, whether group discussion simultaneously reduces uncertainty *and* increases the salience of the expected value on gambling problems remained questionable. However, Marquis and Reitz confirmed the following predictions of the model: (1) With initial *certainty* in the risk-taking situation, the direction of shift after discussion will depend on the expected value of the risky outcome – a positive expected value will produce a risky-shift, a negative expected value will produce a conservative shift, and zero expected value will produce no shift, (2) When *uncertainty* exists, group discussion will produce shifts toward risk when the risky outcome has either positive or zero expected value. When the risky outcome has negative expected value, the direction of shift depends upon which force is stronger, the decrease of uncertainty or the increased salience of negative expected value.

Given this preliminary support, Marquis and Reitz's model appears promising. Its particular appeal is that it can also account for risky-shifts obtained on the Choice-Dilemma items since one can argue that such items are uncertain risk problems (i.e., they do not specify the exact value of success and failure cf., Hubbard, 1963).

In a recent paper, Moscovici and Zavalloni (1969) similarly take an enhancement position to account for risky-shifts. According to their analysis, the various treatments (e.g., familiarization, group discussion, ballotting) that have produced the risky-shift have one thing in common; they involve subjects in the situation in which they are embedded and increase the importance of the common judgmental object – the problem requiring a decision. It is this increase in involvement that presumably causes a polarization of initial risk-taking tendencies. Although the design of their own research does not (and cannot) experimentally confirm this view, there is a substantial literature in attitude change research that makes this position tenable (Miller, 1965; Sherif & Hovland, 1961; Sherif, Sherif, & Nebergall, 1967). The attitude change research shows that persons who are more involved with their position on an issue take a more extreme stand in reporting their position. In addition, when subjects' involvement with a position is experimentally increased, their stand becomes more extreme (Miller, 1965). Other research shows that those who are more certain about their position on an issue hold more extreme views (Suchman, 1950). Moscovici and Zavalloni also suggest that the degree of opinion polarization will be a function of the initial diversity in risk-taking levels. Diversity, they argue, produces interpersonal disagreement, tension, and anxiety. They further assume that anxiety is associated with taking extreme stands. While these suppositions seem reasonable, it remains to be seen if they will be substantiated with empirical data. Particularly important is whether the various treatments that produce the risky-shift do indeed increase subject involvement as Moscovici and Zavalloni contend.

B. SOCIAL COMPARISON MODELS

As indicated in the discussion of the cultural-value explanation, several investigators focus on social comparison processes in group risk-taking. Levinger and Schneider (1969) propose a "conflict-compromise" model to account for the risky-shift. Presumably, prior to discussion, individual choices on Choice-Dilemma items must compromise between two conflicting pressures: (1) personal idealistic considerations and (2) realistic considerations. Similar to the cultural-value interpretation, Levinger and Schneider view discussion as performing an information function. The group provides a reference function by defining a realistic choice. Prior to discussion, a realistic choice is governed by an individual's expectation concerning the choices of others. Upon entering discussion, individuals discover that others are actually riskier than expected. The effect of this discovery is to move the group's reference point to a position closer to the average, ideal level of risk-taking and to result in a risky-shift. Steiner (personal communication) proposes a similar model, which assumes that group discussion corrects the "pluralistic ignorance" of individuals' initial considerations. Group discussion alters initial decisions in the direction of disconfirmation.

Recent research provides instances of evidence consistent with the conflict-compromise model. Pruitt's (1969) interpretation of the "Walter Mitty" effect (*viz.*, that prior to discussion, subjects imagine and represent themselves as "risky" while making objectively cautious decisions) is essentially congruent with a conflict-compromise model. Similar to Levinger and Schneider's analysis, Pruitt explains the latter phenomenon as a compromise between two conflict forces: (1) the desire to be consistent with a cultural value favoring riskiness and (2) fear of putting one's self out on a limb. Pilkonis and Zanna (1969) also provide some evidence for the conflict-compromise explanation in an experiment specifically designed to test predictions derived from it.

In contrast to some researchers' recent focus on social comparison processes, Madaras and Bem (1968) present a version of cultural-value theory that differs from Brown's (1965) position primarily in that it denies the importance of the social comparison mechanism. Their position also resembles the rhetoric-of-risk interpretation in some respects. According to Madaras and Bem, the risky-shift is primarily due to persuasion resulting from individual's inclinations to "generate and favor risky arguments" when seriously considering decision situations that elicit values favoring risk. However, unlike the rhetoric-of-risk interpretation, if a situation elicited values favoring caution, individuals presumably should favor cautious arguments. This version of cultural-value theory possesses several advantages. It can accommodate the familiarization studies of Bateson and Flanders and Thistlethwaite, it explains evidence supporting the leadership hypothesis and finally, it does not rely on a selective (or undirectional) social comparison mechanism. However, since there are problems with their study, their explanation lacks strong empirical support.

VII. Retrospective Wisdom

In many respects a review of the literature on the risky-shift illustrates several of the poor methodological practices that characterize much social psychological research. While prescience is rare, retrospective wisdom is pedestrian. Nevertheless, it is instructive to review and reiterate some of these indiscretions. The major points of criticism can be organized under four headings: (1) theoretical chauvinism (or the folly of particularistic explanations), (2) single testing sites (or problems of generality and demand), (3) oblique experimentation (or indirect tests of hypotheses), and (4) myopic scholarship (or inattention to the work of others).

A. THEORETICAL CHAUVINISM

Much of the available research on the risky-shift (particularly the early studies) implies that a single process or explanation accounts for the observed effect. Discussions of results often seem to presume that if a given explanation receives support, no other competing explanation (even though untested by the design of the particular experiment) conceivably remains true or relevant. This view of social psychological processes is too simplistic and, consequently, naive. Little empirical precedent exists for assuming that any single hypothesis or social psychological process operates to the exclusion of all others in a given setting. There certainly is no *a priori* logical incompatibility between any of the four explanations treated in detail in this chapter. Indeed, the various explanations proposed seem to complement one another rather than imply mutual incompatibility. As previously reiterated, the explanations often make similar rather than competing predictions. In fact, it is not easy to construct critical experimental tests among them. To illustrate the point, there is no reason why the essential truth of the cultural-value explanation must preclude the possibility that the leadership explanation also has a core of truth. For instance, it may be the case that those whose initial decisions are riskier do in fact exert more influence in the group setting, but, further, that this greater influence is restricted to situations in which the value of risk is salient. Or considering things in a slightly different way, the critical proposition of the leadership explanation might require incorporation as a special subproposition or subhypothesis of cultural value theory. More specifically, it is conceivable that the greater influence of those whose initial decisions were most risky is restricted to group discussions of risk-relevant problems. In other spheres of life they may be no more influential than others. Likewise, if cultural-value theory is essentially true, this in no way precludes the additional possibility that group discussion also diffuses responsibility or that the diffusion of responsibility can by itself produce shifts toward risks. Similar arguments could be made for the familiarization explanation and its relation to other hypothesized explanations. Yet many authors unfortunately speak as if support for their own hypothesis disconfirms the other alternative hypotheses explaining shifts toward risks. Given that the various interpretations of the risky-shift seem to complement rather than contradict one another, this type of "disconfirmation" is clearly specious and misleading.

B. SINGLE TESTING SITES

The exclusive focus on a single experimental arena is a second problem with the risky-shift research. This problem has two facets: (*a*) most research on the risky-shift uses a particular type of risk-taking task (or measure), the Choice-Dilemma items and (*b*) material is generally employed within a single type of experimental design: the repeated measures design.[39] Although the hazards are probably well known, some of the issues associated with each of these methodological practices warrant continued airing.

While some researchers rightly argue that some advantages result from loyalty to a single set of materials (or a single apparatus), the long-run disadvantages are greater. The most obvious point is that standardization severely limits the generalizability of any obtained effects. Numerous, unknown subtle characteristics of the specific experimental materials may be essential for producing the observed effects. For example, one possible explanation for the high frequency of risky-shifts obtained with Choice-Dilemma items is that this task requires decisions about hypothetical situations with no personal consequences for the decision-maker. In this respect, one could effectively argue that the research on the risky-shift tells us very little indeed about group risk-taking, although it may provide substantial information about group processes.

Relatedly, the specific content of the Choice-Dilemma items may be responsible for producing the effects obtained with them. The authors of the Choice-Dilemma items defined risk in a particular way (Mackenzie & Bernhardt, 1968) and the items hardly constitute a random sample of all possible risk-taking situations. Consequently, it is impossible to ascertain the extent to which the results of the studies reviewed in this paper depend upon a specific narrow definition of "riskiness." The markedly lower frequency of risky-shift effects on gambling tasks as opposed to Choice-Dilemma items probably reflects a systematic difference in the components of risk-taking behavior being assessed by these two measuring procedures. A lack of correspondence among presumably similar measures of the same behavior or construct is hardly novel in psychology. Investigators of learning once believed that a diverse set of response measures (resistance to extinction, latency, number of trials to criterion) were equivalent. However, subsequent research showed that these measures often correlated poorly with one another. This same problem may exist for the risky-shift literature.

---

[39]The exclusive reliance on the pretest-posttest design in research on the risky-shift is rather curious. The question as to whether or not groups make riskier decisions than individuals in no way requires a design in which members of groups must be the same persons who previously made the same decisions as individuals. Nor is the pretest necessarily indispensable to experimental method (Campbell & Stanley, 1963). Indeed, the alternative "after-only" design offers the additional advantages that (1) the obtained results do not depend upon the sensitizing effect of a pretest and (2) greater generality of the outcome to a nonpretested population of people.

Of course, the specific content of the Choice-Dilemma items is not the only source of difficulty with this material. Clark and Willems (1969) recently illustrated that the instructions accompanying the Choice-Dilemma items can strongly influence group decisions. They noted that the standard instructions characteristically employed with Choice-Dilemma items require subjects to indicate the *lowest* probability or riskiest odds they consider reasonable for a hypothetical person if he were to choose the risky option. Clark and Willems compared the effect of these risk-oriented instructions to neutral instructions which omitted the word "lowest" across four conditions: (1) a test-retest control condition, (2) information-exchange, (3) group discussion, and (4) a listening condition in which subjects heard recorded tapes of an interacting group. The result was an interaction between types of instructions and particular treatments. Specifically, risky-shifts occurred in the information exchange and group discussion situations provided that risk-oriented instructions were employed. Since the subjects in both instruction conditions had equivalent initial pretest scores, the differential effects of the risk-oriented instructions apparently produced their effect in interaction with the treatments.

The outcome above illustrates the kind of interaction that particularly worries those concerned about demand characteristics (*viz.*, a circumstance in which a standard feature of the experiment or the instructions interacts with only certain treatments and thereby produces different outcomes across treatments). Similar problems characterize the within-subject repeated measures or multiple treatments design that dominates the risky-shift research. Perhaps most important, these designs, while sometimes recommended for their sensitivity, create more experimental demand (Orne, 1962) than other types of designs (A. L. Edwards, 1963; Rosenblatt and Miller, 1970). When such designs are used, the subject must inevitably ask himself: "Why am I being measured again?" or "What is the point of this part of the experiment?" The subject's answer to this type of question probably is some interpretation of the intervening experimental treatment. Specifically, subjects in a treatment condition may think: "I am supposed to be influenced." On the other hand, those in a control condition might reason: "Maybe it is my consistency from one measurement to the next that is being assessed. I can create a better image by responding as identically as possible to the way I did the first time and prevent the experimenter from thinking I am wishy-washy." Of course, demand characteristics are less problematic if they are uniformly present in all conditions. The problem becomes critical when different treatments elicit different kinds or amounts of demand. *Post hoc* analysis of the data can never reveal an interaction between demand characteristics and treatments. Instead, an accurate assessment of the role played by demand in any given experiment requires either additional comparison groups or another experiment. Of course, demand is never tied solely to one particular type of design or one set of experimental materials. Any of several facets of an experimental situation can produce demand. For this reason, methodologists generally recommend replication of experiments in a nonstandardized manner as a means of determining the generalizability of any experimental finding.

C. OBLIQUE EXPERIMENTATION

A third problem with the research on group risk-taking is that researchers tend to construct either indirect or weak tests of their hypotheses. The failure to confront predictions directly erodes much of the true power of experimentation. By far, the most glaring disappointment in this respect is the absence of direct experimental analysis of actual group decision-making. Each interpretation of the risky-shift postulates the occurrence of different processes during group discussion. They interpret the risky-shift as mediated by different psychological and/or social psychological changes during group interaction. The implication is obvious: directly observe groups in the process of decision-making and directly assess the presence or absence of these processes and changes. Unfortunately, researchers never directly examine these processes (i.e., characteristics of group interaction) or if they do, they do not present their analysis of them. Instead, they impose some restriction upon discussion and then *assume* that this "manipulation" distorts or alters the essential process(es) occurring during discussion. If subsequent differences coincide with the expected outcomes, they summarily *assume* a successful manipulation of the crucial process without making a direct assessment. As a specific example, the direct observation and recording of events during discussion is clearly the simplest and most straightforward technique for testing the central proposition of the leadership explanation—that the riskiest group member on the premeasure exerts more influence. All we need are judges to rate participants on influence.

Apart from the omission of direct analysis of group process, we frequently had the feeling that a better procedure could have been used to test the experimental hypothesis under consideration. Most researchers agree that if one has a hypothesis to test, one should devise the strongest test possible. Obviously this does not mean using a design or set of materials which, through artifactual means or other experimental blunder, may enable one to obtain one's preferred outcome. What it does mean is that one should use procedures or materials that are most likely to detect a difference when one does indeed exist (in statistical terms this is referred to as "power"). Unfortunately, such considerations have not been adequately emphasized in the risky-shift research. We will present some examples of this below. We do so not to embarrass particular researchers for we do indeed write from the vantage point of retrospective wisdom. Rather, we hope to make explicit for the future that which can be learned from our collective experience in the past.

1. If one takes the stand that the diffusion of responsibility is critical (or at least important) for producing shifts toward risk, it seems logical to use materials or situations in which the outcome of the decisions has real consequences in terms of felt or perceived gains and losses for the decision-makers. From this standpoint, the use of Choice-Dilemma items as the primary arena for testing the responsibility-diffusion explanation seems a poor choice. These materials may best allow other processes besides responsibility-diffusion to affect the outcome. What difference can it make to any one subject whether or not he alone or other group members are responsible for the decision of the group when, in fact, whatever their decision, it hardly matters to anybody?

2. A similar argument can be made about the use of Choice-Dilemma items in studies concerning the familiarization hypothesis. In tests of the familiarization explanation, researchers typically use private study of the Choice-Dilemma items to induce greater comprehension. Indeed, some forms of the familiarization explanation view comprehension as the critical ingredient. This being the case, it makes little sense to use the Choice-Dilemma items since it is not clear what comprehension could mean in terms of these items. What is there to comprehend? The hypothetical situations do not have such a vast array of ingredients that one is pressed to weigh things properly within a short period of time. If one thinks that comprehension is in fact a critical element, one would want to use materials that allow one to assess readily the level of comprehension postexperimentally. This does not seem particularly possible with Choice-Dilemma items, and may explain why researchers are forced to rely on a response measure that has little to do with comprehension, namely, recall of the content of the items.[40]

3. The manipulation of affective bonds provides another illustration of weak hypothesis testing. For example, affective bonds have been manipulated by varying whether or not subjects are allowed to engage in

---

[40]Of course, in fairness, some of the familiarization research was done to test whether previously stated effects with Choice-Dilemma items could be explained via familiarization.

discussion. This seems to be attacking the problem from the wrong direction since the level of affective bonding produced within the interaction period of a single experiment cannot be very great. As a result, such procedures are not optimal since they seem incapable of producing the strong differences in affective feelings that one would like to create between experimental conditions so as to test the effect of this variable. The better procedure might be to use enduring groups, cliques, or families. Here one knows in advance that strong affective feelings exist.

D. MYOPIC SCHOLARSHIP

In reviewing the literature on the risky-shift, one is struck by the frequency with which researchers fail to see or discuss the relevance of their findings to theoretical explanations other than their own. Relatedly, in many cases, researchers could have easily added other dependent measures to test predictions derived from alternative interpretations of the risky-shift. In still other instances, the investigators regrettably failed to note that their data pertained to the competing predictions stemming from a different theoretical explanation. Citing specific instances to document each type of omission needlessly embarrasses us all. Perhaps the mere mention of these problems will be sufficiently corrective.

E. SUMMARY

This paper reviewed the four major explanations of the risky-shift as well as more recent theoretical trends in this research area. At this point, we will recapitulate our overall conclusions. (1) Although the initial data made the diffusion of responsibility explanation appear promising, subsequent empirical disconfirmations led its proponents to revise its underlying assumptions and corollary hypotheses continually — many of which are not necessarily relevant to the basic truth or falsity of a responsibility-diffusion explanation. Furthermore, the relevant disconfirmations may stem from the use of inappropriate tasks (i.e., hypothetical risk-taking situations). In other words, although data suggest that many of the corollary hypotheses associated with the responsibility-diffusion explanation may indeed be false, responsibility-diffusion itself may effectively explain risky-shifts in situations in which persons experience (or are led to anticipate) actual consequences from their decisions. (2) There are few "firm" data to support the leadership interpretation. Most of the supporting evidence consists of correlations between initial riskiness and participant's ratings of one another's relative influence during discussion. Unfortunately, such data do not permit causal inference. For example, they may reflect rationalizations of the fact that the raters themselves did become more risky (for other reasons). That is, after group discussion, participants may say to themselves "Since I became riskier, I must have been influenced, and I was probably influenced most by the guy who was initially riskiest". Thus as we have stressed, the leadership explanation and its underlying assumptions still await critical or direct tests. (3) The familiarization explanation appears to be a theoretical "blind-alley." Unsuccessful attempts to produce a risky-shift with familiarization procedures naturally cast doubt upon its validity. But more importantly, it now seems likely that the risk-inducing features of group discussion do not solely depend on greater familiarity with the materials. (4) As a *single* explanation of the risky-shift, the cultural-value interpretation fares best. It has a solid core of supporting evidence behind some of its basic assumptions. Furthermore, it served as a direct stimulus for recent attempts to develop more precise models of the risky-shift. Lastly, when we reach a complete understanding of group decision-making and risk-taking, it should not surprise us if propositions from several of the competing theoretical positions turn out to be true.

The ultimate significance of risky-shift research rests not with the finding that group decisions are riskier than individual decisions. Indeed, as Flanders (1970) strongly argues, from the standpoint of the dictionary definition or the man in the streets' understanding of the term risk, we may have learned little thus far about group risk-taking. Rather, the constructive stance takes the research on group risk-taking as a window through which one can view the processes of group interaction.

To take one example and elaborate on it a bit, consider the basic question of the effect of the group on individual judgments. As mentioned at the outset of this chapter, the position that groups exert a moderating influence on individual judgments held sway since the time of Allport (1924). In contradiction to this view, the research on the risky-shift suggests a quite different conclusion concerning the impact of group discussion on individual judgments and decisions—namely, that, under some conditions, groups engender greater extremity or polarization in judgments and decisions. Supporting this generalization, recent studies find that groups adopt more extreme positions after they consider political (Moscovici and Zavalloni, 1969) or racial attitudes (Myers & Bishop, 1970). Related to the polarization principle, advocates of the cultural-value interpretation suggest that group discussion enhances any underlying dominant value, e.g., intellectualism, generosity, and altruism (Brown, personal communication; Levinger & Schneider, 1969). Second, some research on the risky-shift (Bem *et al.*, 1965; Wallach *et al.*,

1964) reinforces the conclusion also emerging from such other research areas as bystander intervention — that groups are less responsible than individuals. Finally, some research on the responsibility diffusion interpretation (e.g. Bem *et al.*, 1965) of the risky-shift suggests that group decisions are less rational than those of individuals. In these respects, risky-shift research serves as a general model of group effects and suggests several interesting contrasts between groups and individuals.

Although we have frequently made research suggestions throughout this chapter, most of them consist of alternate procedures for testing hypotheses related to the various explanations of the risky-shift. Before concluding this chapter, it seems important to note that there are other directions in which research might be profitably directed. In particular, more attention might be focused on the characteristics of "real" or "naturally existing" groups. Indeed, the research on group risk-taking attracts the interest of those concerned with organizational psychology in that it is thought to comment on the decision-making process in real ongoing organizations. Yet interestingly, few of the actual characteristics of the decision making groups that exist in organizations are paralleled in those studied in the laboratory. For instance, real groups typically contain status differentiations. Few of the laboratory experiments on group risk-taking experimentally introduce status differences between group members. When groups with status differentiation confront a decision problem we might expect the temporal sequence in which individuals present their initial propensities to be correlated with status. That is to say, group members with high status might be expected to present their initial inclinations about the decision problem at an earlier point in the discussion than those of lower status. If so, this implies the possibility of making predictions about group decisions in a particular organization by combining (1) what we know about the influence of primacy in persuasion with (2) knowledge about the correlation of status and advocacy of a particular position on the dimension in question. Also, if high status persons are more prone to speak early in a discussion, it may lead other group members to associate certain values (other than those implicit in the decision problem itself) with particular decision alternatives about the problem. Another problem that needs to be considered concerns the extent to which sanctions are tied with status differentiation. Surely real groups must differ in the extent to which the status differences among group members also reflect fate control or behavioral control over other group members.

Other questions that have not been systematically studied focus on who the decision affects. Does the decision primarily have consequences for: some large impersonal entity (like a corporation) with which the decision-makers are not directly tied but which must nevertheless be responsive to the decision outcome; an impersonal entity in which the decision makers hold administrative or executive positions; the individual group members making the decision; or lastly, one individual member in the group? While we have commented on the fact that most of the laboratory experimentation on group decision-making uses a task that has no important consequences for the decision-making group, the question above focuses on the comparison or potential difference in outcome depending upon the target of the decision. Additional interesting questions arise from a consideration of surveillance and anonymity. The extent to which the group decision is public or subject to the surveillance of others should enhance the impact of such values as social responsibility. Lastly, none of the research on group decision making has considered the question of whether the group discussion redefines the end points of the scale implicitly used by the decision-makers. In sum, many of the structural and psychological dimensions along which real groups in ongoing organizations vary have yet to be studied. Although the research on group risk-taking is one of the most voluminous of any concerned with group behavior, it is clear that our knowledge of group decision-making remains very modest at this point.

REFERENCES

Allen, V. L. Situational factors in conformity. In L. Berkowitz (Ed.), *Advances in experimental social psychology*. Vol. 2. New York: Academic Press, 1965. Pp. 133-170.

Allport, H. *Social Psychology*. Boston: Houghton Mifflin, 1924.

Atkinson, J. W. (Ed.) *Motives in fantasy, action and society*. Princeton, N. J.: Van Nostrand, 1958.

Atkinson, J. W. *An introduction to motivation*. Princeton, N. J.: Van Nostrand, 1964.

Atkinson, J. W., & Litwin, G. H. Achievement motive and test anxiety conceived as motive to approach success and motive to avoid failure. *Journal of Abnormal and Social Psychology*, 1960, 60, 52-63.

Atthowe, J. M., Jr. Types of conflict and their resolution: A reinterpretation. *Journal of Experimental Psychology*, 1960, 59, 1-9.

Barnlund, D. C. A comparative study of individual majority and group judgment. *Journal of Abnormal and Social Psychology*, 1959, 58, 55-60.

Baron, R. S., Dion, K. L., & Baron, P. and Miller, N. Group norms, elicited values and risk-taking. Unpublished manuscript, University of Minnesota, 1970.

Bateson, N. Familiarization, group discussion, and risk taking. *Journal of Experimental Social Psychology*, 1966, 2, 119-129.

Bem, D. J., Wallach, M. A., & Kogan, N. Group decision-making under risk of aversive consequences. *Journal of Personality and Social Psychology*, 1965, 1, 453-460.

Blank, A. Effects of group and individual conditions on choice behavior. *Journal of Personality and Social Psychology*, 1968, 8, 294-298.

Brown, R. *Social psychology.* New York: Free Press, 1965.

Burns, J. F. An extremity-variance model of risk-taking. Unpublished doctoral dissertation, School of Industrial Management, Massachusetts Institute of Technology, 1967.

Burnstein, E. Decision-making and problem-solving in groups. Unpublished manuscript, University of Michigan, 1967.

Campbell, D. T., & Stanley, J. C. *Experimental and quasi-experimental designs for research.* Chicago: Rand McNally, 1963.

Carter, L., Haythorn, W., Meirowitz, B., & Lanzetta, J. The relation of categorizations and ratings in the observation of group behavior. *Human Relations*, 1951, 4, 239-254.

Clark, R., III, & Willems, E. P. Where is the risky shift? Dependence on instructions? *Journal of Personality and Social Psychology*, 1969, 13, 215-221.

Clausen, G. Risk taking in small groups. Unpublished doctoral dissertation, University of Michigan, 1965.

Cohen, A. R. *Attitude change and social influence.* New York: Basic Books, 1964.

Collins, B. E., & Guetzkow, H. *A social psychology of group processes for decision-making.* New York: Wiley, 1964.

Crowne, D. P. & Marlowe, D. *The approval motive: Studies in evaluative dependence.* New York: Wiley, 1964.

Darley, J. M., & Latané, B. Bystander intervention in emergencies: Diffusion of responsibility. *Journal of Personality and Social Psychology*, 1968, 9, 142-146.

Dion, K. L., Miller, N., & Magnan, M. A. Group cohesiveness and social responsibility as determinants of the risky-shift. Paper presented at the convention of the American Psychological Association, Miami, Florida, September, 1970.

Edwards, A. L. *Experimental design in psychological research.* New York: Holt, Rinehart & Winston, 1963.

Edwards, W. Probability preferences in gambling. *American Journal of Psychology*. 1953, 66, 349-364.

Ferguson, D. A. & Vidmar, N. Familiarization-induced risky and cautious shifts: A replication of sorts. Paper presented at the Midwestern Psychological Association Convention: Cincinnati, Ohio, 1970.

Festinger, L. A theory of social comparison processes. *Human Relations*, 1954, 7, 117-140.

Festinger, L., Pepitone, A., & Newcomb, T. M. Some consequences of deindividuation in a group. *Journal of Abnormal and Social Psychology*, 1952, 47, 382-389.

Flanders, J. P. Research on the risky-shift: Questions asked and unasked. Unpublished manuscript. Walter Reed Army Institute of Research, 1970.

Flanders, J. P. & Thistlethwaite, D. L. Effects of familiarization and group discussion upon risk-taking. *Journal of Personality and Social Psychology*, 1967, 5, 91-97.

Gibb, J. R. The effects of group size and of threat reduction upon creativity in a problem-solving situation. *American Psychologist*, 1951, 6, 324. (Abstract)

Hermann, M., & Kogan, N. Negotiation in leader and delegate groups. *Journal of Conflict Resolution*, 1968, 12, 332-344.

Hinds, W. C. Individual and group decisions in gambling situations. Unpublished master's thesis, School of Industrial Management, Massachusetts Institute of Technology, 1962.

Hovland, C. I., & Janis, I. L. (Eds.) *Personality and persuasibility.* New Haven: Yale University Press, 1959.

Hoyt, G. C., & Stoner, J. A. F. Leadership and group decisions involving risk. *Journal of Experimental Social Psychology*, 1968, 4, 275-285.

Hubbard, J. H. Effects of uncertainty on individual and group risk-taking. Unpublished master's thesis, School of Industrial Management, Massachusetts Institute of Technology, 1963.

Hunt, E. B., & Rowe, R. R. Group and individual economic decision making in risk conditions. In D. W. Taylor (Ed.), *Experiments on decision making and other studies*. Arlington, Va.: Armed Services Technical Information Agency, 1960. Pp. 21-26.

Jellison, J. M., & Riskind, J. A social comparison of abilities interpretation of risk taking behavior. *Journal of Personality and Social Psychology*, 1970, 15, 375-390.

Jones, E. E. & Gerard, H. B. *Foundations of social psychology*. New York: Wiley, 1967.

Kelley, H. H., & Thibaut, J. W. Experimental studies of group problem solving and process. In G. Lindzey (Ed.), *Handbook of social psychology*. Vol. 2. Reading, Mass.: Addison-Wesley, 1954. Pp. 735-785.

Kelley, H. H., & Thibaut, J. W. Group problem-solving. In G. Lindzey & E. Aronson (Eds.), *Handbook of social psychology*. (Rev. ed.) Vol. 4. Cambridge, Mass.: Addison-Wesley, 1968. Pp. 1-104.

King, B. T., & Janis, I. L. Comparison of the effectiveness of improvised versus non-improvised role-playing in producing opinion change. *Human Relations*, 1956, 9, 177-186.

Kogan, N., & Carlson, J. Group risk taking under competitive and noncompetitive conditions in adults and children. *Journal of Educational Psychology*, 1969, 60, 158-167.

Kogan, N., & Doise, W. Effects of anticipated delegate status on level of risk-taking in small decision-making groups. *Acta Psychologica*, 1969, 29, 228-243.

Kogan, N., & Wallach, M. A. The effect of anxiety on relations between subjective age and caution in an older sample. In P. Hoch & J. Zubin (Eds.), *Psychology of aging*. New York: Grune & Stratton, 1961. Pp. 123-135.

Kogan, N., & Wallach, M. A. *Risk-taking: A study in cognition and personality*. New York: Holt, 1964.

Kogan, N., & Wallach, M. A. Effects of physical separation of group members upon group risk taking. *Human Relations*, 1967, 20, 41-49. (a)

Kogan, N., & Wallach, M. A. Group risk taking as a function of members' anxiety and defensiveness levels. *Journal of Personality*, 1967, 35, 50-63. (b)

Kogan, N., & Wallach, M. A. Risk taking as a function of the situation, the person and the group. In G. Mandler, P. Mussen, N. Kogan & M. A. Wallach (Eds.), *New directions in psychology*. Vol. III, New York: Holt, Rinehart & Winston, 1967. Pp. 224-266. (c)

Kogan, N., & Wallach, M. A. The risky-shift phenomenon in small decision-making groups: A test of the information-exchange hypothesis. *Journal of Experimental Social Psychology*, 1967, 3, 75-85. (d)

Lamm, H. Will an observer advise high risk taking after hearing a discussion of the decision problem? *Journal of Personality and Social Psychology*, 1967, 6, 467-471.

Lamm, H., & Kogan, N. Risk-taking in the context of intergroup negotiation. *Journal of Experimental Social Psychology*, 1970, 6, 351-363.

Levinger, G., & Schneider, D. J. Test of the "Risk is a value" hypothesis. *Journal of Personality and Social Psychology*, 1969, 11, 165-170.

Levitt, E. A. The relationship between abilities to express emotional meanings vocally and facially. In J. R. Davitz (Ed.), *The communication of emotional meaning*. New York: McGraw-Hill, 1964. Pp. 87-100.

Levy, P. K. The ability to express and perceive vocal communication of feeling. In J. R. Davitz (Ed.), *The communication of emotional meaning*. New York: McGraw-Hill, 1964. Pp. 43-55.

London, H., Meldman, P. J. & Lanckton, A. Van C. The jury method: How the persuader persuades. *Public Opinion Quarterly*, 1970, 34, 171-183.

Lonergran, B. G. & McClintock, C. G. Effects of group membership on risk-taking behavior. *Psychological Reports*, 1961, 8, 447-455.

Mackenzie, K. D., & Bernhardt, I. The effect of status upon group risk taking. Unpublished manuscript, Wharton School of Finance and Commerce, University of Pennsylvania, 1968.

Madaras, G. R., & Bem, D. J. Risk and conservatism in group decision making. *Journal of Experimental Social Psychology*, 1968, 4, 350-366.

Marquis, D. G. Individual responsibility and group decisions involving risk. *Industrial Management Review*, 1962, 3, 8-23.

Marquis, D. G. Individual and group decisions involving risk. *Industrial Management Review*, 1968, 9, 69-76.

Marquis, G., & Reitz, H. J. Uncertainty and risk taking in individual and group decisions. *Behavioral Science*, 1969, 14, 281-288.

McGuire, W. J. Personality and susceptibility to social influence. In E. F. Borgatta & W. W. Lambert (Eds.), *Handbook of personality theory and research*. Chicago: Rand McNally, 1968. Pp. 1130-1187.

Messick, S., & Damarin, F. Cognitive styles and memory for faces. *Journal of Abnormal and Social Psychology*, 1964, 69, 313-318.

Miller, N. Involvement and dogmatism as inhibitors of attitude change. *Journal of Experimental Social Psychology*, 1965, 1, 121-132.

Miller, N. and Dion, K. L. An analysis of the familiarization explanation of the risky-shift. Paper presented at the convention of the American Psychological Association, Miami, Florida, September, 1970.

Moscovici, S., & Zavalloni, M. The group as a polarizer of attitudes. *Journal of Personality and Social Psychology*, 1969, 12, 125-135.

Myers, D. G. Enhancement of initial risk-taking tendencies in social situations. Unpublished doctoral dissertation, University of Iowa, 1967.

Myers, D. G., & Bishop, G. Discussion effects on racial attitudes. Paper presented at the convention of Midwestern Psychological Association, Cincinnati, Ohio, 1970.

Nordhøy, F. Group interaction in decision-making under risk. Unpublished master's thesis, School of Industrial Management, Massachusetts Institute of Technology, 1962.

Orne, M. T. On the social psychology of the psychological experiment: With particular reference to demand characteristics and their implications. *American Psychologist*. 1962, 17, 776-783.

Pilkonis, P. A., & Zanna, M. P. The choice-shift phenomenon in groups: Replication and extension. Unpublished manuscript, Yale University, 1969.

Pruitt, D. G. The "Walter Mitty' effect in individual and group risk-taking. Paper presented at the convention of the American Psychological Association, Washington, D. C., September 1969.

Pruitt, D. G., & Teger, A. I. Is there a shift toward risk in group discussion? If so, is it a group phenomenon? If so, what causes it? Paper presented at the convention of the American Psychological Association, Washington, D. C., September 1967.

Pruitt, D. G., & Teger, A. I. The risky shift in group betting. *Journal of Experimental Social Psychology*, 1969, 5 115-126.

Rabow, J., Fowler, F. J., Jr., Bradford, D. L., Hofeller, M. A., & Shibuya, Y. The role of social norms and leadership in risk-taking. *Sociometry*, 1966, 29, 16-27.

Rettig, S. Group discussion and predicted ethical risk taking. *Journal of Personality and Social Psychology*, 1966, 3, 629-633.

Rim, Y. Risk-taking and need for achievement. *Acta Psychologica*, 1963, 21, 108-115.

Rim, Y. Interpersonal values and risk-taking. Paper presented at the First International Congress of Psychiatry, London, 1964. (a)

Rim, Y. Intolerance of ambiguity and risk-taking. *Revue Suisse de Psychologie et de Psychologie Apliquée*, 1964, **23**, 253-259. (b)

Rim, Y. Personality and group decisions involving risk. *Psychological Record*, 1964, **14**, 37-45. (c)

Rim, Y. Leadership attitudes and decisions involving risk. *Personnel Psychology*, 1965, **18**, 423-430. (a)

Rim, Y. Dimensions of interpersonal behavior and risk-taking. *Revista de Psicologia General of Applicada*, 1965. (b)

Rim, Y. Machiavellianism and decisions involving risk. *British Journal of Social and Clinical Psychology*, 1966, **5**, 30-36.

Rogers, C. R. A theory of therapy, personality, and interpersonal relationships as developed in the client-centered framework. In S. Koch (Ed.), Vol. 3. *Psychology: A study of a science*. New York: McGraw-Hill, 1959. Pp. 184-256.

Rosenblatt, P. C. & Miller, N. Experimental methods. In C. G. McClintock (Ed.), *Experimental social psychology*. New York: Holt, Rinehart and Winston, 1970.

Schachter, S. Deviation, rejection and communication. *Journal of Abnormal and Social Psychology*, 1951, **46**, 190-207.

Schachter, S. *The psychology of affiliation: Experimental studies of the sources of gregariousness*. Stanford, California: Stanford University Press, 1959.

Secord, P. F., & Bachman, C. W. *Social psychology*. New York: McGraw-Hill, 1964.

Sherif, C. W., Sherif, M., & Nebergall, R. E. *Attitude and attitude change; the social judgment-involvement approach*. Philadelphia: Saunders, 1965.

Sherif, M., & Hovland, C. I. *Social judgment: Assimilation and contrast effects in communication and attitude change*. New Haven: Yale University Press, 1961.

Siegel, S., & Zajonc, R. B. Group risk-taking in professional decisions. *Sociometry, 1967*, **30**, 339-350.

Singer, J. E., Brush, C. A., & Lublin, S. C. Some aspects of deindividuation: Identification and conformity. *Journal of Experimental Social Psychology*, 1965, **1**, 356-378.

Stoner, J. A. F. A comparison of individual and group decisions involving risk. Unpublished master's thesis, School of Industrial Management, Massachusetts Institute of Technology, 1961.

Stoner, J. A. F. Risky and cautious shifts in group decisions: The influence of widely held values. *Journal of Experimental Social Psychology*, 1968, **4**, 442-459.

Suchman, E. A. The intensity component in attitude and opinion research. In S. A. Stouffer et al. (Eds.), *Studies in social psychology in World War II*. Vol. 4. *Measurement and prediction*. Princeton: Princeton University Press, 1950. Pp. 213-276.

Teger, A. I., & Pruitt, D. G. Components of group risk taking. *Journal of Experimental Social Psychology*, 1967, **3**, 189-205.

Teger, A. I., Pruitt, D. G., St. Jean, R., & Haaland, G. A re-examination of the familiarization hypothesis in group risk-taking. *Journal of Experimental Social Psychology*, 1970, **6**, 346-350.

Thibaut, J. W., & Kelley, H. H. *The social psychology of groups*. New York: Wiley, 1959.

Vidmar, N. Group composition and the risky-shift. *Journal of Experimental Social Psychology*, 1970, **6**, 153-166.

Wallach, M. A., & Kogan, N. Aspects of judgment and decision making: Interrelationships and changes with age. *Behavioral Science*, 1961, **6**, 23-36.

Wallach, M. A., & Kogan, N. The roles of information, discussion, and consensus in group risk taking. *Journal of Experimental Social Psychology*, 1965, **1**, 1-19.

Wallach, M. A., Kogan, N., & Bem, D. J. Group influence on individual risk taking. *Journal of Abnormal and Social Psychology*, 1962, **65**, 75-86.

Wallach, M. A., Kogan, N., & Bem, D. J. Diffusion of responsibility and level of risk taking in groups. *Journal of Abnormal and Social Psychology,* 1964, **68**, 263-274.

Wallach, M. A., Kogan, N., & Burt, R. Can group members recognize the effects of group discussion upon risk taking? *Journal of Experimental Social Psychology,* 1965, **1**, 379-395.

Wallach, M. A., Kogan, N., & Burt, R. Group risk taking and field dependence-independence of group members. *Sociometry,* 1967, **30**, 323-339.

Wallach, M. A., Kogan, N., & Burt, R. Are risk takers more persuasive than conservatives in group decisions? *Journal of Experimental Social Psychology,* 1968, **4**, 76-89.

Wallach, M. A., & Wing, C. W., Jr. Is risk a value? *Journal of Personality and Social Psychology,* 1968, **9**, 101-107.

White, R. W. *The abnormal personality.* New York: Ronald Press, 1964.

Whyte, W. H., Jr. *The organization man.* New York: Simon & Schuster, 1956.

Witkin, H. A., Dyk, R. B., Faterson, H. F., Goodenough, D. R., & Karp, L. A. *Psychological differentiation.* New York: Wiley, 1962.

Witkin, H. A., Lewis, H. B., Hertzman, M., Machover, K., Meissner, P. B., & Wapner, S. *Personality through perception.* New York: Harper & Row, 1954.

Zajonc, R. B., Wolosin, R. J., Wolosin, M. A., & Loh, W. D. Social facilitation and imitation in group risk taking. *Journal of Experimental Social Psychology,* 1970, **6**, 26-46

Zajonc, R. B., Wolosin, R. J., Wolosin, M. A., & Sherman, S. J. Individual and group risk taking in a two-choice situation. *Journal of Experimental Social Psychology,* 1968, **4**, 89-107.

Zajonc, R. B., Wolosin, R. J., Wolosin, M. A., & Sherman, S. J. Group risk taking in a two-choice situation: Replication, extension, and a model. *Journal of Experimental Social Psychology,* 1969, **5**, 127-140.

Zimbardo, P. G. The human choice: Individuation, reason, and order versus deindividuation, impulse, and chaos. In W. J. Arnold and D. Levine (Eds.), *Nebraska symposium on motivation.* Lincoln, Nebraska: University of Nebraska Press, 1969. Pp. 237-308.

A Questionnaire in Search of a Theory[1]

Norman Miller

UNIVERSITY OF SOUTHERN
CALIFORNIA

When we first reviewed the literature on group decision making and the risky-shift, we focused on four alternative explanations. Our analysis, as well as most of the studies that we reviewed, primarily concerned the Choice-Dilemma Questionnaire (CDQ) developed by Kogan and Wallach (1964). This instrument asks subjects to make a choice between two hypothetical decision alternatives, a less desirable alternative with a high likelihood of success and a more desirable alternative with a lower likelihood of success. The basic finding, termed the "risky-shift," was that individuals (or groups) tended to prefer more risky alternatives after they had discussed the decision alternatives with others. Among the four major explanations we examined—diffusion of responsibility, leadership or rhetoric-of-risk, familiarization, and cultural value—we concluded that the last explanation fared best.

Almost a decade has passed since we reviewed this literature. At that time we concluded that there was certainly a body of facts to be explained but that they may well have had relatively little to do with group risk-taking. Instead, we suggested that there was greater profit in thinking of this research "as a window through which one can view the process of group interaction [p. 292, this volume]." As we shall see, this continues to be the major value of the research on group decision

[1] The author would like to thank Linda Collins for assistance in the preparation of this chapter.

301

making; more specifically, it provides a paradigm for studying social influence processes within group settings.

In recent years a number of other fine reviews have also been published (Cartwright, 1971; Cartwright, 1973; Clark, 1971; Myers & Lamm, 1976; Pruitt, 1971a,b; Sanders & Baron, 1977; Vinokur, 1971). Cartwright has pointed out that groups are not invariably more risky than individuals on CDQ items. Instead, the effects of group discussion depend on the particular content of the items and the distribution of initial choices within the group. An analysis of the effect of group discussion on individual items in the CDQ reveals considerable variation. For some items, group discussion leads to a tolerance for a 10% to 15% reduction in the likelihood that the risky alternative will be successful, that is, to an increased tolerance for risk. For other items the tolerance for risk is only increased by about 6% or 7%, and, for still others, group discussion *decreases* the tolerance for the risky choice. In consonance with this last finding, other researchers have been able to successfully construct items that consistently yield conservative shifts after group discussion (Nordhøy, 1962; Rabow *et al.*, 1966; Stoner, 1968).

Cartwright also points out that the average magnitude of shift per item across the 12 contained in the CDQ is roughly from a tolerance for 6 in 10 to one of 5 in 10 chances for success. Thus he argues that the shift is not large. Perhaps more importantly, he notes that existing research provides little information about the way subjects perceive the riskiness of choices, their initial levels of risk, or their assessment of the values of outcomes. In other words, group discussion can produce shifts by changing the perceived riskiness of the choices, by altering the ideal level of risk, or by modifying the value of the outcome implied by each of the decision alternatives. Which of these processes occurs, or whether they all occur, still remains unknown.

Beyond these problems, Cartwright raises the issue of how to best conceptualize the process by which group discussion alters prior individual decisions. Subsequent reviews almost uniformly call attention to two major processes whereby group discussion might alter prior individual decisions: (*a*) the presentation or knowledge of others' positions and the social comparison of one's own opinion with that of others; and (*b*) the presentation of new information or arguments concerning the decision alternatives (Baron & Sanders, 1977; Clark, 1971; Myers & Lamm, 1976; Pruitt, 1971a,b; Vinokur, 1971). The first emphasizes normative influence processes, whereas the second, by pointing instead to information beyond that implicit in the positions of others, emphasizes persuasion processes. Recent empirical work has also tended to focus on the role of one or the other of these two mechanisms in producing group influence. Although our original article discussed each of these two processes, the emphasis and elaboration on them in recent empirical work makes it important to consider them in further detail.

I. The Social Comparison Explanation

The roots of the first explanation lie in Festinger's social comparison theory (Festinger, 1954). In order to predict the extremity shifts that seem to follow from group discussion, however, the theory must be modified to include the notion that the poles of attitudinal and belief dimensions, like those of ability dimensions, can often be clearly labeled as positive and negative. Indeed, this notion forms the core of Brown's (1965; 1974) cultural value explanation of the risky-shifts that occur on CDQ items after group discussion; namely, for attitudes about the dimension "riskiness," risky is typically the good end, and conservatism or caution the unfavorable end, of the continuum. If attitude dimensions, like abilities, are often evaluatively unambiguous, social comparison might operate in a number of ways to produce group polarization effects (extremity shifts).

In one account, a person may value a particular direction of opinion but, for fear of being labeled extreme, deviate, or unreasonable, may express a more moderate position than that which he personally prefers. Group discussion, however, reveals to him that some other group members espouse positions closer to his personal ideal or even more extreme than that he had expressed on the premeasure. The concomitant release from fear of negative evaluation enables such a person to subsequently agree with and express more extreme positions. Since this process presumably operates in varying degrees for several of the group members, it thereby produces the group polarization effect (Levinger & Schneider, 1969).

An interpretation with only a slightly different emphasis stresses a more positive aspect of the individual's behavior; it points to concerns about the positive aspect of impression management or self-presentation, rather than the "release" from fears. It sees group members as competing to express more extreme views, as vying with one another to come closer to expressing the more admired (extreme) position (Jellison & Arkin, 1977). Initially, Jellison argued that the evaluative anchoring of the "risk" dimension is mediated by perceptions of ability (Jellison & Riskind, 1970). In their view, observers attribute greater competence and ability to those who advocate a more risky course of action. This interpretation converts the attitudinal dimension to a unidirectional evaluative dimension as a consequence of its attributional implications for ability. It thereby maintains to some degree Festinger's original distinction between opinions, which supposedly lacked clear evaluative anchors, and abilities, which clearly possessed them.

Alternatively, as argued earlier, the attitude dimension itself may directly possess an evaluative component. From this latter standpoint, since ability is a dimension with clear evaluative poles, the attribution of ability to those who express an extreme attitude may be a generalization of an evaluative judgment made directly on the basis of the person's attitudinal stand. Though Jellison and Riskind prefer the first interpretation, these subtleties of interpretation have not been clearly resolved. Regardless of which of these interpretative nuances will ultimately be shown to be more accurate, either one fits comfortably with the view of Jellison and Arkin (1977) that the fulfillment of self-presentational concerns underlies group polarization effects.

II. The Persuasion Explanation

The persuasion explanation argues that mere knowledge of others' positions per se is not the critical ingredient for group polarization effects. Instead, it is the information that is exchanged during the course of the discussion that plays the critical role. According to this interpretation, the group discussion produces a preponderance of arguments that favor a more extreme position. Though individual group members may have been aware of some of these supporting elements, most were not aware of all of them. Thus, the net effect is a shift toward greater extremity as a result of the new persuasive information to which group members are exposed. Of course, the persuasive-arguments explanation must also account for why it is that most of the discussion contains information that supports a particular pole of the attitudinal dimension, specifically, in the case of the CDQ items, the risky pole. Ultimately, then, this explanation must introduce some cultural value notion to account for the direction of the preponderance of persuasive arguments. Vinokur (1971) and Burnstein and his other co-workers have favored this interpretation as a complete explanation (Burnstein & Vinokur, 1973; Burnstein & Vinokur, 1975; Burnstein, Vinokur, & Pichevin, 1974; Burnstein, Vinokur & Trope, 1973), although numerous other researchers who do not endorse this explanation exclusively have also provided additional support (e.g., Clark, Crockett, & Archer, 1971; Myers, Wong, & Murdoch, 1971).

III. Why Neither Social Comparison Nor Persuasion
Is a Complete Explanation

In concluding our earlier review we raised a number of criticisms about prior research. First among them was "theoretical chauvinism," the tendency to argue that only a single explanation accounts for the observed effect. Today, it is both perplexing and discouraging to find the same tendency in action among those in this reasearch area. Researchers continue to act as if the discovery of empirical support for their own preferred hypothesis must mean that another alternate hypothesis, though distinct from their own and neither logically nor psychologically competitive with it, is nevertheless untenable. Interestingly, advocates of both of the major positions outlined above have taken such stands. Burnstein, on the one hand, has argued that the persuasive argument explanation is the necessary and sufficient explanation. In contrast, Jellison and Arkin argue in a similar fashion that social comparison for the purpose of managing or manipulating one's image is the necessary and sufficient explanation of group polarization effects, as well as of numerous other social psychological findings. It will be instructive to examine each of these positions in more careful detail.

In attempting to demonstrate that social comparison is neither necessary nor sufficient to explain group polarization effects, Burnstein and his collaborators have tried to confirm two points: First, that predictable shifts occur when group members do not know one another's actual positions and therefore cannot engage in any social comparison; and, second, that if the development and presentation of persuasive arguments within the group is somehow prevented, mere knowledge of others' positions does not by itself produce polarization effects. Sanders and Baron (1977), in a critical analysis, show quite clearly that neither of these two propositions finds support. First, they present evidence indicating that when researchers arrange for group members to discuss CDQ items either without presenting their own position (Clark, 1971) or under circumstances designed to make it difficult to correctly infer others' positions (Burnstein & Vinokur, 1973), group members nevertheless continue to make inferences about others' positions and use these inferences as bases for social comparisons.

In one seemingly ingenious attempt to prevent group members from knowing one another's positions and thereby to rule out any possibility of comparison, Burnstein and Vinokur (1973, Study 1) told subjects that some group members would be arguing for and stating their own positions, but others, "in order to ensure a good discussion that would cover an array of positions," would be defending positions assigned to them. Actually, in one condition each group member was assigned his own position to advocate, whereas in another each was assigned a position other than his own. Since shifts on the CDQ were obtained in the first condition, and since subjects in this condition (as well as in the second) could not be sure of others' true positions, Burnstein and Vinokur concluded that social comparison was not necessary for the polarization effect.

Sanders and Baron (1977) point out, however, that certainty about others' true positions is not a critical ingredient in order for social comparison to occur. Group members might still make inferences about others' positions and change their own responses as a result of these inferences. Burnstein and Vinokur's own data suggest that this was the case. Only 14% of the group members responded "do not know" when later asked to make inferences about the positions of the other group members. When Sanders and Baron replicated Burnstein and Vinokur's procedure, they found that, on the average, subjects in this first condition felt that two among the four group members were "truth tellers." More importantly, only on the CDQ item for which the positions of the persons designated as "truth tellers" were more extreme than the subject's own opinion did significant shifts occur. Contrary to the persuasive argument explanation and in accord with a comparison interpretation, this suggests that group members not only make inferences about others' positions but also relied on them as a basis for social comparisons that in turn triggered their own shifts toward a more risky position.

As Sanders and Baron note, however, the critical uncertainty surrounding Burnstein's position concerns whether comparison per se of others' positions in the absence of any subsequent exposure to persuasive information and discussion that supports those positions is sufficient to produce group shifts. Apart from their arguments showing that Burnstein's subsequent research (Burnstein & Vinokur, 1975; Burnstein, Vinokur, & Pichevin, 1974; Burnstein, Vinokur, & Trope, 1973) does not test and refute the sufficiency of social comparison, they show how other research does demonstrate this sufficiency. In the autokinetic situation, subjects in a very dark room give estimates of the distance a pinpoint of light travels. When these subjects are led to believe that large estimates imply valued traits, knowledge of others' estimates increases the magnitude of their own (Baron & Roger, 1976). Here subjects partook in no group discussion. Blascovich and his co-workers (Blascovich & Ginsberg, 1974; Blascovich, Ginsburg, & Veach, 1975; Blascovich, Veach, & Ginsburg, 1973) report similar findings on blackjack gambling; knowledge of others' bets increased the subject's own bets. Furthermore, the opportunity for discussion did not add to the effect. If the social comparison made possible from mere knowledge of others' distance estimates or size of bets produced shifts in the direction of the new information, it seems very plausible that the same kind of information—knowledge of others' positions—will produce shifts on the CDQ.

Turning to the other extreme, Jellison's viewpoint is that knowledge of others' position per se is the *only* relevant ingredient for polarization effects. Other information is thought to be irrelevant. All that operates in the group setting is normative influence. The sole purpose of this responsiveness to group norms is to gain others' approval and acceptance. According to this view, knowledge of others' positions forms the foundation for an inference about how to behave in order to maximize social rewards.

Myers and Lamm (1976), in their comprehensive and balanced review of group polarization effects, contend that there is a core of information available arguing that supporting arguments do play some role. Of particular interest are analyses of the details of group discussion that examine the distribution of individual positions prior to discussion and the nature of the arguments that are expressed. First, the direction of arguments elicited by a CDQ item predicts the shift that is obtained (correlations are about .9). The initial mean predicts the prediscussion and discussion arguments that can be elicited from group members, and these in turn accurately predict the mean postdiscussion shift. "The predictive power of the initial response mean is thus captured almost entirely by the content of the subsequent discussion, suggesting that it is that nature of the expressed arguments which mediates the relationship between initial mean and mean shift [p. 616]." Furthermore, novel arguments have more impact than do unoriginal arguments.

Of course, Jellison and Arkin can argue that the only important function of number, direction, quality, and/or originality of arguments is to further clarify and accurately specify the actual positions of the group members and thereby to facilitate social comparison. There are two important problems, however, for Jellison's position. To lay the groundwork for these problems, it must first be made clear that the self-presentation interpretation argues against the concept of attitude. In its extreme form, it sees all behavior as immediate instrumental action designed to maximize social rewards in the existing situation. Viewing all behavior as responsive to situational pressures and reward structures, the self-presentation interpretation finds no role for stored residues of prior experience (attitudes, learning, values, etc.). If this is indeed the case, one must now ask why it is that particular CDQ items so reliably elicit the initial and postdiscussion levels and direction of endorsement that they do. Why do some items consistently elicit cautious shifts, whereas others reliably yield risky shifts? These elementary and consistent findings argue that to whatever extent social comparison and compliance effects do occur, to whatever extent people do try to maximize social rewards, there are nevertheless values or attitudes elicited by the individual CDQ items that exert their own contribution apart from self-presentational concerns.

A second major problem for the view that self-presentation is the sole and complete explanation arises from the literature on recidivism—the tendency for individuals to revert toward their initial positions after leaving the group context. The Barons and their associates (Baron, Baron, & Roper, 1974; Baron, Monson, & Baron, 1973) show that the lack of recidivism in earlier research with the CDQ stems from methodological characteristics of the experimental designs that were used, namely, pretest–posttest designs. When they eliminated the pretest, significant recidivism was found. Likewise, when anonymity was guaranteed on the posttest, recidivism was strongest (Baron *et al.*, 1974).

Jellison and Arkin (1977) and Arkin (1976) correctly point out that these results and others (Paicheler & Bouchet, 1973; Blank, 1968) support the self-presentation or compliance view, namely, that subjects in the group setting are responding to the normative position of the group. Although they can be viewed as also supporting the self-presentation explanation, they cannot support the more extreme notion that self-presentation is the sole and complete explanation of group polarization effects and that attitudes or values are superfluous and irrelevant. In order to buttress this latter view, Jellison and Arkin would have to show that recidivism is complete. In other words, in the absence of the group and with the guarantee of anonymity the self-presentation interpretation should predict *no* residual effects of the group experience. According to the self-presentational view, with no persons to observe one's response and with no potential social rewards available, there is no reason to advocate a position other than one's initial stand. This should be particularly true if persons in fact have no "true" new attitude about the issue. They should return completely to their premeasure position. The very research that demonstrates recidivism effects, however, does not uniformly support this latter requirement. Indeed, Arkin (1976), although he adopts this strong self-presentational position, fails even to test for incomplete recidivism.

Although the preceding discussion aimed to show that neither the social comparison nor the persuasive argument explanation can by itself account for existing data, it should not be inferred that either of them are mistaken hypotheses. Instead, it is most likely that both operate in concert within the group discussion setting. As we noted when we first reviewed that literature, Brown (1965) incorporated both of these elements in his original discussion. Other more recent theoretical treatments also offer more balanced presentations (Myers & Lamm, 1976; Sanders & Baron, 1977). Finally, both hypotheses have long been recognized as components of the social influence process (Deutsch & Gerard, 1955). Sanders and Baron correctly put the supposed conflict between the two interpretations into its broader perspective by recasting it in terms of Jones and Gerard's (1967) later discussion of the conformity process, in which they draw a distinction between effect dependence and information dependence. In their model of conformity behavior, Jones and Gerard see one influence of the group over its members as stemming from individuals' desires to receive approval and avoid rejection from other group members. They achieve this by adopting positions that are normative within the group. This notion obviously corresponds closely to the function of social comparison within the group setting, as presented in the preceding discussion. The substantial evidence that exists for effect dependence makes it difficult to imagine that it does not operate in the context of group discussions of CDQ items as well as in the myriad of other situations in which its operation has been observed. Information dependence, on the other hand, parallels the persuasive arguments explanation. It too emphasizes the role of supporting information in producing conformity among those group members unsure of the "correct" position or the proper decision.

Our texts and experimental design courses tend to stress the discovery of simple cause and effect relations, and this emphasis seems to occlude the fact that the independent and dependent variables that we study in laboratory isolation more typically operate as parts of complex interdependent systems in which there are intricate circular feedback connections. Consequently, what in one instance is a dependent effect is in the next a causal impetus. In group discussions, the relation between aspects of effect dependence and information dependence undoubtedly exemplifies these system properties.

IV. Conclusion

Now, more than a decade and a half since the first research using the CDQ, our own initial view, that the major value of the research it generated lay in our increased understanding of group processes, seems even more strongly upheld. Myers and Lamm echo this view (1976, p. 610). Despite Cartwright's astute analysis, there has been little additional work on the meaning of risk within the context of research on group decision. Indeed, it seems likely that the vast array of studies using the CDQ have little, if anything, to say about risk-taking. In large part this reflects the hypothetical nature of the CDQ items, but it also reflects problems of generalizability as well as inadequate definition and analysis of the meaning of risk. A tendency to ignore related literatures also continues to characterize recent as well as earlier discussions of group decision making as concerns the CDQ items. For instance, the research and theorizing on coalition formation and bargaining seem relevant to analysis of the effect of different distributions of initial positions on the CDQ items among the members of the group. Similarly, they seem relevant to discussions of group decision rules or schemes. Other literatures, such as jury decision research, should seemingly invoke the same group processes as the group risk-taking situation. Finally, the most positive aspect of recent work and perhaps most important among the theoretical contributions of this research paradigm is the impetus it has provided for a renewed interest in Festinger's social comparison theory and the need to reconsider and modify some of its propositions (e.g., Jellison & Arkin, 1977; Baron *et al.*, 1977).

REFERENCES

Arkin, R. M. Self presentation: The effects of anticipated approval and public commitment on attitudes. Unpublished doctoral dissertation, University of Southern California, 1976.

Baron, P. H., Baron, R. S., & Roper, G. External validity and the risky shift: Empirical limits and theoretical implications. *Journal of Personality and Social Psychology*, 1974, **30**, 95–103.

Baron, R. S., Monson, T. C., & Baron, P. H. Conformity pressure as a determinant of risk taking: Replication and extention. *Journal of Personality and Social Psychology*, 1973, **28**, 406–413.

Baron, R. S., & Roper, G. Reaffirmations of social comparison views of choice shifts: Averaging and extremity effects in an autokinetic situation. *Journal of Personality and Social Psychology*, 1976, **33**, 521–530.

Baron, R. S., Roper, G., & Baron, P. H. Group discussion and the stingy shift. *Journal of Personality and Social Psychology*, 1974, **30**, 538–545.

Baron, R. S., Sanders, G. S., & Baron, P. H. Social comparison reconceptualized: Implications for choice shifts, averaging effects and social facilitation. Unpublished manuscript, University of Iowa, 1977.

Blank, A. D. Effects of group and individual conditions on choice behavior. *Journal of Personality and Social Psychology*, 1968, **8**, 294–298.

Blaskovich, J., & Ginsburg, G. P. Emergent norms and choice shifts involving risk. *Sociometry*, 1974, **37**, 205–218.

Blaskovich, J., Ginsburg, G., & Veach, T. A pluralistic explanation of choice shifts on the risk dimension. *Journal of Personality and Social Psychology*, 1975, **31**, 422–429.

Blaskovich, J., Veach, T. L., & Ginsburg, G. P. Blackjack and the risky shift. *Sociometry*, 1973, **36**, 42–45.

Brown, R. *Social psychology*. New York: Free press, 1965.

Brown, R. Further comment on the risky shift. *American Psychologist*, 1974, **29**, 468–470.

Burnstein, E., & Vinokur, A. Testing two classes of theories about group induced shifts in individual choice. *Journal of Experimental Social Psychology*, 1973, **9**, 123–137.

Burnstein, E., & Vinokur, A. What a person thinks upon learning he has chosen differently from the others: Nice evidence for the persuasive-arguments explanation of choice shifts. *Journal of Experimental Social Psychology*, 1975, **11**, 412–426.

Burnstein, E., Vinokur, A., & Pichevin, M. F. What do differences between own, admired, and attributed choices have to do with group shifts in choice? *Journal of Experimental Social Psychology*, 1974, **10**, 428–443.

Burnstein, E., Vinokur, A., & Trope, Y. Interpersonal comparison versus persuasive argumentation: A more direct test of alternative explanations for group-induced shifts in individual choice. *Journal of Experimental Social Psychology*, 1973, **9**, 236–245.

Cartwright, D. Risk taking by individual and groups: An assessment of research employing choice dilemmas. *Journal of Personality and Social Psychology*, 1971, **20**, 361–378.

Cartwright, D. Determinants of scientific progress: The case of the risky shift. *American Psychologist*, 1973, **28**, 222–231.

Clark, R. D. III. Group-induced shift toward risk: A critical appraisal. *Psychological Bulletin*, 1971, **76**, 251–270.

Clark, R. D. III. Crockett, W. H., & Archer, R. L. Risk-as-value hypothesis: The relation between perception of self, others, and the risky shift. *Journal of Personality and Social Psychology*, 1971, **20**, 425–429.

Deutsch, M., & Gerard, H. G. A study of normative and informational social influence upon individual judgment. *Journal of Abnormal and Social Psychology*, 1955, **51**, 629–636.

Festinger, L. A theory of social comparison processes. *Human Relations*, 1954, **7**, 117–140.

Jellison, J. M., & Riskind, J. A social comparison of abilities interpretation of risk taking behavior. *Journal of Personality and Social Psychology*, 1970, **15**, 375–390.

Jellison, J., & Arkin, R. Social comparison of abilities: A self-presentation approach to decision making in groups. In J. M. Suls & R. L. Miller (Eds.), *Social comparison processes: Theoretical and empirical perspectives*. New York: Hemisphere Publishers Inc., 1977, 235–259.

Jones, E. E., & Gerard, H. B. *Foundations of social psychology*. New York: Wiley & Sons, 1967.

Kogan, N., & Wallach, M. A. *Risk taking: A study in cognition and personality*. New York: Holt, Rinehart, and Winston, 1964.

Levinger, G., & Schneider, D. J. Test of the "Risk is a value" hypothesis. *Journal of Personality and Social Psychology*, 1969, **11**, 165–170.

Myers, D. G., & Lamm, H. The group polarization phenomenon. *Psychological Bulletin*, 1976, **83**, 602–627.

Myers, D. G., Wong, D. W., & Murdoch, P. H. Discussion arguments, information about others' responses and risky shift. *Psychonomic Science*, 1971, **24**, 81–83.

Nordhøy, F. Group interaction in decision-making under risk. Unpublished master's thesis, School of Industrial Management, Massachusetts Institute of Technology, 1962.

Paicheler, G., & Bouchet, J. Attitude polarization and group polarization. *European Journal of Social Psychology*, 1973, **3**, 83–90.

Pruitt, D. G. Choice shifts in group discussion: An introductory review. *Journal of Personality and Social Psychology,* 1971, **20,** 339–360. (a)

Pruitt, D. G. Conclusions: Toward an understanding of choice shifts in group discussion. *Journal of Personality and Social Psychology,* 1971, **20,** 495–510. (b)

Rabow, J., Fowler, F. J., Jr., Bradford, D. L., Hofeller, M. A., & Shibuya, Y. The role of social norms and leadership in risk-taking. *Sociometry,* 1966, **29,** 16–27.

Sanders, G. S., & Baron, R. S. Is social comparison irrelevant for producing choice shifts? *Journal of Experimental Social Psychology,* 1977, **13,** 303–314.

Stoner, J, A. F. Risky and cautious shifts in group decisions: The influence of widely held values. *Journal of Experimental Social Psychology,* 1968, **4,** 442–459.

Vinokur, A. A review and theoretical analysis of the effects of group decisions involving risk. *Psychological Bulletin,* 1971, **76,** 231–250.

COMMUNICATION NETWORKS

Marvin E. Shaw
DEPARTMENT OF PSYCHOLOGY
UNIVERSITY OF FLORIDA
GAINESVILLE, FLORIDA

I. Introduction

Communication lies at the heart of the group interaction process. No group, whether an informal or formal organization such as an industrial unit, governmental body, or military group, can function effectively unless

Reprinted from *Advances in Experimental Social Psychology*,
Volume 1, 111–147.

its members can communicate with facility. One major function of a chain of command is to provide channels of communication extending from the top downward throughout the group structure. The free flow of information (factual knowledge, ideas, technical know-how, feelings) among various members of a group determines to a large extent the efficiency of the group and the satisfaction of its members.

Administrative personnel often assume that the optimum pattern of communication for a given group or organization can be derived from the requirements of the task. Bavelas (1948, 1950) noted this assumption and raised several questions about the effects of fixed communication patterns upon group process. Do some communication networks have structural properties that limit group efficiency? What effects can such structural properties have upon problem-solving effectiveness, organizational development, leadership emergence, the ability of the group to adapt successfully to sudden changes in the environment? Bavelas also suggested a technique for investigating these questions in the laboratory. As a consequence of his work, extensive research has been carried out to analyze the relationships among structural properties of groups (communication networks) and group process variables.

This chapter reviews several of these studies and attempts an integration by means of certain theoretical constructs. Section II reviews the methodology employed in the research on communication networks and considers some of the structural properties of these networks. Section III summarizes the major findings of experimental investigations of the effects of networks on group process. Section IV explicates theoretical constructs advanced to explain network effects. In Section V an attempt is made to relate these theoretical constructs to specific experimental results and to concepts that have been proposed by other experimenters. Section VI summarizes the present state of knowledge and suggests some areas where further research is badly needed.

II. Research Methodology

A. Methods of Imposing Communication Patterns

Although the experimental method suggested by Bavelas is simple, it allows for maximum control of the communication structure. Group members are placed in cubicles which are interconnected by means of slots in the walls through which written messages can be passed. Slots may be closed to create any selected communication structure. Each cubicle is fitted with a silent switch which controls a signal light and a timer located at the experimenter's desk. The most common procedure has been to permit free (continuous) communication within limits imposed by the network. However, some investigators (Christie *et al.*, 1952; Schein, 1958) have used an "action quantization" procedure that restricted each subject to single, addressed messages transmitted at specified times. Also, some investigators substituted an intercom system for the slots—written messages system (Heise and Miller, 1951).

B. Network Characteristics

Figure 1 shows communication networks that have been studied experimentally. The dots represent persons or positions in the network, and the lines represent communication channels (slots) between positions. Most channels are symmetrical (two-way); asymmetrical (one-way) channels are indicated by arrows. The labels are arbitrary designations, intended only to facilitate identification. It will be noted that the same label is used for similar networks for groups of different sizes, although there has been some criticism that networks of different sizes are not comparable. For

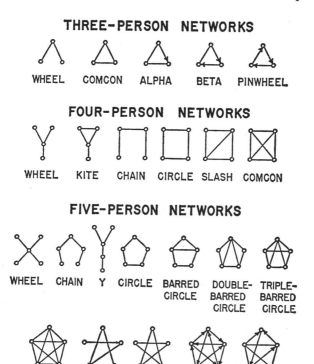

Fig. 1. Communication networks used in experimental investigations. Dots represent positions, lines represent communication channels, and arrows indicate one-way channels.

example, Glanzer and Glaser (1961) asked why the three-person "wheel" could not be called "chain" and compared to the four- and five-person chains. They apparently could see no difference between the three-person wheel and the larger chains. Actually, there is good reason to label it "wheel" and compare it with the larger wheels rather than chains. The essential characteristic of a wheel network is that one person communicates with all others, whereas all other members communicate only with this central person.

The three-person wheel in Fig. 1 clearly fits this description. In a chain, on the other hand, there are two isolates (end-persons) who communicate with only one other (but different) person in the group and two or more persons in the group must serve as message relayers in order to disseminate information throughout the group. The three-person wheel does not have these features and, therefore, is clearly more comparable to the larger wheels than to the larger chains. Similar considerations are involved in the labeling of other networks. When the group size is decreased, different networks do not necessarily coalesce into a single pattern, as Glanzer and Glaser suppose; rather, certain networks found with larger groups cannot be formed with smaller groups.

One further comment regarding the networks shown in Fig. 1. According to the Bavelas analysis, the spatial arrangement of positions is of no consequence; it is the relationships among positions that is important. Two of the networks shown in Fig. 1 depart from this conception: the chain (X) and the circle (X) used by Christie *et al.* (1952).

C. STRUCTURAL INDICES

It has seemed desirable to quantify the structural properties of networks to facilitate the analysis of their effects on group behavior. Bavelas (1950) suggested a "centrality index" as a measure of the differences between networks and positions within networks, based upon the distance, in communication links, between any two positions in the network. Other suggested indices include "relative peripherality" (Leavitt, 1951) and "independence" (Shaw, 1954b). Each of these indices merits consideration.

1. Centrality Indices

In formulating the centrality measures, Bavelas (1950) first defined the *sum of distances* $(d_{x,y})$ for a given position as the minimum number of communication links that must be crossed in order for that position to communicate with all other positions in the group. The individual sums of distances in a network may be summed to obtain a sum of distances for the network $(\Sigma d_{x,y})$. Comparisons among networks may then be made on the basis of $\Sigma d_{x,y}$. However, to make comparisons among positions within a network, a relative measure (called relative centrality) was suggested. This measure is computed as the ratio of the network sum of distances to the sum of distances for the given position. Thus,

$$\text{Relative centrality} = \frac{\Sigma d_{x,y}}{d_{x,y}}$$

Leavitt (1951), working with Bavelas, computed a centrality index for the group by summing the relative centralities of all positions in the network.

In general, total centrality for networks has been found to correlate poorly with group performance and satisfaction (Shaw, 1954a,b), although relative centrality has been found to account moderately well for positional differences in performance and satisfaction (Leavitt, 1951; Shaw, 1954a). However, relative centrality does not reflect differences among positions in similar networks of different sizes, nor does it permit easy comparisons among positions in different networks. Leavitt (1951) noted the first inadequacy and proposed an index of relative peripherality as an alternate to relative centrality; Shaw (1954b) pointed to the second difficulty and attempted to overcome it by an independence index. These indices are described below.

2. Peripherality Indices

The relative peripherality of any position in a network is the difference between the relative centrality of that position and the relative centrality of the most central position in that network. A total peripherality index for the network may be computed by adding all the individual peripheralities in the network.

Leavitt believed peripherality is related to group behavior variables via differences among positions in answer-getting potentials which structure group members' perceptions of their roles in the group. The relative peripherality index reflects differences in position independence, which in turn determine behavioral differences.

Since relative peripherality and relative centrality are perfectly correlated (negatively) within a given network, the two indices relate equally well to positional differences in performance and satisfaction. The advantage of peripherality over centrality lies in the greater comparability among positions in networks of different sizes. However, two positions in different networks having the same relative peripherality index do not necessarily produce the same behavior. Nor do two positions having different peripherality indices necessarily give rise to different behaviors. Likewise, the total peripherality index does not adequately reflect differences among networks. (The same statements hold, of course, for the centrality indices.) The independence index was developed in an attempt to overcome some of these shortcomings.

3. The Independence Index

Since the centrality-peripherality indices are imperfectly correlated with behavioral measures, there clearly are some important characteristics of networks and positions that are not reflected by these measures. Shaw (1954b) proposed that it is necessary to determine what characteristics do contribute to position independence. On a logical basis these features seemed to be (a) the number of channels available to a given position, (b) the total number of channels in the network, and (c) the number of positions for which a given position must relay information. The independence index (I index) for any one position was designed to reflect the weighted contributions of these various characteristics, as shown by the formula:

$$I = n + \left[n \left(1 - \frac{n}{N} \right) \right] + \log R_d + \log R_i$$

where

$n =$ number of channels available to the given position

$N =$ number of channels in a completely interconnected network of the same size

$R_d =$ number of positions for which the particular position must serve as a direct relayer (i.e., directly connected to the position served)

$R_i =$ number of positions for which the position must serve as an indirect relayer (i.e., one or more links removed from the position served)

The I index has been shown to be better than either the centrality or the peripherality index in the sense that it makes possible direct comparisons among positions in different networks (Shaw, 1954b, 1955a). However, no satisfactory method of computing the total independence in a network has been found, since the mere summation of positional values does not correlate highly with behavioral differences among networks.

In summary, the various structural indices described thus far have some explanatory value with regard to differences among positions within networks, but are inadequate to predict or explain differences among different networks.[1] Consequently, more general processes have been invoked as explanatory concepts. Although several such concepts have been described by various writers, this chapter will attempt to explain group behavior in communication networks in terms of two underlying processes

---

[1] For other mathematical analyses, see Flament (1958c), Glanzer and Glaser (1959), and Luce et al. (1953).

which have been labeled "independence" and "saturation." Before describing these concepts, however, a consideration of the effects of networks upon group process is helpful.

III. Effects of Networks upon Group Process

This section reviews the initial studies conducted in the Group Networks Laboratory at MIT and selected follow-up experiments to demonstrate that communication networks are related to group behavior in systematic ways.

A. THE MIT EXPERIMENTS

The initial studies are reported in articles by Bavelas (1950), Bavelas and Barrett (1951), Leavitt (1951), and Christie et al. (1952). The experiment reported by Leavitt is representative of the approach described in the first three articles. He carried out his investigation in order to explore the relationship between the behavior of small groups and the patterns of communication in which the groups operate. Leavitt examined the effects of the five-man circle, chain, Y, and wheel (Fig. 1) on problem-solving effectiveness, satisfaction, and organization characteristics of the group. The tasks were extremely simple symbol-identification problems. Each member in the group was given a card containing a number of symbols such as a square, a diamond, an asterisk, etc. Only one symbol appeared on each and every member's card. The task was to identify this commonly held symbol. Measures of performance were time taken to solve the problem, errors, and number of messages sent. Networks did not differ in average time to solve, but the circle was slower than the other patterns when the measure was the single, fastest, correct trial. The circle groups made the greatest number of errors (16.6), the Y the smallest number (2.6), and the chain and wheel an intermediate number (9.8 in each case). The circle required considerably more messages than did the other networks.

Satisfaction of members was determined by means of a questionnaire. Circle members reported greatest satisfaction, wheel members least, with chain and Y members intermediate. Leadership emergence was also measured by questionnaire. The total frequency of persons named as leaders and the unanimity of opinion as to who was the leader, increased in the order: circle, chain, Y, wheel. For example, 23 of the 25 persons in the wheel named a leader and all agreed that he was the person in the most central position, whereas only 13 of the 25 persons in the circle named a leader, and those named were scattered among all positions in the circle.

Operational methods (i.e., organizational patterns) used by the wheel, Y, and chain were such that the information was funneled to the central position, where the decision was made and sent out to the peripheral positions. (Hereafter this organizational pattern will be referred to as a centralized organization.) The circle showed no consistent, operational organization.

With regard to individual positions in the network, persons in the more central positions generally required less time to solve the problem, sent more messages, made fewer errors, were better satisfied, and were named leader more often than persons in the more peripheral positions.

Leavitt's findings clearly demonstrated a systematic relationship between the communication network imposed upon the group and the behavior of the group members. The experiments reported by Bavelas (1950) and by Bavelas and Barrett (1951) were carried out for essentially the same purpose, and their findings generally agree with those obtained by Leavitt. The experiments conducted by Christie et al. (1952, 1956), however, had a rather different purpose. They were concerned with the effects of networks upon the information-handling process, learning in networks, and the testing of mathematical models.

Christie and associates (1952) investigated the effects of five-person networks: circle, circle (X), chain, chain (X), pinwheel, barred circle, wheel, comcon, and alpha (see Fig. 1). The tasks assigned to the groups were number-identification problems, similar to the symbol-identification task used by Leavitt. An action-quantized procedure was used; that is, all subjects prepared messages simultaneously and transmitted them at specified times. On a given exchange, each group member was permitted to send a single message to one other group member. In this experiment, the investigators were interested primarily in testing certain mathematical models; hence, not all data were presented. However, it is clear that the geometrical arrangement of the circle and chain networks had little or no effect on the group process; no differences in time scores and other behavioral measures between the circle and circle (X) or between the chain and chain (X) were observed. With regard to time, networks were ordered (from fastest to slowest) as follows: chain, pinwheel, circle, comcon. The probability of minimum solution (task completion with the smallest number of message exchanges possible in a given network) was found to be greatest in the chain and smallest in the pinwheel. The alpha, barred circle, and wheel networks were about the same on this measure, while the circle was somewhat better and the comcon worse than this group of networks.

In a separate publication, Christie (1954) reported data concerned with the effects of learning in the comcon, circle, chain, and pinwheel networks. The task was reconstruction of a number list, and the performance criterion was number of communication acts (i.e., message exchanges) required for solution. All networks did better than chance from the beginning, but only the chain and the circle showed significant learning. Since task solution required a minimum of five message exchanges in the chain, its absolute performance was poor in comparison with each of the other networks. The circle groups, on the other hand, achieved a high level of efficiency in comparison with other networks.[2]

These early experiments demonstrated quite clearly that the pattern of communication imposed upon a group is an important determinant of the behavior of that group. However, the findings also indicated that the particular relationship between communication pattern and group behavior depends in part upon other variables. The MIT studies stimulated a considerable amount of research designed to examine the effects of these other variables upon network-group behavior relationships, and to test various theoretical interpretations of the underlying processes. A review of selected experiments will reveal the general conclusions that may be drawn from these follow-up investigations.

B. Selected Follow-up Investigations

The experiments described in this section were selected to demonstrate two general conclusions that can be drawn from the many communication network studies: (a) the major differences in group performance and satisfaction are between the centralized (wheel, Y, chain) and decentralized (comcon, circle) networks; and (b) the direction of these differences in group performance depends upon the kind of task assigned to the group.

Leavitt's experiment demonstrated differences among networks on the completion of one kind of task (symbol identification) under noise-free conditions. Heise and Miller (1951) extended this design by varying the intelligibility of the message and the type of problem given the group. They studied the three-person networks shown in Fig. 1, using as tasks word-construction problems, sentence-construction problems, and anagrams. Each problem was attempted in each network under three conditions of intelligibility. Speech, instead of written messages, was the method of communication. Intelligibility of the message was manipulated by controlling the relative intensities of speech and noise. Their results showed that for the word-construction problems the comcon was the most efficient, the wheel was intermediate, and the pinwheel was by far the least efficient network. For the sentence-construction problems, the results were similar except that the wheel replaced the comcon as the most efficient network. There were no marked differences in efficiency for the anagram problems. Noise generally accentuated differences for the first two kinds of tasks but not for the anagram problems.

Unfortunately, Heise and Miller used only three undergraduate sub-

---

[2] It is interesting to note that Leavitt and Knight (1963) recently concluded, on purely theoretical grounds, that the circle should be the most efficient network.

jects who went through all conditions for the word problem. Two of these subjects continued through the sentence problem with a new recruit as the third member, while the subjects for the anagram problems consisted of two groups of three graduate students each. Therefore, no statistical tests of the reliability of the obtained differences was possible. Since there are individual differences in group performance, we do not know to what extent the observed differences were due to the experimental variables or to individual differences. The large differences in the relationship between networks and group-effectiveness measures (time and errors) as a function of the task, however, suggest strongly that the kind of network that is most efficient depends upon the kind of task faced by the group.

This conclusion is also indicated by the findings in two experiments by Shaw (1954a,b). Leavitt, it will be recalled, interpreted his results in terms of centrality-peripherality indices which were supposed to reflect each position's accessibility to information. The first experiment reported by Shaw was designed to test this interpretation. On the basis of Leavitt's argument, it seemed reasonable to suppose that increasing the information input from external sources (i.e., the experimenter) should have the same effect upon a position as increasing the centrality of that position. For this purpose, the effects of four-person wheel, slash, and circle networks (see Fig. 1) on group effectiveness in solving arithmetic problems were examined. Although quite simple, these problems required more than the mere collation of information required by the symbol-identification task used by Leavitt. For half of the groups in each network the necessary information was distributed equally among group members (as Leavitt did), whereas for the other groups, the information was distributed unequally. In the unequal distribution condition, one of the most peripheral positions in each network was given five units of information, whereas all other positions were given one unit of information each.

Leavitt's measure of performance, the single fastest correct trial, indicated no differences among networks. However, the total time required to complete the task did reveal significant differences, although in a complicated manner. Analysis of variance revealed a significant trials × networks × distribution interaction; analysis of the interaction means showed that significant differences among networks occurred only on the third (and last) trial. The circle was fastest, the slash next fastest, and the wheel the slowest of the three patterns, with the difference between the circle and the other patterns being greatest with unequal distribution of information. For overall situational conditions the order was circle, slash, wheel—just the opposite of that expected from Leavitt's results. There were no differences among networks in number of errors, although the ability to correct errors (corrective power of the network) was greatest for the circle, next for slash, and least for the wheel network—again contrary to expectations.

Findings with respect to number of messages, satisfaction, and leadership emergence were in general agreement with Leavitt's results.

All groups used either the centralized organization or "each-to-all" organization (all information sent to all group members and then each member solved the problem independently). Centralized organization was used by 73% of the wheel, 7% of the slash, and 7% of the circle groups. Each-to-all organization was used by 27% of the wheel, 93% of the slash, and 93% of the circle groups.

The most striking result of the unequal-distribution-of-information experiment was the reversal of effectiveness of the wheel and circle networks as compared with Leavitt's results. The most obvious difference in the two experiments was the greater complexity of the problems employed in the later study. Two observations suggested that task differences probably accounted for the reversal. First, the central person in the wheel was overloaded by the many communication demands of the situation, and second, persons in the peripheral positions were unwilling merely to accept a solution offered by the central person. Both of these effects presumably were more likely to occur with complex than with simple tasks. However, since there were several other differences between the Shaw and Leavitt experiments, an experiment (Shaw, 1954c) was carried out to demonstrate the effects of the task variable in which the three-person wheel and comcon (Fig. 1) were compared using symbol-identification and arithmetic problems as tasks. Although the difference was not statistically reliable, the wheel groups required less time than did the comcon with the relatively easy symbol identification problems, whereas the comcon required less time than did the wheel with the more complex arithmetic problems. There were no differences in number of errors on the identification problems, but the wheel made more errors than did the comcon on the arithmetic problems.

Numerous other investigators have reported results which support the general conclusions that the major differences are between the centralized and decentralized networks, and that the direction of such differences is contingent upon the kind of task. First, let us consider studies employing fairly simple group tasks. Using symbol identification problems, Guetzkow and Simon (1955) investigated the five-person wheel, circle, and comcon networks shown in Fig. 1. They were interested primarily in the effects of free communication between trials for organizational purposes (discussed later), but they also reported that the wheel was faster than the circle, with the comcon intermediate in speed.

Many other investigations using the symbol identification task might also be mentioned. Cohen et al. (1961) found that five-person wheel groups took less time to solve identification problems, made fewer answer changes and fewer final errors, sent fewer messages, and recognized a leader more frequently than did circle groups. Both groups improved with practice, but wheel groups improved more than circle groups did. Networks did not produce overall differences in satisfaction, but central positions were better satisfied than peripheral positions. Hirota (1953) repeated Leavitt's experiment using Japanese subjects. The wheel required less time to solve, followed by the Y, circle, and chain, in that order; however, differences were not statistically reliable. Hirota also reported positional differences in frequency of communication and leadership emergence which agreed with Leavitt's findings. In a cross-cultural study, Mohanna and Argyle (1960) repeated the portion of Leavitt's experiment that made use the wheel and circle networks. Leavitt's results were confirmed in that the wheel was found to be superior to the circle in time required, number of messages required, and errors made.

Studies using more complex problems usually found the decentralized networks more effective. In Holland, Mulder (1960) examined the effects of four-person wheel and circle networks on the solution of arithmetic problems. Overall, but especially in their early trials, the circle groups required less time per problem than the wheel groups. However, contrary to findings reported in this country (Shaw and Rothschild, 1956), with practice the wheel groups became relatively more efficient than the circle networks. The studies using sentence- and word-construction tasks (Heise and Miller, 1951) and those using noisy marbles (Macy *et al.,* 1953) also showed that more decentralized networks are more efficient than centralized networks when solving even moderately complex problems (Flament, 1956, 1958b; Shaw *et al.,* 1957).

C. Some General Conclusions Concerning the Effects of Networks upon Group Process

The evidence strongly supports the generalization that the major network difference is between centralized (e.g., wheel, chain, Y) and decentralized (e.g., circle, comcon) networks, and that the direction of this difference is determined in part by the degree of complexity of the task. It is instructive to tabulate the number of comparisons showing specified differences between centralized and decentralized networks with simple and with more complex problems. The results of one such tabulation are shown in Table I. The "simple problems" classification includes tasks that require the mere collation of information (symbol-, letter-, color-identification tasks). The "complex problems" classification includes tasks that require some data operation procedures after the information has been collected in one place (arithmetic, word arrangement, sentence construc-

TABLE I

NUMBER OF COMPARISONS SHOWING DIFFERENCES BETWEEN CENTRALIZED
(WHEEL, CHAIN, Y) AND DECENTRALIZED (CIRCLE, COMCON)
NETWORKS AS A FUNCTION OF TASK COMPLEXITY

	Simple problems[a]	Complex problems[b]	Total
Time			
Centralized faster	14	0	14
Decentralized faster	4	18	22
Messages			
Centralized sent more	0	1	1
Decentralized sent more	18	17	35
Errors			
Centralized made more	0	6	6
Decentralized made more	9	1	10
No difference	1	3	4
Satisfaction			
Centralized higher	1	1	2
Decentralized higher	7	10	17

[a] Simple problems: symbol-, letter-, number-, and color-identification tasks.

[b] Complex problems: arithmetic, word arrangement, sentence construction, and discussion problems.

tion, and discussion problems).[3] A comparison is a single difference in means (as reported in the literature) between a centralized and a decentralized network, without regard to level of significance. For example, if an experiment involved three centralized and one decentralized network (as did Leavitt's study), three comparisons were made.

Examination of Table I shows that with simple problems the more centralized network required less time than the decentralized network on 14 of the 18 comparisons, whereas, with more complex problems, the decentralized required less time than the centralized network on every comparison made.[4] Errors show the same pattern, but differences are less

[3] This is a rough classificaion, and admittedly does not do justice to the complexity variable. The categorical grouping of tasks is not intended to deny the essential continuity of the complexity dimension.

[4] Eighteen different experiments were involved in this tabulation. Since a single experiment sometimes contributed more than one comparison, there is some lack of independence among scores in Table I. To overcome this, one comparison was drawn from each experiment. The results of this procedure showed the centralized network faster than the decentralized in 6 of 8 experiments using simple problems, whereas the decentralized network was faster in 10 of 10 experiments using complex problems. This difference is highly reliable ($\chi^2 = 11.26$, $p < .001$).

consistent than for time scores. With very few exceptions, individuals in decentralized networks are more active (send more messages) and are better satisfied than are persons in centralized networks, regardless of kind of task. Differences in activity are probably due merely to the fact that centralized have fewer channels than decentralized networks, as suggested by Glanzer and Glaser (1961). Organizational differences were not included in Table I because relatively few investigators examined this aspect in any systematic way. In most instances, however, centralized networks develop centralized organizations (i.e., all members send their information to one member who solves the problem and sends the answer to other members). Decentralized networks develop either each-to-all or centralized organizations about equally.

Numerous other variables have been examined in relation to group behavior in communication networks. The effects of these other variables generally raised or lowered the behavioral level of groups across all networks, either equally or in a manner which increased the differences among networks. For this reason, it is believed that these effects can be interpreted in terms of the same explanatory concepts used to explain the results presented thus far. Therefore, the independence and saturation constructs will be explained more fully before considering the effects of additional variables.

IV. Explanatory Concepts

It has been suggested (see Section II, C, 3) that the various effects of communication networks upon group behavior can be accounted for by two general processes labeled "independence" (Leavitt, 1951; Shaw, 1954b) and "saturation" (Gilchrist et al., 1954).[5] Further, it is believed that the various explanations advanced by other investigators can be subsumed under one of these two more general processes.

A. Independence

The concept of independence was introduced by Leavitt (1951) to account for differences among network positions. He pointed to the differences in answer-getting potential among positions and suggested that the group members' perceptions of these differences structure their perceptions of their own roles in the group. In the wheel, for example, group members readily perceive the degree of information accessibility and the nature of their own roles. The central person is autonomous and controls the group.

[5] The labels applied to these two classes of intervening processes are unimportant. Independence and saturation were chosen because of the author's familiarity with them; however, "autonomy" or "self-realization" would be just as acceptable as independence, and "vulnerability" or "demands" as acceptable as saturation.

In the circle, on the other hand, any given group member is not exclusively dependent upon anyone else in the net, and his role is not clearly different from anyone else's role. Thus, his action is not as greatly controlled by others. Morale is higher with greater independence because independence permits the gratification of the culturally supported needs for autonomy, recognition, and achievement. Independence bears a more direct relationship to group performance via its organizational influences upon the group. Leavitt concluded, "In summary, then, it is our feeling that centrality determines behavior by limiting independence of action, thus producing differences in activity, accuracy, satisfaction, leadership, recognition of pattern, and other behavioral characteristics" (Leavitt, 1951, p. 49).

While the general notion of independence of action is useful, it has become clear that the original formulation is too limited. As stated earlier, structural indices such as centrality, relative peripherality, and the I index do not appear to be highly valid measures of independence as reflected by measures of performance and satisfaction. As a consequence of the experimental results obtained from many studies, it is now clear that the concept of independence must be expanded to include freedom from all restrictions on action. We shall use the term "independence," then, to refer to the degree of freedom with which an individual may function in the group. A person's independence of action may be influenced not only by accessibility of information but also by the actions of others in the group, by situational factors (such as communication "noise," reinforcement, kind of task), and by the person's own perceptions and cognitions regarding the overall situation. The concepts of "autonomy" (Trow, 1957) and "exercise of power" (Mulder, 1958, 1959a) are similar to independence.

The author's own view is that independence, as defined above, is related to both performance and satisfaction; however, independence probably has a greater effect on satisfaction than on performance. We agree with Leavitt (1951) that its effect on member satisfaction probably is due to the fulfillment of culturally approved needs for autonomy, recognition, and achievement. This chapter will try to show, however, that the effect of independence upon performance is due, not to its organizational influences, but to the individual's willingness and ability to perform under the more autonomous conditions. That is, lowered independence not only directly limits the possibilities for action (hence performance), but also reduces the person's willingness to perform at his optimum level.

B. Saturation

In addition to independence, a second process operates in group situations to influence group performance and satisfaction. This process, called "saturation," was first described by Gilchrist *et al.* (1954). They observed that when the number of required messages for a given position passed a certain optimal level, communication requirements began to counteract the effects of position centrality. Two kinds of saturation were distinguished: "channel saturation," which refers to the number of channels with which a position must deal, and "message unit saturation," which refers to the number of messages the position must handle. These two kinds of saturation, of course, are correlated. Each of these main classes may be broken down further into "input" and "output" saturation. Total saturation experienced by a position is the sum of all the input and output requirements placed upon that position.

Like the original formulation of independence, this notion of saturation is too limited. The requirements placed upon the position, from whatever source, call for action by the individual who occupies that position. Therefore, the total saturation of a position is the result not only of the communication requirements, but also of the other requirements in the situation, such as data manipulation procedures that are necessary for task completion. Requirements of this sort are essentially those referred to as "task demands" by Lanzetta and Roby (1956a,b, 1957). Saturation, then, refers to the total requirements placed upon an individual in a given position in the network. It varies with communication demands and task demands. Communication demands are determined by the number of channels available to the position, the task information to be transmitted, and the demands imposed by the vagaries of the other members who have access to that position's channels. Task demands are determined by the requirements of the task, per se, and by interferences that must be overcome in the process of task solution.

Several explanatory concepts have been suggested by other investigators that are similar to saturation. The notion of "vulnerability" (Mulder, 1959b, 1960) is essentially the same as saturation; organizational arrangements (Guetzkow and Simon, 1955; Guetzkow and Dill, 1957), task demands (Lanzetta and Roby, 1956a,b, 1957; Roby and Lanzetta, 1956), and "inadequation" (Flament, 1958a,b) may be regarded as special cases of saturation.

Group effectiveness varies inversely with saturation. The greater the saturation the less efficient the group's performance.

C. Network and Task Effects in Terms of Independence and Saturation

It is proposed, then, that independence and saturation processes jointly determine group behavior; variables such as the communication network and kind of task influence group performance and satisfaction through their effects upon these two underlying processes. In Section III, C, evidence was presented indicating that centralized networks are more effective for simple problems, whereas decentralized networks are more effective for complex problems. To what extent can the concepts of independence and saturation help explain these results?

The effects of networks on independence have already been discussed. Persons in peripheral positions in centralized networks have limited freedom of action (hence low independence), while persons in central positions have relatively great freedom (hence high independence). In decentralized networks, by contrast, all positions have approximately equal freedom of action (hence moderate to high independence). Generally speaking, independence is greater in decentralized than in centralized networks. Task complexity has relatively little effect upon independence.

Saturation is determined in part by the communication channels available to a position. Therefore, the central position of a centralized network is more vulnerable to saturation than any position in the decentralized network. Whether this potential saturation occurs, however, depends upon task demands and member demands. When the task is a complex one, the demands upon the central person are greater than when the task is less complex, not only because the communication demands are increased by complexity, but also because data manipulation procedures are more demanding with the more complex tasks. Therefore, the probability that the network will become saturated is greater with complex than with simple tasks, and greater for the centralized than for the decentralized network.

Further, if our assumptions about the effects of independence are correct, this difference in saturation as a function of task complexity should be enhanced by the unwillingness of group members to accept the dictates of the central person when the task is a complex one. With simple identification problems, needs for achievement and recognition can hardly be satisfied by the simple act of noting that the same symbol appears on each of several cards; hence, subjects in a centralized network are typically willing to accept the report of the central person. This should actually reduce saturation in centralized networks relative to decentralized networks because of decreased message requirements. When the task requires data manipulation, however, needs for achievement and recognition can attain some degree of satisfaction through the problem solution process; members therefore are less likely to accept the solution arrived at by the central person. For example, they ask for the information upon which his solution was based. This increases communication demands and, hence, saturation. In decentralized networks the willingness or unwillingness of members to accept the solution of another has relatively little effect on saturation, since each person typically has all the information for solution and often achieves the solution through his own efforts.

In summary, independence is greater in the decentralized than in the centralized network, regardless of the kind of task. Saturation should be less in the centralized than in the decentralized network with simple tasks, but greater in the centralized network with complex tasks. Therefore, member satisfaction should be greater in more centralized positions, but overall satisfaction should be higher in decentralized networks. With simple tasks, the centralized should be more effective than the decentralized networks; whereas with complex tasks, the decentralized should be more efficient than the centralized networks. As we have seen, the experimental data generally agree with these expectations.

The author's feeling is that the effects of network and task variables upon group behavior are adequately accounted for by independence and saturation processes. Section V attempts to interpret the effects of other variables upon group behavior in different networks, again using the concepts of independence and saturation.

V. Independence and Saturation in Relation to Other Experimental Variables

Since it was well known from theoretical analyses and experimental observations that group behavior is influenced by many other variables, in addition to the communication network and kind of task, the question naturally arose whether these other variables influenced behavior differentially in different networks.[6] The numerous studies concerning this general problem may be classified under three broad headings, in terms of whether they deal with: (1) network-related variables, (2) information-input variables, or (3) group-composition variables.

Network-related variables are those that cannot be manipulated without simultaneous modification of some aspect of the communication network. These include size of the group, change of network, and organizational arrangements. Information-input variables are variables relating to amount and kind of information that is available to group members. This class includes noise, information distribution, and reinforcement variables. Group composition variables are concerned with variations in the personality and behavior characteristics of group members. Ascendance, authoritarianism, popularity, and leadership style are group composition variables

[6] Contrary to the conclusion drawn by Glanzer and Glaser (1961), the extension of communication network studies to include other (often nonstructural) variables indicates the vitality of this research; few variables produce a monotonic effect upon group process, regardless of variations in other known determinants of group behavior.

that have been studied in communication networks. These investigations will now be reviewed and the findings interpreted in terms of independence and saturation.

A. NETWORK-RELATED VARIABLES

1. Group Size

Most investigators have compared communication networks having similar topological characteristics without regard to the size of the group. For example, it seemed reasonable to assume that a network in which one person can communicate with everyone else in the group, who are themselves unable to communicate directly with each other (the wheel), would have similar effects upon group process regardless of whether three, four, or five persons are in the group. But what do the notions of independence and saturation lead us to expect with regard to differences in group size?

For a given network increased size of the group should decrease the independence of each member, since the mere presence of additional positions would limit, to some extent, each person's freedom of action. On the other hand, the larger the group the greater the saturation, since a larger group means more channels, more messages, and greater member demands. There is no apparent reason to suppose, however, that group size will affect independence and saturation differentially in different networks. Therefore, we would expect that increasing the size of the group would, within a given network, decrease satisfaction and efficiency (as measured by time and error scores). However, these effects should be the same for all networks.

The experimental evidence concerning the effects of group size is limited but consistent. Walker (1954) compared three-, four-, and five-person wheel networks with comcon networks of the same size. Arithmetic problems were assigned as group tasks. The results showed that as size increased: (a) group efficiency, as measured by problem solution times and errors, decreased; (b) group morale, as measured by ratings of satisfaction and by members' sociometric rejections of their own positions, decreased; (c) number of messages increased; and (d) unanimous selection of a leader decreased. Efficiency, satisfaction, and messages transmitted tended to be higher in the comcon than in the wheel, but there was no significant interaction between network and group size.

The findings are similar if we compare the results of several experiments that differed primarily in the size of the groups employed. Lawson (1963a) examined four-person wheel, comcon, and circle networks using the symbol-identification task. He found that the wheels were fastest in time and made fewer errors than the other networks. Circles were slowest and had the highest member satisfaction. These results agreed with those reported by Leavitt (1951) for five-person networks solving symbol-identification problems, and with those reported by Shaw (1954c) for three-person networks with symbol-identification tasks.

It seems clear that size of the group does have an effect on group problem-solving, but this variable does not interact with the communication network variable (at least, this is true for three-, four-, and five-person groups). It is also clear that the findings are in accord with expectations derived from independence and saturation.

2. Change of Network

Randomly assembled groups are composed of individuals with widely different past experiences in naturally occurring communication networks. Variations in behavior due to this factor are usually treated as errors of measurement. For theoretical as well as practical reasons, however, a number of researchers have thought it worthwhile to study the effects of past experience upon group behavior in communication networks.

It is difficult to make specific predictions about the consequences of changing from one network to another in terms of independence and saturation, since the expected effects vary with the particular change introduced. In general, independence and saturation should be unaltered if the new network permits the group to continue operating in the same manner as it did in the old network. However, if the new network requires a change in operational procedure, independence should be decreased and saturation increased by the network change.

This general principle was well demonstrated by Flament (1956) in an experiment designed to study the effects of imposed networks that were either congruent or incongruent with emergent structures. In the first part of the experiment, five-person groups solved seven symbol-identification problems in the comcon network. Those groups that developed a centralized organization (only two did not) were selected for the second part and assigned to one of four conditions: (1) imposed wheel network with emergent leader in the central position; (2) imposed wheel with the emergent leader in a peripheral position; (3) imposed chain with emergent leader in the most central position; and (4) imposed chain with emergent leader in one of the most peripheral positions. Groups were required to solve seven more symbol-identification problems under the new conditions. On the first problem after the new networks were imposed, the wheel groups required less time and fewer messages than the chain groups, and groups with emergent leaders in the central position required less time and fewer messages than groups with the emergent leader in a peripheral position. These differences tended to disappear with practice in the new situation. Satisfaction was higher in both the wheel and the chain when the emergent leader was in the central position than when he was in a peripheral position. In short, when the new network prevented the group from functioning as it had in the previous network, independence presumably was decreased (leading to lowered satisfaction). Saturation apparently was increased (leading to reduced efficiency).

The results of several other studies also supported this interpretation. Lawson (1961) evaluated the effects of training a group in one network, shifting to a different network, and returning to the original network. Four experimental and two control conditions were established: wheel-comcon-wheel-wheel (WCWW), wheel-wheel-comcon-wheel (WWCW), comcon-comcon-wheel-comcon (CCWC), comcon-wheel-comcon-comcon (CWCC), wheel-wheel-wheel-wheel (WWWW), and comcon-comcon-comcon-comcon (CCCC). In the WCWW condition, for example, groups worked in the wheel on day 1, in the comcon on day 2, and in the wheel on days 3 and 4. The problems to be solved were arithmetic problems. In general, we would expect that changes in networks would be accompanied by changes in independence and saturation, thus producing variations in satisfaction and efficiency. The present analysis suggests, further, that the change from the comcon to the wheel would have greater effects on these intervening processes than changing from the wheel to the comcon network. (The comcon would permit the group to continue functioning as before, whereas the wheel would not necessarily allow this.) The results generally supported these expectations. Group performance and satisfaction improved when the group was changed from the wheel to the comcon, and dropped when the change was from comcon to wheel. It is interesting to note that satisfaction dropped to its lowest point in the WCWW groups on day 3. Presumably, perceived independence in the wheel was reduced by contrast to the relative freedom experienced in the comcon on day 2.

The extensive work of Cohen and associates (summarized by Cohen, 1961, 1962) also lends credence to the independence-saturation hypothesis. These experiments demonstrated that when groups are changed from one network to another they continue the same operational procedures developed in the old network if the new network makes this possible. In one experiment, Cohen and Bennis (1961) explored the effects of changes in networks upon leader continuity. Five-person groups were assigned to one of the following conditions: wheel-circle, circle-wheel, wheel-wheel, and circle-circle. Each group attempted 30 symbol-identification problems in the first network, followed by 30 in the second network. After network change, groups in the wheel-circle condition developed chain problem-solving systems; groups in the circle-wheel and wheel-wheel conditions developed centralized problem-solving systems; and circle-circle groups developed each-to-all problem-solving systems. In the wheel-circle groups, the central person in the wheel remained in the central position in the emergent chain in only one of eight groups. This latter finding led to a second experiment in which groups in the wheel-circle condition were permitted to choose the person to occupy the central position in the wheel after having completed 15 symbol-identification problems. Organizational development after network change was the same as in the first experiment; however, the elected center of the wheel continued in the central position in the emergent chain in eight of the ten groups tested.

In another study using the same general design, Cohen and associates (1962) found similar results with regard to organization and leadership continuity. They reported, in addition, that circle groups with prior experience in wheel networks were better satisfied than those groups that had been in circle networks throughout the experiment. Groups that had been in the wheel throughout the experiment were better satisfied than groups that were changed from circle to wheel networks. Cohen *et al.* were surprised by these findings, since they had supposed that groups in a changed network would show greater similarities in satisfactions to those of prior networks. A consideration of independence as perceived by group members would have eliminated this surprise. When a group is changed from a restrictive network like the wheel to a less restrictive network like the circle, the amount of experienced freedom of action is undoubtedly greater (by contrast with the previous situation) than when the group has been in the less restrictive network throughout the experiment. Their perceived independence is therefore greater, and higher satisfaction is expected. The reverse is true for the circle to wheel groups. This interpretation is essentially the same as that offered by Cohen *et al.* to account for their findings, although they used the term "relative deprivation" rather than "perceived independence."

In a final report, Cohen and Bennis (1962) described an experiment comparing five-person groups working in the comcon network throughout the experiment with groups that were changed to the comcon after thirty trials in the wheel network. Symbol-identification problems were used. The results revealed that the groups that had had prior experience in the wheel developed centralized problem-solving systems when changed to the comcon, whereas groups working only in the comcon network developed each-to-all problem-solving systems. They concluded that the problem-solving systems that are adopted in antecedent networks are continued after network change if they are the more efficient systems, and if the new network permits the adaptation of the old system to the new situation.

The author cannot agree that only efficient problem-solving systems will be continued after network change. However, the saturation concept assumes only that continuation of a prior problem-solving system will not affect saturation, whereas forced change to a new system will increase saturation. The Cohen studies are important in demonstrating that a prior problem-solving system will be continued after network change if the new network permits.

In summary, when changes in the network suggest changes in independence and saturation, the expected variations in group satisfaction and performance are observed. This further demonstrates the explanatory value of these concepts.

3. Organizational Opportunity

The variable labeled "organizational opportunity" might be considered one aspect of network change. However, it differs from that variable in that all problem solving takes place in one network. Differential opportunity for organization is provided by varying the opportunity to organize between problem-solving trials. This technique was developed by Guetzkow and associates (Guetzkow and Simon, 1955; Guetzkow and Dill, 1957) to test their hypothesis that network patterns effect the efficiency of groups only indirectly by governing the members' ability to organize themselves for efficient task performance. They assumed that circle and comcon networks are less efficient than the wheel for symbol-identification problems because the people in the former communication nets have greater difficulty developing a hierarchical organization. Theoretically, then, added opportunity for organizational activity should eliminate differences among networks.

Again we might ask, what would be predicted from a consideration of the effects of organizational opportunity upon saturation? The writer thinks agreement among group members regarding organizational procedures would reduce saturation, regardless of the kind of organizational procedure agreed upon. With simple tasks, this reduction should be greater in the circle and comcon than in the wheel where agreement is already evident (see Section IV, C). In other words, the saturation concept would lead to the same prediction as the Guetzkow hypothesis, although the effect would not be expected to be as great as his hypothesis suggests. Let us examine the evidence.

In order to test their hypothesis, Guetzkow and Simon (1955) tested five-person groups in the wheel, circle, and comcon networks, using symbol identification tasks. Groups were given periods of not more than two minutes between successive task trials to provide an opportunity to solve the organizational problem. The results showed clearcut differences among nets, in agreement with previous findings; i.e., the intertrial periods did *not* eliminate network effects as expected. However, Guetzkow and Simon examined the organizational arrangements actually used by their groups, and compared networks using only those groups that developed hierarchical organizations. Since the different networks did not differ when only the selected groups were compared, they argued that their hypothesis was correct.

In a second report, Guetzkow and Dill (1957) tested twenty five-person groups in the circle network, again using the symbol identification task. This time, all communication restrictions were removed during the intertrial planning periods. These groups were then compared with the circle groups in the first experiment. The circle groups with complete freedom of communication during the intertrial periods did *not* perform any better than did the circle groups in the initial experiment. Once again, groups were divided into organized and unorganized, and the organized groups were found to be more efficient than the unorganized groups.

Similar findings are reported by Mulder (1959b, 1960), who proposed that groups with more centralized "decision-structures" (defined as "who makes decisions for whom") will perform better than groups with more decentralized decision structures. He required four-person groups to solve arithmetic problems in the wheel and circle networks. Groups were divided into more-and-less-centralized structures for the wheel and circle considered separately. Within each network, groups with the more centralized decision-structures required less time and fewer messages than the less centralized groups.

At first thought, it appears that all of these results support the Guetzkow hypothesis. Such a conclusion would also be consistent with the saturation hypothesis, since the predictions are the same. Further, it might be argued that saturation is the preferred interpretation, since it accounts for other findings as well as those presented above. Unfortunately, there are some serious questions about the soundness of the methodology employed in these studies. It is clear that opportunity to plan for problem-solution improves performance, as demonstrated by an excellent study by Shure *et al.* (1962). But it is not at all clear that degree of organization accounts for the obtained network differences. The evidence presented by Guetzkow *et al.* and by Mulder merely shows that efficiency and organization are correlated; it does not show that organization *causes* the efficiency. In a series of experiments dealing with this question, Schein (1958) traced the achievement of organization and efficiency across trials. Although he found efficiency and organization perfectly correlated at the end of the experiment, the achievement of efficiency developed *earlier* than organization. Thus it is evident that organization is not a prerequisite of efficiency. It could be just the opposite: more efficient groups tend to become organized, perhaps because the same abilities are required for both efficiency and organization development.

In conclusion, the evidence concerning organizational development is consistent with the saturation hypothesis (and the Guetzkow-Mulder hypothesis), but this evidence rests upon a comparison of selected groups. We have seen that the soundness of this approach is questionable.

B. INFORMATION INPUT VARIABLES

Group behavior may be influenced either positively or negatively by the amount and kind of information that is available to the group. If the information is noise free, for example, the demands upon the group are not great, and the performance should be better than under noisy conditions. Several aspects of this problem have been investigated, including the effects of noise, distribution of information, and reinforcement.

1. Noise

Noise may be introduced into the information available to the group in several ways. Some common types of noise include *channel noise* (transmission of messages is interfered with), *coding noise* (encoding and decoding processes are ambiguous), and *information noise* (apparently task-relevant information is interspersed among task-irrelevant information). Noise, regardless of type, increases saturation relative to noise-free conditions, and, therefore, should decrease performance and, to some extent, satisfaction. Centralized networks should be affected to a greater extent than decentralized networks because the burden of handling the noise falls upon the central positions.

Channel noise was studied by Heise and Miller (1951) in the experiment described in Section III, B. The five three-person networks shown in Fig. 1 were created by means of an intercom system. White noise was introduced into the system at three speech-to-noise ratios: $+6$ db, -2 db, and -10 db. For those problems requiring communication (word arrangement and sentence construction) lowering the signal-to-noise ratio increased errors, time, and messages required to complete the task, and accentuated the superiority of decentralized over centralized networks.

Coding noise was studied by Macy *et al.* (1953). Coding processes were made ambiguous by requiring group members to write and interpret descriptions of colors that are not easy to describe, in this case mottled and streaked marbles. Varying degrees of coding ambiguity were compared in five-person wheel, chain, circle, and pinwheel networks. Coding noise greatly increased the errors made by all networks. However, certain networks learned to reduce their errors to a previously determined noise-free level, while others did not. The circle showed rapid learning and a good error reduction; the wheel and chain showed no learning and no error reduction; and the pinwheel showed some initial learning and poor error reduction. In general, the results were consistent with predictions from the saturation hypothesis.

The effects of noisy information were studied by Shaw (1958) in the four-person wheel and comcon. The group tasks were arithmetic problems, each requiring twelve items of information for solution. Each group member was given three of these items. Half of the groups in each network were given only this task-relevant information (noise-free condition) and for the other half, each member was given two additional items that appeared relevant but were actually irrelevant to the task (noisy condition). The comcon required significantly less time than did the wheel under both conditions. Noise increased time for both networks, and more in the wheel than in the comcon, but this latter finding was not statistically reliable. Noise significantly increased the number of messages required by the wheel, but not by the comcon. (Incidentally, this finding supports the view that noise increases saturation more in centralized than in decentralized networks.) Noise significantly reduced satisfaction in both the wheel and the comcon. The direction of the findings are consistent with the saturation interpretation, but not all differences are significant.

The results of the studies of the various kinds of noise upon performance in communication networks are unusual in that all agree in showing that noise tends to reduce the effectiveness of groups in centralized, restrictive networks more than groups in decentralized networks, in agreement with expectations based upon the saturation process.

2. Information Distribution

Several investigators have suggested that the way in which the information is distributed among network positions will influence the performance of the group, and that the distributional effects will be different for different networks. These expectations are based upon the undisputed fact that information availability and number of communication channels are positively correlated; hence the conclusion that at least part of the differential effects among network positions can be accounted for by differences in availability of information. It is also obvious that independence and saturation are influenced by variations in information distribution, although the precise effects will depend upon the particular distributional variable under consideration. The distributional variables that have been studied include uniformity of distribution, organization of distributed information, and the temporal aspects of distribution.

Uniformity of information distribution refers to the degree to which available information is distributed equally among group members. Relative to equal distribution, unequal distribution of information should decrease saturation when a peripheral position has the most information and increase saturation when the most central person has the most information. Independence should be increased for the person who has the most information and decreased for those that have the least information, since additional information reduces an individual's dependence on others in the group.

The initial study of distribution uniformity (Shaw, 1954a) compared equal and unequal distribution of task-relevant information in four-person wheel, slash, and circle networks. In the equal distribution condition, the eight items of information needed to solve the assigned arithmetic problems were randomly, but uniformly, distributed among group members; whereas in the unequal condition, one group member (one of the most peripheral positions) in each group was given five items and the other three members only one item of information. Information distribution had no effect upon group time scores in any of the networks. However, in all networks the position having five units of information required less time to solve than the corresponding position in the equal condition. Unequal distribution reduced errors in all networks. Satisfaction was not influenced by the distribution variable, but the position having the most information was valued more highly than other positions in all networks. Messages varied with information distribution only in the wheel and the slash; the position having the most information in the unequal condition sent more messages than the corresponding position in the equal condition. These findings only partially supported the independence-saturation analysis.

In a follow-up study, Gilchrist et al. (1954) compared three information distribution conditions in the four-person wheel network: task-relevant items were distributed equally (four items to each person), unequally peripheral (five items to a peripheral position, one item to all others), or unequally central (five items to the most central position, one item to all others). Additional information to the peripheral position significantly decreased the time taken and increased messages and satisfaction for that position relative to comparable positions in the equal condition; whereas, additional information to the central position increased time, messages, and satisfaction as compared with the central position in the equal condition. It was this finding that led to the formulation of the saturation concept.

The uniformity of distribution studies cited above varied the amounts of information input to each position, but information input may also vary in content. In most network studies, the items of information were distributed randomly among positions, so that the content of the items given a position was matter of chance. In most extant groups, however, the information content available to a given position is determined by the particular functions of the position. For example, the sales manager has information about markets, demands, composition, etc. that derives from the activities associated with his position. This systematic distribution of information should increase independence (hence raise satisfaction) and decrease saturation (hence improve performance), relative to the random distribution situation.

This problem was attacked by Shaw (1956), who suggested that systematic distribution should allow for selective communication and that this effect should be increased by prior knowledge of the distribution. Systematic versus random distribution of information was studied in the four-person wheel and comcon networks. For half of the groups in each network, the information needed to solve arithmetic problems was distributed so that a given position had all the information concerning one particular aspect of the problem, whereas for the other half, the relevant information was randomly distributed. Similarly, half of the groups in each condition were told the nature of the distribution and half were not. Knowledge of distribution did not significantly influence the group process. Systematic distribution, as compared with random distribution, decreased time and errors and increased satisfaction scores.

For time and error scores, this effect was stronger in the wheel than in the comcon. These findings were interpreted as not supporting the selective communication hypothesis; rather, systematic distribution appeared to impose a useful kind of organization upon the content of transmitted messages. That is, each person transmitted all his information on a single message card so that related information was grouped at the time it was received. This led to a reduction in saturation, since the task demands were not as great as when the information had to be grouped before the problem could be solved.

An experiment by Shelly and Gilchrist (1958) represents the third approach to information distribution. They were concerned with the effects produced by variations in the amount of information the group must deal with at any given time when the group must handle a given amount of total information. Eight arithmetic problems, each requiring four items of information for solution, were formulated, so that it was possible to combine some or all of these problems into larger problems which also would have a single answer. These eight problems were given to all groups, but were presented in four different ways: singly in sequential order, sequentially in groups of two as a single problem, sequentially in groups of four as a single problem, and all eight as a single problem. Groups of four persons each were run in the wheel and comcon networks. Under these conditions, saturation effects would be expected to increase and independence to decrease with increasing work load per unit time. The effects of increased work load should of course be greater in the wheel than in the comcon. The results generally supported these expectations.

Although not statistically reliable, the comcon was faster than the wheel under all information conditions, and there was some indication of an increasing divergence with increased communication demands. Questionnaire results indicated that as the number of items to be handled simultaneously increased, group members became less satisfied, and perceived levels of performance and cooperation dropped off. Group members in the wheel were significantly less satisfied with the way they had to operate, whereas those in the comcon were less satisfied with the performance of the group. This suggests that wheel group members blamed the structure for their information-handling difficulties, whereas those in the comcon blamed themselves.

In general, it appears that performance and satisfaction are depressed when information is distributed so that the central person's work load is increased, when the input information is randomly distributed, or when the total work load must be handled at one time. Although the results are not always significant, it appears that the detrimental effects of such adverse information distributions are greater in the more centralized and restricted networks than in the more decentralized and unrestricted networks. Similarly, persons in favorable positions for information accessibility and processing (for example, a position that has most of the information needed to solve the problem, or a position that has many channels to persons who have such information) are more efficient and better satisfied than are persons in less favorable positions. These findings are consistent with the independence-saturation hypothesis.

3. Reinforcement

In addition to information that may be needed to complete the task, other kinds of information may be available to the group. Information about the group's performance, for example, may affect the group's behavior in systematic ways. This information may be knowledge of results, or it may only be information that the performance was good or bad (i.e., positive or negative reinforcement). Only this latter form of information has been studied systematically.

The effects of reinforcement on independence and saturation are not easy to predict. Based on findings by Berkowitz and Levy (1956) that evaluations given to the group as a whole promoted feelings of group interdependence, perceived independence might be supposed to decrease by group reinforcement. Further, Lawson's (1963a,b) conclusions that reinforcement imposes additional pressure upon group members would suggest that reinforcement increases saturation. Evidence on the effects of reinforcement was provided by two experiments. The first experiment (Lawson,

1963a) examined the effects of reinforcement in four-person groups in wheel, circle, and comcon networks. Symbol identification tasks were used. Reinforced groups received randomly positive reinforcement (a chime) at the end of 50% of the trials, and randomly negative reinforcement (a raucous buzzer) at the end of the remaining 50% of the trials. The meaning of the two types of reinforcement was explained to the subjects before the first trial. Reinforcement had a significant effect on group performance only with respect to time scores in the circle. Reinforced circle groups required less time to solve problems than nonreinforced groups. Reinforcement effects were in the same direction in the wheel and in the opposite direction in the comcon, but neither difference was significant. There were no differences in satisfaction of reinforced and nonreinforced groups.

The second experiment (Lawson, 1963b) was similar to the first one except that groups were assigned arithmetic problems. The results showed that under nonreinforced conditions the comcon was fastest, the wheel slowest, and the circle intermediate in speed. Reinforcement exaggerated these differences: the reinforced comcon groups were faster, but the reinforced wheel groups were slower than their controls. Circle groups showed the least change as a consequence of reinforcement. Morale scores tended to be higher in the nonreinforced than in the reinforced groups.

The effects of reinforcement upon performance and morale are consistent with the independence-saturation interpretation except for the improved performance of the comcon groups under reinforcement.

C. Group Composition Variables

Group composition, defined as the individual characteristics of group members, has long been considered one of the major variables in group behavior. Cattell (1948) and Carter (1954) suggested that member personality is one of three broad classes of variables needed to understand group behavior, and Schutz (1958) built an entire theory of group behavior based upon member characteristics. It is obvious that an adequate analysis of network effects must include a consideration of the kinds of individuals who compose the group. The characteristics of group members surely affect independence and saturation, but the direction of these effects will of course depend upon the particular characteristics under consideration. The characteristics that have been studied in networks are ascendance, authoritarianism, popularity, and leadership style.

1. Ascendance

Ascendance may be defined as the tendency to dominate others (i.e., to speak, resist opinion change, interrupt, etc., more often than others). Ascendant persons in groups may be expected to elicit negative attitudes and uncooperative behavior in other group members; especially if they do not have legitimate authority over other group members. Therefore, we would expect ascendant persons to decrease independence and increase saturation effects in networks. Only one study relative to the effects of ascendance has been reported.

Berkowitz (1956) selected high, moderate, and low ascendant persons on the basis of their scores on the Guilford-Zimmerman Ascendance scale and certain measures obtained from their participation in a group mechanical assembly task. Four-person groups, each composed of one high, one low, and two moderates, solved arithmetic problems in the wheel network. In one condition, the high ascendant person was in the central position, while in another condition the low ascendant person was in the central position. Highs in the central position were faster and more active than lows in the central position on the first trial, but these differences did not persist. Highs in the periphery had faster rates of communication than lows on all trials; however, highs in the periphery tended to become more passive with practice. Both highs and lows rated satisfaction higher than did moderates, even when in a peripheral position. Although Berkowitz did not make group comparisons, analysis of his data revealed that groups with a high ascendant person in the central position required more time (59.2 vs. 51.6) and rated satisfaction lower (4.1 vs. 4.5) than did groups with low ascendant persons in the central position. These findings are in accord with the assumption that ascendant behavior decreases independence and increases saturation. When the ascendant person is in the periphery he is unable to affect the group to the same extent that he can in the central position.

2. Authoritarianism

Authoritarianism is in some ways similar to ascendance. However, the high authoritarian may be either dominant or submissive, depending upon the position in which he finds himself. When he is in an authority (or power) position, he tends to use his authority to control others; when he is in a subordinate position, he accepts this status and submits to the person or persons in superior positions. When in a leadership position, for example, he issues orders to others in the group, and tries to have things done according to the accepted rules, but otherwise he presumably communicates very little. The other group members probably learn fairly quickly what to expect of him. Unlike the high ascendant who communicates a great deal to others, his actions should reduce communication demands. Therefore, we would expect that a high authoritarian assigned to a leadership position in the center of a centralized network would decrease both saturation and independence. Similar effects should occur when a high authoritarian is assigned a leadership position in a decentralized network, but the magnitude of these effects should be attenuated because the communication structure does not permit the leader to exert tight control over the group.

The effects of member authoritarianism upon group behavior was examined by Shaw (1959). The degree to which an individual accepts authority, as measured by the Bales Acceptance of Authority (AA) scale (Bales, 1956), was varied for members of the four-person groups in wheel and comcon networks. Group members were selected to have either high or low scores on the AA scale. One-fourth of the groups in each network was assigned to each of the following conditions: high scoring leader, high scoring followers (HH); high leader, low followers (HL); low leaders, high followers (LH); and low leader, low followers (LL). The leader was always assigned to one of the most central positions in the network.

The fact that assigning a high authoritarian leader to a central position reduces saturation relative to the low leader condition is demonstrated by the results of an analysis of messages transmitted. In the wheel, the mean number of messages by H-leader groups was 15.6 as compared with 25.6 for the L-leader groups; in the comcon, the means were 25.7 and 24.7, respectively. The results also suggested that this reduction in saturation improved performance. Leader AA scores and time scores correlated $-.29$ in the wheel and $+.38$ in the comcon, with the effects of intelligence partialled out. The results are not so clear with regard to independence, since ratings of satisfaction were a function of the follower AA score. With H-followers, satisfaction varied only with the network; with L-followers, satisfaction was greater in the comcon than in the wheel, and greater with L-leaders than with H-leaders.

In summary, authoritarian leaders appear to decrease saturation in the centralized network, but not in the decentralized network, and these effects are reflected in performance scores. Authoritarian leaders also tend to reduce independence, but this effect is reflected only in the satisfaction ratings of nonauthoritarian followers. Again, the results are generally consistent with the independence-saturation hypothesis.

3. Leadership Style

The study of leadership styles in communication networks may be regarded as simply another way of investigating the effects of authoritarianism, since leaders of laboratory groups are often asked to behave in either an autocratic, authoritarian manner, or a democratic, nonauthoritarian manner. The difference is that the laboratory variations in leader behavior are produced by instructions rather than by differences in personality characteristics. Since the outward behavior is the important variable, however, the effects on independence and saturation should be similar.

A study by Shaw (1955a) was designed specifically to test predictions based on the independence and saturation concepts. It was hypothesized that authoritarian leadership should decrease independence for followers (and hence decrease satisfaction) and should decrease saturation for all group members (and hence improve performance). Nonauthoritarian leadership should increase both independence (hence increase satisfaction) and saturation (hence lower performance). Leaders were assigned to four-person groups in wheel, kite, and comcon networks. For half of the groups in each network, the leader was instructed to behave in an autocratic manner; for the other half, he was instructed to behave in a democratic manner. Autocratic leadership improved group efficiency (shorter times and fewer messages) relative to democratic leadership in all networks, but differences in the kite were not significant. Similarly, groups were better satisfied under democratic, nonauthoritarian leadership in all networks. Thus, predictions were generally verified.

4. Popularity

Popularity might be viewed as the manifestation of certain personality attributes. But generally, popularity refers to the acceptance of the individual by fellow group members rather than to his behavioral characteristics. A popular group member should be better able to elicit the cooperation of others than an unpopular person, and this should reduce the demands on the group. Thus, a popular person in the central position of a centralized network should decrease saturation relative to an unpopular person in this position. Independence, on the other hand, should be increased since members should feel that they are free to act in other ways if they wish. A study by Mohanna and Argyle (1960) throws some light on this question. The central person in a five-person wheel network was selected sociometrically to be either popular or unpopular. Solving symbol identification problems, wheel groups with a popular center were more efficient (in time taken and messages used) than were groups with unpopular centers. This finding is in agreement with the expected effects on saturation, but no evidence was obtained relative to independence and satisfaction.

In summary, studies of group composition involving the ascendance, authoritarianism, leadership style, and popularity of group members demonstrated the importance of considering the composition of the group. The findings generally supported the independence-saturation hypothesis in that characteristics which may reasonably be expected to influence independence and saturation are related to satisfaction and performance in the predicted direction. However, knowledge of the effects of group composition is woefully limited, and much additional research is badly needed.

VI. Summary and Conclusions

This section summarizes what is known about the effects of communication networks upon group process, and indicates some areas where further research is needed.

A. THE PRESENT STATE OF KNOWLEDGE

The experimental evidence reviewed in this paper clearly demonstrates that the communication network imposed on the group influences its problem-solving efficiency, communication activity, organizational development, and member satisfaction. The major network difference is between centralized (e.g., wheel, chain, Y) and decentralized (e.g., circle, comcon) networks. The direction and magnitude of these effects are modified by the following variables: kind of task, noise, information distribution, member personality, reinforcement, and the kind of prior experience the members have had in networks. Of all those reviewed here, the variable having the most pronounced effect is the kind of task the group must perform. Centralized networks are generally more efficient when the task requires merely the collection of information in one place. Decentralized networks are more efficient when the task requires, in addition to the information collection process, that further operations must be performed on the information before the task can be completed. Decentralized networks are more satisfying to group members regardless of the kind of task.

The experimenter has tried to show that these several effects can be interpreted using the concepts of independence and saturation. The hypothesis that independence is positively related to satisfaction and saturation is negatively related to efficiency is generally supported by the experimental evidence.

B. UNFINISHED BUSINESS

The experiments conducted to date have taught us a great deal about the effects of communication networks, but the precise nature of many of the relationships among variables still remains unclear. In particular, the following variables need much clarification.

1. Network Characteristics

The several attempts to develop quantitative measures of network characteristics (centrality, peripherality, independence) have met with only partial success. Many measures appear adequate for positional differences, but fail to represent network differences adequately. In general, these measures seem to reflect extreme differences, such as differences between the highly centralized networks (wheel, chain, Y) and the highly decentralized networks (circle, comcon). They do not, however, distinguish between networks having smaller pattern differences, such as the Y and chain or the circle and comcon. Perhaps such quantification must await more extensive theoretical developments, or perhaps greater differences than, for example, those between the wheel and chain are necessary to produce consistent effects on group behavior.

2. Kind of Task

The evidence shows unequivocally that network differences vary with the kind of task. So far, the analysis has involved only a crude, dichotomous classification of tasks. Even so, the evidence is strikingly consistent in showing directional differences in favor of centralized networks for simple problems, but in favor of decentralized networks for complex problems. A more careful analysis of tasks is called for. At this stage the hypothesis may be ventured that the magnitude of differences in effectiveness between centralized and decentralized networks will be a positive, increasing function of the differences in task complexity.

3. Group Composition

Despite the obvious relevance of member characteristics for the functioning of groups, the relationship between such characteristics and network performance has been much neglected. Only a few studies have considered this variable, and these have varied only one "trait" of the group members. How effective will a network be when the critical member (e.g., the central person in the wheel) is the least intelligent group member? Can other characteristics of members, such as cooperativeness, need for affiliation, etc., compensate for lack of ability? Are "compatible" or "cohesive" groups relatively more effective in some networks than in others? Many other questions might be asked, but perhaps these will suffice to indicate the problems yet to be solved with reference to group composition variables.

4. Reinforcement

Lawson's (1963a,b) experiments showed that random reinforcement produces differential effects in networks which vary with the task. This is only a beginning. Still to be investigated are the effects of contingent reinforcement, schedules of reinforcements, magnitude of reinforcement, kind of reinforcement, and so on.

5. Network Embeddedness

Experimentation has been limited to isolated networks; however, small groups generally function as subgroups—parts of larger organizations. It may be that the most efficient network in isolation is quite different from the most efficient network embedded in a larger group. Furthermore, a given network may function differently depending upon the structure of the larger organization. These and other related problems demand attention, although no experimental techniques are readily available for their solution.

This brief resume lists only some of the more obvious research needs. Many more problems might be listed and many additional ones no doubt will arise in the course of future experimentation. The communication network studies have provided a great deal of information regarding structural effects upon group behavior. Much more remains to be done.

REFERENCES

Bales, R. F. (1956). "Factor Analysis of the Domain of Values in the Value Profile Test," Mimeo. Rept., Laboratory of Social Relations. Harvard Univ., Cambridge, Massachusetts.

Bavelas, A. (1948). *Appl. Anthropol.* **7,** 16–30.

Bavelas, A. (1950). *J. Acoust. Soc. Am.* **22,** 725–730.

Bavelas, A., and Barrett, D. (1951). *Personnel* **27,** 366–371.

Berkowitz, L. (1956). *Sociometry* **19,** 210–222.

Berkowitz, L., and Levy, B. I. (1956). *J. Abnorm. Soc. Psychol.* **53,** 300–306.

Carter, L. F. (1954). *Personnel Psychol.* **7,** 477–484.

Cattell, R. B. (1948). *Psychol. Rev.* **55,** 48–63.

Christie, L. S. (1954). *J. Operat. Res. Soc. Am.* **2,** 188–196.

Christie, L. S., Luce, R. D., and Macy, J., Jr. (1952). "Communication and Learning in Task-Oriented Groups," Tech. Rept. No. 231, Research Laboratory of Electronics. MIT.

Christie, L. S., Luce, R. D., and Macy, J., Jr. (1956). *In* "Operations Research for Management" (J. F. McCloskey and J. M. Coppinger, eds.), Vol. II, pp. 417–537. Johns Hopkins Press, Baltimore, Maryland.

Cohen, A. M. (1961). *J. Communication* **11,** 116–124 and 128.

Cohen, A. M. (1962). *Admin. Sci. Quart.* **6,** 443–462.

Cohen, A. M., and Bennis, W. G. (1961). *Human Relat.* **14,** 351–367.

Cohen, A. M., and Bennis, W. G. (1962). *J. Psychol.* **54,** 391–416.

Cohen, A. M., Bennis, W. G., and Wolkon, G. H. (1961). *Sociometry* **24,** 416–431.

Cohen, A. M., Bennis, W. B., and Wolkon, G. H. (1962). *Sociometry* **25,** 177–196.

Flament, C. (1956). *Annee Psychol.* **56,** 411–431.

Flament, C. (1958a). *Bull. Centre Etudes Rech. Psychotechn.* **7,** 97–106.

Flament, C. (1958b). *Annee Psychol.* **57,** 71–89.

Flament, C. (1958c). *Annee Psychol.* **58,** 119–131.

Gilchrist, J. C., Shaw, M. E., and Walker, L. C. (1954). *J. Abnorm. Soc. Psychol.* **49,** 554–556.

Glanzer, M., and Glaser, R. (1959). *Psychol. Bull.* **56,** 317–332.

Glanzer, M., and Glaser, R. (1961). *Psychol. Bull.* **58,** 1–27.

Guetzkow, H., and Dill, W. R. (1957). *Sociometry* **20,** 175–204.

Guetzkow, H., and Simon, H. (1955). *Mgmt. Sci.* **1,** 233–250.

Heise, G. A., and Miller, G. A. (1951). *J. Abnorm. Soc. Psychol.* **46,** 327–335.

Hirota, K. (1953). *Japan J. Psychol.* **24,** 105–113.

Lanzetta, J. T., and Roby, T. B. (1956a). *J. Abnorm. Soc. Psychol.* **53,** 307–314.

Lanzetta, J. T., and Roby, T. B. (1956b). *Sociometry* **19,** 95–104.

Lanzetta, J. T., and Roby, T. B. (1957). *J. Abnorm. Soc. Psychol.* **55,** 121–131.

Lawson, E. D. (1961). "Change in Communication Nets and Performance." (Paper read at the 1961 Eastern Psychological Association convention.)

Lawson, E. D. (1963a). "Reinforced and Non-Reinforced Four-Man Communication Nets." (Paper read at the 1963 Eastern Psychological Association convention.)

Lawson, E. D. (1963b). "Reinforcement in Group Problem-Solving with Complex Problems." (Paper read at the 1963 Canadian Psychological Association convention.)

Leavitt, H. J. (1951). *J. Abnorm. Soc. Psychol.* **46,** 38–50.

Leavitt, H. J., and Knight, K. E. (1963). *Sociometry* **26,** 260–267.

Luce, R. D., Macy, J., Jr., Christie, L. S., and Hay, H. D. (1953). "Information Flow in Task-Oriented Groups," Tech. Rept. No. 264, Research Laboratory of Electronics. MIT, Cambridge, Massachusetts.

Macy, J., Jr., Christie, L. S., and Luce, R. D. (1953). *J. Abnorm. Soc. Psychol.* **48,** 401–409.

Mohanna, A. I., and Argyle, M. (1960). *J. Abnorm. Soc. Psychol.* **60,** 139–140.

Mulder, M. (1958). "Groepsstructuur, Motivatie en Prestatie." Nederlands Instituut veer Praeventieve Geneeskunde, Leiden.

Mulder, M. (1959a). *Acta Psychol.* **16,** 178–225.

Mulder, M. (1959b). *Acta Psychol.* **16,** 356–402.

Mulder, M. (1960). *Sociometry* **23,** 1–14.

Roby, T. B., and Lanzetta, J. T. (1956). *Sociometry* **19,** 105–113.

Schein, E. H. (1958). "The Development of Organization in. .Small Problem-Solving Groups," Final Rept. on Sloan Project No. 134. MIT, Cambridge, Massachusetts.

Schutz, W. C. (1958). "FIRO: A Three-Dimensional Theory of Interpersonal Behavior." Rinehart, New York.

Shaw, M. E. (1954a). *J. Abnorm. Soc. Psychol.* **49,** 547–553.

Shaw, M. E. (1954b). *J. Psychol.* **38,** 139–149.

Shaw, M. E. (1954c). *J. Exptl. Psychol.* **48,** 211–217.

Shaw, M. E. (1955a). *J. Abnorm. Soc. Psychol.* **50,** 127–134.

Shaw, M. E. (1956). *J. Pers.* **25,** 59–69.

Shaw, M. E. (1958). *J. Soc. Psychol.* **47,** 33–37.

Shaw, M. E. (1959). *J. Pers.* **27,** 196–210.

Shaw, M. E., and Rothschild, G. H. (1956). *J. Appl. Psychol.* **40,** 281–286.

Shaw, M. E., Rothschild, G. H., and Strickland, J. C. (1957). *J. Abnorm. Soc. Psychol.* **54,** 323–330.

Shelly, M. W., and Gilchrist, J. C. (1958). *J. Soc. Psychol.* **48,** 37–44.

Shure, G. H., Rogers, M. S., Larsen, I. M., and Tassone, J. (1962). *Sociometry* **25,** 263–282.

Trow, D. B. (1957). *J. Abnorm. Soc. Psychol.* **54,** 204–209.

Walker, L. C. (1954). "The Effects of Group Size and Group Structure on Problem Solving Behavior in Small Groups." Unpublished doctoral dissertation, University of Wisconsin, Madison, Wisconsin.

Communication Networks
Fourteen Years Later

Marvin E. Shaw

UNIVERSITY OF FLORIDA

It is always hazardous to read what one has written in the past. Far too often one finds little correspondence between earlier assertions and current beliefs. Consequently, when I read again the report I had written on communication networks some 14 years ago, I was pleasantly surprised to discover that I still believe most of what I said at that time. Perhaps this only means that I have given too little thought to the problem during the intervening years, but I suspect that my unchanging view can be attributed to the relative lack of research on communication networks. A search of the literature revealed an average of only about one article a year devoted to this topic. Nevertheless, it may be worthwhile to consider the current status of research and theory in this area.

In this brief update I shall present a summary of new research, reconsider theoretical issues, and propose potentially important directions for future research.

I. New Research on Communication Networks

In the 1964 review of communication networks I suggested several areas in which further research and theory might be helpful, namely, network characteristics, kinds of tasks, group composition, reinforcement, and network embeddedness. A review of the literature

Group Processes

published since 1965 on communication networks reveals that indeed the major portion of the work has been devoted to one or more of these variables, although there is no convincing evidence that these studies were a consequence of my suggestions. Five studies were devoted to improving mathematical descriptions of network characteristics, four studies concerned kinds of tasks, and two studies dealt with the effects of reinforcement in communication networks. Two studies were at least peripherally related to group composition, but no one seems to have dealt with the effects of embeddedness. In addition to research related to variables mentioned in the 1964 review, several persons have attempted to show that networks are relatively unimportant variables in group behavior. Research in each of these areas will be briefly reviewed.

A. NETWORK CHARACTERISTICS

The research on network characteristics has been limited almost entirely to attempts to improve indices of centrality. Beauchamp (1965) attempted to improve Bavelas' (1948) centrality index by using graph theory. His formula was based upon the minimal distance between nodes within a given network and represented a definite improvement over the Bavelas measure, although it still is limited to completely connected networks of a given size. Mackenzie (1966a) also attempted to improve the Bavelas index of centrality. He developed an index based on the incidence matrix of actual communications rather than of possible communications. This index can vary from zero for the completely connected (comcon) network to unity for the most centralized network (wheel). In a later paper Mackenzie (1966b) demonstrated that this index is best interpreted as a total expected participation index.

Moxley and Moxley (1974) attempted to improve upon the Bavelas measure of relative centrality. They developed a concept of point centrality that applies to the relative position of each point in a network rather than to an overall summary index for the entire network. This procedure has the great advantage that it can be applied to social systems containing isolates and separate clusters as well as to completely connected networks. They also presented a computer algorithm for computing the index.

The most recent attempt to improve the measure of centrality was proposed by Freeman (1977), who also based his work upon the early intuitions of Bavelas (1948). He introduced a family of new measures of point and graph centrality that define centrality in terms of the degree to which a point falls on the shortest track between others and therefore has a potential for control of communication. This index also has the advantage of being applicable to any large or small network of symmetrical relations, whether connected or unconnected.

In summary, major advances have been made in the mathematical description of centrality indices in communication networks. The Freeman index for net centrality and the Moxley and Moxley indices for point centrality can be recommended for future research.

B. KINDS OF TASKS

The task variable has been attacked most directly by Morrissette and his associates. In the first paper (Morrissette, Pearson, & Switzer, 1965), information theory was used to develop a measure of difficulty of Leavitt type tasks. Using this measure, three tasks were developed, with difficulty indices of 1.6, 2.0, and 2.4, in which larger indices mean greater difficulty. Using only the difficulty = 2.0 task, they compared the Y and circle networks. As in other studies using this type of task, they found that the Y network was faster than the circle. In a second part of the study, all three tasks were used, but only in the Y network. In this communication pattern, performance varied inversely with difficulty.

In a second study (Morrissette, Switzer, & Crannell, 1965), two Leavitt type tasks with difficulties = 1.6 and 2.4 were used in the wheel and circle networks with 4- and 5-person groups. Size of group affected performance only in the circle, where the 4-person groups were faster than the 5-person groups. However, both communication structure and task produced significant differences in performance. The wheel was faster than the circle in all cases, and the time required was less for the less difficult task.

In a final paper, Morrissette (1966) compared 3-person groups with the 4- and 5-person groups from the previous study. Again the tasks were of the Leavitt type and were either easy or difficult. The communication networks were the circle and the wheel. He found that those in the circle structures made more errors than those in the wheel, and that as size of group increases, in the circle performance deteriorates, whereas in the wheel there is no relation between group size and performance. Furthermore, he noted that as group size decreases, the effect of structure decreases.

All of these findings are entirely consistent with previous data, which generally show that simple tasks of the Leavitt type require more time in decentralized communication networks than in centralized ones.

C. GROUP COMPOSITION

The studies that I have classified under this heading are only peripherally related to group composition. For example, Pool (1976) studied coalition formation, using 6- and 7-person groups with variable communication networks. He found that West German adolescent subgroups with more communication channels became the winning coalition more often than other subgroups. It should be noted that Pool did not compare different kinds of networks.

A study by Snadowsky (1974) was also concerned with group composition in a peripheral sort of way. His study involved the comcon and wheel communication networks, autocratic versus democratic leadership, and complex versus simple tasks. He separated organizational and operational phases of group interaction and measured satisfaction with several aspects of the group situation. Significant differences between networks were found only with regard to satisfaction with other group members. However, it is worth noting that the mean rating of satisfaction by group members in the concom was higher than the corresponding mean rating in the wheel on 26 of 28 comparisons. The groups with democratic leaders were generally better satisfied than were the groups with autocratic leaders. Snadowsky argued that we should measure the many aspects of satisfaction, and his finding indicates that the communication network is not as important as leadership style.

In summary, the research relevant to group composition is meager and is at best only indirectly related to the variable.

D. REINFORCEMENT

The two studies that have been conducted relevant to the effects of reinforcement upon performance in communication networks were conducted by Burgess. In the first experiment (Burgess, 1968a) the comcon and wheel networks were compared as to performance on a modified version of the Leavitt common symbols task (specifically, the task was to determine which one of several lights was not lit on every group member's panel). Groups were allowed to reach a "steady state," and performance was then evaluated following this achievement. During this phase, groups were given a positive incentive (reduction of time in the experiment) for success and punishment (a raucous buzzer) for failure on a trial. The wheel was faster than the comcon but the introduction of reinforcement reduced the difference to nearly zero. It should be noted that it required an average of 500 trials to reach steady state, and group members may have become fatigued before that time. In the second study Burgess (1968b) argues that the power function of the form $Y = aX^b$ can describe the results precisely. Thus, this study is an attempt to describe the results of research on communication networks rather than a presentation of new evidence.

E. ANTI-COMMUNICATION NETWORK STUDIES

The studies that are cited in this section are labeled anti-communication network studies because the authors seem to have as their major purpose the demonstration that communication networks have no effect upon group behavior or, at best, the network is a relatively insignificant variable. Two of the studies cited in earlier sections might be classified as anti-network studies. Burgess (1968a) was able to demonstrate that no significant differences existed between the wheel and circle communication networks after several hundred trials on a common symbol task with contingent reinforcement. He suggested the conclusion that previously asserted differences between communication structures were a function of experimental artifacts (i.e., failure to include reinforcement contingent on performance and failure to study groups after they had achieved a "steady state").

Snadowsky (1974), in his study of the effects of networks, leadership, and types of tasks on group members' satisfaction also suggested that the communication network variable is relatively unimportant. He concluded that "the potential or lack of potential for independence in a given communication network does not seem to be the primary factor determining members' satisfaction [p. 52]."

The most concerted attempts to demonstrate the relative unimportance of communication networks were investigations by Moore, Johnson, and Arnold (1972) and by Coleman (1975). Moore *et al.* compared decentralized status homogeneous networks (four graduate students or four freshmen) with centralized status congruent networks (one graduate in the central position and three freshmen in peripheral positions), centralized status homogeneous networks (four graduate students or four freshmen), and centralized status incongruent networks (one freshman in the central position and three graduate students in the peripheral position). The task assigned to the groups was a word construction task (scrabble) that was judged to be complex. Each group was given 15 trials on this task. An efficiency index was computed as the ratio of number of letters used to time in seconds times 100, summed over the 15 trials. The mean efficiency index was 74.74 for the decentralized homogenous network, 62.41 for the centralized homogeneous network, 60.51 for the centralized incongruent network, and 80.14 for the centralized status congruent network. These differences persisted when only the last 10 trials were used to compute the efficiency index. Thus, it can be seen that the decentralized network was more efficient on the word construction task than the centralized network in all cases except the status congruent centralized network. Moore *et al.* concluded that " . . . network centralization or decentralization is not in itself the primary predictor of group efficiency [p. 530]." Moore *et al.* also found that the number of communications to the central person decreased when a graduate student was in the central position (as compared with homogeneous group composition), but increased when a freshman was in the central position. This demonstrates quite clearly that saturation effects varied with efficiency in the centralized communication networks. They also failed to demonstrate network effects on ratings of satisfaction by group members.

Coleman (1975) conducted two experiments in which he attempted to separate the effects of the communication networks from the effects of organizational structures. In order to accomplish this, he superimposed the organizational structure upon the communication network. That is, he specified the kind of organization that the group must use regardless of the kind of communication network. Using this procedure he compared a relay-organized circle with a three-level hierarchy chain and a two-level hierarchy wheel. He found that the communication network per se did not influence time for solution but that the organizational structure did influence performance. He argued that these findings demonstrated that the communication network was not the significant variable determining group performance.

II. Theoretical Considerations

As stated initially, the research that has been conducted since 1964 does little to alter my views regarding theoretical explanations of communication network effects. However, before considering the degree to which new research findings are consistent with the concepts of independence and saturation, it may be well to evaluate the research that purportedly demonstrates that the communication network is a relatively insignificant variable in group behavior. If such critics are correct, there is nothing to explain.

First, let us consider the research reported by Burgess (1968a). Burgess concluded that the communication network is relatively unimportant because network effects decreased with increased number of trials and disappeared after groups had reached a steady state on the common symbols task and were subjected to contingent reinforcement. It is highly questionable that the conclusion drawn by Burgess is justified. Despite Burgess' argument that natural groups typically achieve a steady state, it seems highly unlikely that very many groups are subjected to several hundred trials on identical tasks in a short period of time. At the very least, such groups must have been very bored with the process and possibly fatigued as well. Under such circumstances it does not seem surprising to this reviewer that there would be little effect of almost any kind of manipulation. Perhaps the most valid conclusion that can be drawn from the Burgess study is that groups who are fatigued by several hundred trials on an essentially boring task are not affected by the particular communication network imposed upon them. The conclusion that the findings of previous studies are due to experimental artifacts is certainly not justified.

Little can be said about the Snadowsky (1974) study. His failure to find significant differences between centralized and decentralized communication networks on several measures of group satisfaction does not seem to warrant his conclusion concerning the effects of communication networks. A basic question is whether the negative findings presented by Snadowsky should be accepted in preference to the positive findings of numerous other studies on network differences. It should also be noted that the independence interpretation is not based upon potential or lack of potential for independence; instead, it is argued that the *actual* amount of independence afforded the individual is a determinant of member satisfaction.

Several objections might be directed toward the Moore *et al.* (1972) study. The most serious criticism is that the network variable was at least partially confounded with group composition. The only conditions that are different only with respect to communication networks are the decentralized status homogeneous and the centralized status homogeneous. When these two are compared, the findings are precisely consistent with previous research. In order to draw the conclusion that Moore *et al.* drew concerning the relative importance of the communication network variable, they should have included a decentralized network with one graduate student and three freshmen.

Comparing this with a centralized network with similar group composition might have demonstrated something concerning the relative importance of group composition versus communication networks. It might be noted in passing that the data presented by Moore *et al.* are strongly supportive of the saturation concept proposed earlier.

Finally, it might be noted that the Coleman (1975) study merely demonstrated that if one eliminates the effects of a network on the ways that group members interact with each other, then no network effects are observed. This is not a surprising result and certainly does not demonstrate that the communication network has no effect upon group efficiency.

All of the researchers whose work has been cited above apparently have misinterpreted the viewpoint presented by most previous researchers on communication network effects. It is doubtful that anyone would argue that the pattern of communication imposed upon a group is the primary variable influencing group behavior or even the most powerful one. What has been argued is that the communication network is one variable that has a demonstrable effect upon group efficiency and group member satisfaction. Considered in this light, even if the studies cited above were without blemish their findings still would not pose a great difficulty for communication network research and theory. That is, a demonstration that some other variable may be so powerful as to override the effects of the communication network variable does not negate the significance of a communication network variable when other more powerful variables are not acting in a counterdirection.

Having demonstrated, it is hoped, that the anti-communication network studies are not fatal to previous research findings, we may now turn to a consideration of the degree to which new research data require revision of earlier theoretical interpretations.

It will be recalled that in the 1964 review two processes were proposed to explain communication network effects: saturation and independence. Saturation was defined as the total requirements placed upon an individual in a given position in the network, and it was suggested that saturation has its primary influence upon group performance. Independence was defined as the degree of freedom with which an individual may function in the group, and it was suggested that independence primarily influences satisfaction in the group.

The new research most relevant to the saturation hypothesis is that dealing with complex versus simple tasks in centralized versus decentralized communication networks. The data from these studies (Morrissette, 1966; Morrissette, Pearson, & Switzer, 1965; Morrissette, Switzer, & Crannell, 1965) are entirely consistent with earlier findings and require no modification of the saturation hypothesis.

The work on the effects of reinforcement in communication networks is at least minimally related to the saturation interpretation. However, the new evidence concerning the effects of reinforcement in communication networks is difficult to interpret. There are two differences between the Burgess (1968a) and the earlier studies by Lawson (1964): Lawson randomly reinforced groups, whereas in the Burgess study reinforcement was contingent upon performance; and Burgess introduced reinforcement only after groups had achieved a steady state with respect to performance, whereas Lawson's groups did not have this previous history. It is therefore very unclear what the saturation effects of reinforcement might have been in the Burgess studies, and he made no attempt to determine them. It is possible that contingent reinforcement reduces saturation effects. If so, the data presented by Burgess are consistent with the saturation hypothesis.

The strongest support for the saturation hypothesis is provided by Moore et al. (1972). They found that the number of messages sent to the central position in each centralized group correlated in the expected way with several measures of group efficiency (+.59 with mean time to solution, +.47 with total number of time failures, and +.67 with total number of errors in final solutions).

There is little new evidence concerning the effects of independence on group member satisfaction. The two experiments that purport to show that networks do not affect satisfaction are of questionable validity. It has already been noted that Snadowsky's (1974) data are negative only in the sense that he failed to find significant differences between networks on some dimensions of satisfaction. The data reported by Moore et al. (1972) are questionable because the network comparisons were not comparable to comparisons in previous studies. That is, they failed to make an independent comparison between central and peripheral members in the wheel or between all members of the circle and all members of the wheel. Furthermore, status characteristics of the members were confounded with network position. Nevertheless, the support for the independence hypothesis provided by these studies is weak at best and suggests that the hypothesis should be reevaluated and possibly modified. Unfortunately, the new data provide no clues regarding the direction any such modification should take.

III. Summary

In summary:

1. The most significant contributions to the communication networks research since 1964 are those concerned with network characteristics. The new indices of centrality appear especially promising.
2. Attempts to demonstrate that communication network effects are insignificant or are due to experimental artifacts have not produced convincing results.
3. There is continued support for the saturation hypothesis as an explanation of the effects of communication networks upon group performance.
4. There are some new data suggesting that a reconsideration of the independence hypothesis as an explanation of the effects of communication networks on member satisfaction may be in order, but they provide as yet no suggestions for type of modification.

There has been little research on group composition effects and apparently no work on network embeddedness. Research on these aspects of group behavior in communication networks would seem to be desirable new directions.

REFERENCES

Bavelas, A. A mathematical model for group structures. *Applied Anthropology*, 1948, **7**, 16–30.

Beauchamp, M. A. An improved index of centrality. *Behavioral Science*, 1965, **10**, 161–163.

Burgess, R. L. Communication networks: An experimental reevaluation. *Journal of Experimental Social Psychology*, 1968, **4**, 324–337. (a)

Burgess, R. L. An experimental and mathematical analysis of group behavior within restricted networks. *Journal of Experimental Social Psychology*, 1968, **4**, 338–349. (b)

Coleman, G. A microstudy of communication network behavior under conditions of imposed organization. Doctoral dissertation, University of Sydney, 1975.

Faucheux, C., & Mackenzie, K. D. Task dependency of organizational centrality: Its behavioral consequences. *Journal of Experimental Social Psychology*, 1966, **2**, 361–375.

Freeman, L. C. A set of measures of centrality based on betweenness. *Sociometry*, 1977, **40**, 35–41.

Lawson, E. B. Reinforced and nonreinforced four-man communication nets. *Psychological Reports*, 1964, **14**, 287–296.

Mackenzie, K. D. Structural centrality in communication networks. *Psychometrika*, 1966, **31**, 17–25. (a)

Mackenzie, K. D. The information of theoretic entropy function as a total expected participation index for communication network experiments. *Psychometrika*, 1966, **31**, 249–254. (b)

Moore, J. C., Jr., Johnson, E. B., & Arnold, M. S. C. Status congruence and equity in restrictive communication networks. *Sociometry*, 1972, **35**, 519–537.

Morrissette, J. O. Group performance as a function of past difficulty and size and structure of groups, II. *Journal of Personality and Social Psychology*, 1966, **3**, 357–359.

Morrissette, J. O., Pearson, W. H., & Switzer, S. A. A mathematically defined task for the study of group performance. *Human Relations*, 1965, **18**, 187–192.

Morrissette, J. O., Switzer, S. A., & Crannell, C. W. Group performance as a function of size, structure, and task difficulty. *Journal of Personality and Social Psychology*, 1965, **2**, 451–455.

Moxley, R. L., & Moxley, N. F. Determining point centrality in uncontrived social networks. *Sociometry*, 1974, **37**, 122–130.

Pool, J. Coalition formation in small groups with incomplete networks. *Journal of Personality and Social Psychology*, 1976, **34**, 82–91.

Snadowsky, A. Member satisfaction in stable communication networks. *Sociometry*, 1974, **37**, 38–53.

EXPERIMENTAL STUDIES OF FAMILIES[1]

Nancy E. Waxler and Elliot G. Mishler

HARVARD MEDICAL SCHOOL
AND
THE MASSACHUSETTS MENTAL HEALTH CENTER

[1]This work was partially supported by the National Science Foundation, grant #1225, and the National Institute of Mental Health Research Scientist program, award #K2-38,842.

Reprinted from *Advances in Experimental Social Psychology*,
Volume 5, 249–304.

I. Introduction: Methodologies and the Need for Complex Models

There has been growing interest during the past twenty years in the application of experimental methods for investigating small groups to the study of families. A variety of theoretical perspectives are included in these studies and the problems dealt with range from evaluating the effectiveness of family therapy to testing hypotheses derived from general sociological and social psychological theory. It seems useful at this point to review these studies and to compare them with those from the more traditional body of experimental studies of *ad hoc* groups.

In the course of making these comparisons we will demonstrate the ways in which the experimental method strengthens the study of families. In some instances an extension of experimental methods to families opens up new substantive areas of family study. In others, experimental investigation serves to clarify both family and group process theories. Finally, the application of experimental methods to families requires that attention be paid to assumptions about structure and process that have in the past remained unexamined and untested. Each of these issues will be examined through a selective review of the experimental literature.

Experiments with *ad hoc* groups are reported as far back as the early 1900's. In these early studies the major concern was with the conditions under which high productivity could be reached (e.g., see Gates, 1924). In the 1930's, experimental work expanded both in size and in scope, and some researchers became interested in natural groups and particularly in groups of children (Beaver, 1932; Greenberg, 1932). However, it was not until the period after World War II that parent-child and husband-wife pairs were conceived of as a type of small group and were subjected to experimental investigation using methods already available (Bishop, 1951; Strodtbeck, 1951). Since that time, experiments using family members as subjects have multiplied, and both method and theory have become more sophisticated.

Family experimentation in the short period since the early 1950's seems to have recapitulated the history of small group studies. Historical trends are similar, with family studies showing a lag perhaps ten years behind issues that are the foci of other small group studies. A brief outline of historical trends in family experimentation will provide an overview of the major issues in family research at the present time.

A. TRENDS IN FAMILY RESEARCH

The first apparent trend reflects a historical change in much small group experimentation. There has been a shift from theories and studies centered on dyadic relationships to those in which family structure is central. The early dyadic approach examined ways in which one family member interacted with another (usually the mother and the child, as represented by Bishop, 1951); if there were any interest in the family as a whole, it was assumed that the family consisted of summed dyadic relations. Questions that followed logically from this perspective had to do with the quality of interaction, under specified conditions, between each pair of family members. Recent family studies have shifted from theoretical focus on the individual toward concern with the structure of the family group using sociological concepts such as family role, power structure, norms, and sanctions as guiding concepts. In fact, at present, not only are aspects of family structure measured but they are also experimentally manipulated and used as major independent variables.

A second change in family experimental work—a methodological one—also follows historical developments in small group studies. The first family experiments relied on observation of family interaction, without experimental intervention. Thus, in Kenkel's studies of married couples' discussions (Kenkel, 1957), the experimenter intervened only in the sense that a discussion task and an observer were imposed. Recent studies involve some degree of experimental intervention in natural family groups. For example, families have been asked to communicate in highly structured ways such as transmitting written messages and using abstract symbols (Reiss, 1967), experimenters have trained parents to play specified roles with their own children (Patterson, 1965) and artificial "families" have been constructed in order to measure effects of certain structural patterns (Leik, 1963).

These changes in theoretical focus and experimental method are directly related to the third historical development. Early family experiments measured only relatively undifferentiated qualities of the family's behavior while recent studies test precisely defined hypotheses derived from abstract theory. The earlier approach, in which the "whole family" was investigated, is currently represented largely by the group of researchers interested in family therapy and problems of pathology who argue that a method that abstracts one aspect of family structure from the family system is useless in understanding real families (see Framo, 1965, for this point of view.) However, for other investigators, theoretical questions have moved from the global (for example, Does the interaction of parents affect the personality development of the child?) to the specific (How is the distribution of power between the parents related to the child's choice of an object of imitation?). With increased experience with experimental intervention, this latter specific hypothesis is open to clear experimental test by creating artificial families with a variety of "parental" power structures and examining the child's behavior (Bandura, Ross, & Ross, 1963). The recent trend is, therefore, toward a series of well-controlled experiments each designed to shed light on one specific aspect of family structure or process and each related to a broader theory of families or groups.

While recent developments in the area of family experimentation have tended toward concern with family structure, toward the testing of explicit hypotheses related to a theory, and the introduction of experimental interventions and manipulations, the content of these experimental studies is relatively limited. A large proportion of family experiments focuses on the family groups as the socialization agent for the child; there is much less interest in the economic or religious functions of the family or even in the husband-wife role relations. The explanation of relationships between variables is usually in terms of how a particular kind of family may have produced a child with a certain characteristic. Within the set of studies that focus on the family as a socialization agent are a large number of experiments concerned with the part a family may have had in the development of a deviant child, for example, a delinquent child, one with schizophrenia, or with some other "maladjustment."

Other questions asked by family experimenters are less directly related to the general concern with socialization. Some studies have focused, at a rather descriptive level, on relationships between dimensions of family structure. They attempt to provide concrete descriptive material about how families are organized rather than data on some "outcome" of the family's interaction. For example, Kenkel's studies on role patterns are concerned with such things as whether the high power member is also the expressive member in most family groups (Kenkel, 1957.)

Some attention has also been given to questions about relationships between the family's social and cultural context and its internal structure. Most family experimenters control for these context variables by sampling families with homogeneous backgrounds. The few that have measured the influence of cultural context on family structure and organization have shown clearly the enormous effects of such variables as ethnicity, religion, social class, or race.

A final question asked by family experimenters has to do with family process, that is with the changing patterns of interaction within the family that may occur as a result of the changing family situation. While there are relatively few experiments of this type, the ones that predominate examine changes in interaction occurring over a short period of time — perhaps the experimental hour. Only a few experimenters have been concerned with changes in family processes that may be related to the full life history of the family.

B. The Concept of "Family Structure"

Through these questions runs a common concern with the structure of the family. Basic to the concept of "family structure" is the idea that the family group is an organized system in which there are two or more status levels, roles are differentiated, and norms or internalized rules for behavior have developed about who may take each status position and what kind of role behavior is appropriate. The organization or structure of the family is maintained not only through adherence to the norms but also because family members believe that it is appropriate and right to apply certain sanctions to members who deviate from the norms.

The idea of family structure and the concepts of status, role, norm, and sanction will be used to organize our discussion of issues in family and small group experimentation. First, we will examine studies in which family structure is the dependent variable, explained by the social and cultural background of family members. It is here that individual personality characteristics or cultural context of the family are used to predict the internal organization of the group. Second, we will look at experiments in which one dimension of structure (role, status, norm) is the independent variable that predicts to another dimension of structure, or to the behavior or development of the child. Most family and small group experimentation falls within this general area. Finally, we will look at experiments in which family structure is less central and there is greater concern for explaining process and change.

These three general issues—background, structure, and change—serve an organizing function for our discussion of family experiments. In many instances we have imposed the concepts of family structure upon a study that uses another conceptual framework; we do not intend that the problem and theory original to the study should be ignored. Instead, the ideas about family structure are used only to permit comparisons between experiments that, on the surface, appear to be relatively dissimilar yet at a more abstract level attack the same general problem.

II. Personality and Cultural Background: Relationships to Family Structure

No new group is compelled to negotiate every detail of its own organization. Instead members bring to the group certain conceptions about how a group should or could be organized, certain experiences in role-taking, and certain understandings of the behavior of other group members. These expectations and experiences form the basis upon which the new group is built. For example, past experience may predispose a member to feel comfortable in a particular role or to value one form of decision making over another. An individual's personality needs may also push him to behave in a specified way or to perceive others idiosyncratically. Thus, the structure that develops in a new group is to some extent related to or a result of the cultural background and personalities of individual members.

Is the same true for families? Family theories and more general sociological theories predict that it is. Personalities of parents and the subcultural context of the family are both central to theoretical formulations of family structure and particularly to theories about the socialization of the child. These theories have been extensively investigated using nonexperimental techniques, for example, self-report questionnaires, interviews of children or parents, or psychological tests. Yet, as we will see in our examination of the experimental family literature, the nature of the effects of personality and cultural context on structure remains largely an assumption rather than an empirical fact. The same conclusion holds for most of the experimental studies of *ad hoc* groups. It is assumed that personality and cultural background explain many aspects of group structure even though there are relatively few experiments designed to investigate this issue.

A. PERSONALITY AND FAMILY ORGANIZATION

Small group experiments provide some evidence that personalities of group members are predictive of their roles, or that certain combinations of individual personalities are predictive of certain group structures or outcomes. For example, Schutz (1958) has shown that a group whose members all have similar levels of need for interpersonal affection (either a need to give and receive a high level of affection or a need to give and receive a low level of affection) has a high rate of productivity in contrast to a group in which the need for affection is mixed. Haythorn, Couch, Haefner, Langham, and Carter (1956) demonstrated that groups in which all members have authoritarian personalities develop cultures different from those composed of equalitarian personalities. These experiments and the few others that examine personality as an independent variable, suggest personality dimensions that are closely related to aspects of group structure and outcome.

Theories of the family and socialization also provide guide lines for experimental investigation of the relationship between personality and family organization. Several socialization theories that have developed out of psychoanalytic theory rest on the assumption that the quality of parental personalities is vital to the development of the child's identity. While these theories vary in their specification of the mechanisms of this relationship, all claim that the personality of the parents is significant for the child's development. Parental personality factors are also predicted to be related through specific child training practices to the development of the child's morality, to the ways in which the child expresses aggression, and so forth (see Hoffman & Hoffman, 1964, for reviews of this material). Other theoretical perspectives on the family take personality into account as well; for example, Winch (1958) explains mate selection and, further, marital success, by using personality variables.

1. The Importance of Intervening Behaviors

With personality of parents as a major theoretical focus, one might expect that this variable would have been extensively examined experimentally. This is not true. Instead, empirical work has largely been in the form of interviews, questionnaires, or tests of parents that are then statistically correlated with some quality of the child. Seldom has the behavior of parent with child been observed directly. Therefore, most conclusions from empirical studies about the relationship between parental personality and child development have assumed, as Hoffman (1960) points out, "that it is sufficient to obtain parent data at the personality level since this underlies overt behavior" [p. 140]. The assumption that personality measures are identical to behavioral measures is often not supported, however, when parent-child interaction is examined empirically. For example, Hoffman predicted that a parent with an authoritarian personality would be most likely to use "unqualified power assertions" in disciplining his child and that this form of discipline would be related to certain behavior of the child. Yet he showed (in a working class sample) that it is not the personality of the mother (measured by the F scale) but, instead, the personality of the father that predicts the mother's reported behavior toward her child. In order to explain this relationship between "personality of father" and "behavior of mother toward child" one must postulate certain behavior relationships between the parents, thus adding an intervening explanation that can be tested by direct observation and measurement. The Hoffman findings show clearly the necessity of examining family interaction empirically in order to specify the relationship between personality and child development or other "outcome" variables.

Direct observation and measurement of family behavior alone is not the solution to this problem. Techniques of data analysis must also be carefully selected in order for observations of parental behavior to be made relevant to the question about personality. Hoffman shows that direct correlations between parental personality and child development leave out the complexities of the intervening behavioral process. In order to understand this process, even after the behavior is actually observed, one must examine the empirical relationships between all three sets of variables: personality, family behavior, and outcome. Instead, many family experimenters follow the analytic model used by nonexperimental investigators, that is, they relate all variables (personality or behavior), taken one at a time, to outcome. Stabenau, Tupin, Werner, and Pollin (1965) in their examination of interaction in families having schizophrenic, delinquent or normal children, serve as an example of this methodological problem. Individual personality measures were obtained for each family member using a variety of tests; also, families' discussions were recorded and coded. The analysis of data took the form of comparing the three types of families in average scores on personality tests or average behaviors in the interaction situation.[2] For example, the individual's ability to abstract was measured by the Object Sorting Test. Implicit in the theoretical approach is the hypothesis that unclear conceptualization on the part of parents may be related to the conceptualization difficulties found in schizophrenic patients. Stabenau shows that, indeed, the schizophrenic patients and their parents (along with the delinquent children and their parents) have significantly impaired ability to abstract. The questions raised by the Hoffman example may be raised here: Is impaired conceptualization evident in interaction between parent and child? Do the parents with the most impaired ability interact in a disorganized or fragmented (or "impaired") way with their children? and, if so, Are these children more impaired than others? or, Is there an even more complex relationship between parental impairment, parental interactions, and child qualities?

When an experimenter asks the questions dealing with the way in which personality is related to behavior, if Kenkel's (1961) findings are representative, he is likely to discover that the relationship is a complicated one. Kenkel observed married couples in their homes (no children were present) who were asked to discuss how to spend a gift of $300; interaction was classified with Interaction Process Analysis categories (Bales, 1951) and a measure of influence on the final decision made. A personality inventory developed by Brim was also administered individually to obtain measures of dominance, persistence, and self-confidence.

In contrast to the prediction that there is a high linear correlation

[2]Essentially, this technique of analysis makes the same assumptions as does the method of ecological correlation. It has the same pitfall as well. While mean scores of groups are compared, it is tempting and not legitimate to conclude that the correlation holds at the level of individuals as well (see Riley, 1963, for a discussion of this and related issues).

between personality and behavior, Kenkel found that this relationship depended on the sex of the family member. Husbands who had "dominant" personalities were indeed more influential in interaction with their wives than were the "submissive" husbands. However, wives with high "dominance" scores were less influential than wives who were "submissive" personalities. Of the latter wives, 80% were more influential than were their husbands while only 53% of the "dominant" personality wives were more influential than their husbands. Kenkel found the personality quality of "persistence" to be unrelated to behavior of either husband or wife while differences in level of self-confidence were related to behavior only for wives, with the self-confident woman being more influential than their husbands. Kenkel pointed out in his discussion the limitations of sample size and mode of analysis, particularly the fact that the effects of combined personality qualities of husbands and wives were not examined. Even with these limitations, however, his study provides one model for investigating "personality" as an independent variable and his findings provide evidence that concepts such as "role expectations" (in this instance, "sex role expectations"), "cultural values," or "interpersonal behavior" must be introduced in order to explain the relationship between personality and family structure or outcome.

The statistical interaction between personality, social expectations, and family role, exemplified by Kenkel's findings, raises a general question implicit in almost all experimental work on families. Which came first? Did the personality of the individual determine his role? Did his role alter his personality? Or was there a circular and constantly shifting relationship between the two? When families who have a long history of intimate interaction are the object of study this question is both serious and difficult to answer. However, there are a few experiments on *ad hoc* groups that provide models applicable to family groups.

Berkowitz's experiment is one of these (Berkowitz, 1956). He was concerned with the relative effect of personality and social role on a member's behavior, the same question that Kenkel's findings on families raises. The study design is such that personality and role are varied independently, so that the effects of each can be examined separately as well as in combination. Members, classified as either high or low on "ascendence," the personality variable, were placed either in central or peripheral role positions (determined by the number of people with whom it was possible to communicate). Dependent variables included the member's rate of communication, time taken to complete a task, amount of information transmitted, as well as post-meeting questions regarding satisfaction with the member's role. Berkowitz (1956) reports that, for the first group task, the personality variable is most important in determining behavior; members low in ascendence who find themselves in a central role behave passively, as their personality scores would suggest. However, by the third group task, demands of the central role seem to dominate and "low ascendence" personalities are "significantly more 'active' than the high ascendence subjects in the periphery" [p. 221]. Berkowitz shows, for *ad hoc* groups, that over a short period of time role effects come to predominate over personality effects, yet the effects of personality are not completely lost since, in peripheral positions, ascendent personalities continue to be more active communicators than those low in ascendence.

Berkowitz provides a model for investigating the relative effects of personality and other variables through manipulation of group membership. Naturally, this model is not directly applicable to real families because selection into roles has occurred in the past. However, since there is now evidence that it is possible for an experimenter to compose "artificial families," or groups having some of the abstract qualities of a family, this provides the opportunity to manipulate "parental personality" experimentally in order to test the effects of personality on parental roles and on child behavior.

As yet, no family experimenter has used this technique to examine the effects of parental personalities. However, following from the questions raised by Kenkel's findings and using the experimental group composition technique the relationship between "parental" personalities, "parental" behavior and "child" behavior might be examined. Let "expected family role" be indicated by the sex and age of the group member; let "personality" be indicated by a score for dominance on an individually administered personality test. Then compose four types of groups:

		Female Member	
		Dominant	Submissive
Male Member	Dominant		
	Submissive		

Place a child with each of these pairs of adults and measure the behavior of all three along relevant dimensions, e.g., influence, participation, expressiveness, imitation, and deviation. Following from one theoretical perspective, one might expect that the child will imitate (or "identify" with) the dominant "parent," without regard for his sex role. The socialization theories that include personality qualities as central dimensions might also be examined using the same experimental model. For example, Lidz and Fleck (1965) suggest that the degree to which an adult has resolved the Oedipal conflict is related to his ability to carry out the parental role. This quality of personality might be used as an independent variable for composition of "parental" pairs who are then required to interact with a child. Lidz and Fleck might predict "schizophrenic-like" behavior on the part of the child placed with the "conflicted" parents. The strength of these designs comes from control over the independent variables and the assumption that other important variables have been randomized; neither of these is possible when real families are used.

B. CULTURAL BACKGROUND AND FAMILY STRUCTURE

Everyone knows that families belonging to particular subcultures have qualities in common that set them off from families with other affiliations. While the "Jewish mother" or the "Irish father" or the "middle-class family" are stereotypes, the fact that such stereotypes have developed suggests that differences are real, if more complex, than indicated by these phrases. Our examination of studies of *ad hoc* groups and family groups will show how powerful subcultural membership is in predicting individual and group behavior.

While the mechanism through which personality affects family behavior is not clearly specified in most of the family theories, there is somewhat greater agreement on the way in which cultural background affects family structure. The intervening process is assumed to be one in which the member of the subculture (religious, ethnic, class) is socialized very early in his life to value certain family organizations, to expect that particular people will take particular roles, to learn that some behavior is right. These learned values lead him to interact in his own family and in other groups according to the subcultural expectations.

Experimental studies of *ad hoc* groups point to cultural background variables that are clearly related to group structure and they also suggest aspects of structure most likely to be influenced by culture. All of these experiments have similar designs; a cultural background variable (or two of them) is treated as the independent variable and the behavior of members, or the outcome of the group interaction is examined. Thus, Strodtbeck, in his series of jury studies, has shown that social class membership is a predictor of the role a member will take on a jury (Strodtbeck & Mann, 1956). For example, many more members of the highest occupational levels are elected jury foremen than would be expected by chance. This may be partially explained through the concept of "expectations" brought by members to the jury from earlier socialization experiences, specifically the expectation by lower class members that a higher class member will lead. Katz, Goldston, and Benjamin (1958) have demonstrated the differential effects of race on group structure, in this instance in groups composed of two whites and two Negros. Here the pattern of speech reflected a high-participation coalition between the white pair, with each of the Negro members talking more to white members than to each other. Again, expectations about role seem to be carried into new groups and the subcultural experience re-enacted. We would also expect that other subcultural affiliations such as religion might predict to the group behavior of a member. Milgram (1964) reported that Roman Catholics, assumed to value a dependent role, were more likely than members having other religious affiliations to be influenced by a group to carry out behavior that they did not personally value, in this case giving another person an electric shock.

If the subcultural expectations about group structure and roles are originally learned within the family then one would expect even clearer differences between family groups of differing subcultural background. Yet, there has been relatively little interest in measuring the effects of these differences; instead, many family experimenters control or match on these variables, some ignore the issue entirely.

We know from nonexperimental research on families that subcultural membership is related to socialization practices, to the nature of parental roles, to values about the child's future, to ways children are expected to behave (see N. W. Bell & Vogel, 1960). Experimental examination of these families provides the additional opportunity to show if and how verbally stated values are actually carried out in interaction in the family. When the cultural affiliation variable has been experimentally examined, its power to predict structural and interactional differences is apparent. The prototype of this approach is Strodtbeck's study of husband-wife decision making in three cultures (Strodtbeck, 1951). Three sets of husband-wife pairs were selected to represent three subcultures differing in values about the role of men and women. As expected, Protestant-Texan couples showed relative equality in influencing the final decision on the revealed differences task; Navajo women, from a matriarchal society, took precedence over their husbands; Mormon men, from a subculture having written prescriptions about male dominance, were more powerful than were their wives. While the Strodtbeck experiment provides a model, both in design and task, the data were not analyzed so as to show the internal structural patterns; for example, we are not told how a Navajo wife interacted with her husband in order to insure her influence. This limitation was overcome in a related experiment (Strodtbeck, 1958) on Jewish and Italian family triads in which adolescent sons were included in the decision making, and the relationships between internal family structure and the sons' need for achievement were of major importance. Family power relationships, indicated by a "who won the decision" measure, were related to cultural expectations; the Italian father was more powerful than both wife and son while the Jewish parents were equally powerful and more powerful than their son. But, in contrast to predictions, there were no differences between Italian and Jewish families in "expressive" role taking (nor on any of the other measures derived from Interaction Process Analysis). This latter finding provides concrete evidence that the observation of family behavior is important in understanding how cultural values are implemented; support for the stereotype of the "smothering" Jewish mother is clearly not evident in the data.

Since the early Strodtbeck work, other experimenters have examined ethnic/religious variation in family interaction. Haley's experiment is particularly interesting, not only because it looks at Japanese-American families but also because the dependent variable measures depart from the usual "who influences whom," "who speaks to whom," or "who is instrumental" (Haley, 1967b). Instead, the only data collected are the sequence of speakers in the experimental session; the only question asked is the following: Does the sequence of speakers deviate from a random one? While we will be concerned in a later section with this measure of "family process" the focus here is on ethnic family patterns. Haley found that the Japanese-American families deviated more from a random sequence than did American families examined in an earlier study (Haley, 1964, 1967a). This suggests somewhat greater rigidity or formality in the Japanese-American family, perhaps consistent with verbally stated values about how family members should behave. However, no pair of family members followed each other more often than other possible pairs. This lack of differences is important since some students of the Japanese family might have predicted a low rate of father-child interactions, consistent with the value on social distance between the two.

Straus' family experiment exemplifies a more complex experimental design in which two social classes are examined within each of three cultural groups: Indian families, Puerto Rican families, and American families (M. A. Straus, 1968). The extent to which families differ across cultures is fully apparent in Straus' reports that the experimental task originally planned was ". . . too strange and difficult for the [Indian] families . . ." [M. A. Straus, 1968, p. 421]. This task, a puzzle in the form of a game in which family members were asked to discover the rules of the game by playing it, had to be greatly simplified for use with both middle- and working-class Indian families. Here cultural differences between families are so great that a task requiring communication and the sharing of cognitive skills is apparently not within the range of experience or mode of organization of the Indian family. Social class differences, however, are strong and consistent across cultures. Middle-class families are more able to solve the problem and the author traces this to greater communication among members of these families than among working-class family members. While cultural and class differences are evident in the Straus experiment, no data are presented on internal family structure, for example, on communication patterns between husband and wife or parent and child within the middle-class families or on the quality of these communications.

One experiment that does examine the internal family structure in order to understand subcultural differences was carried out in Hong Kong by Liu (1966). He compared local Chinese families with those that had recently emigrated from the Chinese mainland. The effects of the acculturation process are shown in his findings that in the refugee family the father has a powerful position which is supported by deference rather than agreement by his son. The local families presented a set of relationships similar to those expected in Western families; the father is most powerful, yet he is also high on warmth; the family consensus also appears to be based on true agreement rather than simply on respect.

These investigations of the effects of cultural background have indicated, first, the importance that subcultural membership has for family structure. There are clearly different power relationships, rates of communication, and agreement patterns that depend on cultural expectations. Second, the experimental method gives the investigator the opportunity to examine specific ways in which cultural expectations are actually carried out in family members' behavior. It is evident from several of the studies cited that there is no clear positive correlation between cultural expectations about family structure and the family structure indicated by interaction measures, just as there is no clear correlation between personality of parents and the parents' behavior.

C. SUMMARY

Both personality and cultural background have been examined here as causal variables that are assumed to predict to aspects of family structure and outcome for the child. There is, however, relatively little experimental evidence supporting this assumption, particularly in the personality area. When there is experimental evidence, that is, when the family's behavior or modes of interaction are measured along with personality/culture variables and other dependent variables, two conclusions may be drawn. First, the theoretical model of the family must be stated in a much more complex form than most family investigators originally stated it. Neither personality nor cultural background of the parents fully explains empirical differences in family structure; instead, the intervening behavioral process must also be accounted for and theory revised so as to represent the statistical interactions between personality/background, behavior, and outcome variables.

Secondly, as soon as an investigator opens up the question of relationships between background and family structure by examining how family members interact with each other, the assumption of causality is also open to question. When families are observed in interaction in order to link parental personality to child's behavior through parental behavior-with-child, questions about which came first, and the relative effects of each, become important. For this reason an experimental design in which the causal variable is controlled by the experimenter becomes useful. Establishing synthetic or artificial family groups by selecting members to represent the major control variables makes the time order of variable effects clear and allows for a more precise statement of the relative importance of these background variables for family structure.

III. Aspects of Family Structure

Most family experimenters have been concerned, either explicitly or implicitly, with the internal structure of the family. "Structure" here refers to the organization of the family group, that is to the different statuses and roles in the family and to the set of norms that govern this organization. We will use these three general structural concepts — status, role, and norm — to organize the large body of experimental work that falls in this area.

One source of variation in these experiments is whether the structural dimension is taken as the independent or dependent variable. In some studies, one aspect of family structure is the independent variable that is experimentally manipulated and then related to a "family outcome" measure such as the child's identification with his parents (Bandura et al., 1963; Hetherington & Frankie, 1967). In other studies, one aspect of family structure is treated as the independent variable and another aspect as the dependent variable; for example, the organization of family statuses may be experimentally controlled and related to family role-taking (Scott, 1962). In the latter approach the experimental goal is to understand the interrelationships between dimensions of family structure at one point in time. In a third type of experiment, the structural variable is the dependent variable and a quality of the child is controlled, thus treating this quality in the experiment as the independent variable (Lennard, Beaulieu, & Embrey, 1965). We will be concerned in this section only with studies that use a structural dimension as an independent variable, and will discuss the third approach in a later section on outcome studies.

A. FAMILY STATUS STRUCTURE

A central proposition in sociological theories about group structure is that all groups differentiate into two or more status levels, each status having different prestige and responsibilities (Bales, 1951), just as a *sine qua non* of family theory is the assumption of status differentiation based on generation.

A status in a group, or in a family, is a position that is relatively higher or lower than other positions in terms of its prestige, rewards, and responsibilities. In contrast to the hierarchical conception of family status, the concept of "role" will be defined here in terms of the quality of behavior expected of the role player. In a family one would expect several roles to be associated with one status level; theoretically both parents, for example, take a high status position *vis-a-vis* their children yet the role behavior expected of each is quite different in quality.

A large number of experiments on *ad hoc* groups have investigated the division of labor and status differentiation. Some have looked at the process through which status differentiation takes place; for example, Hopkins (1964) showed that initial high participation rate is related to high status which, in turn, is associated with high power role-taking. It is important to note that these are not one-to-one relationships, and particularly that some high status members are not powerful. In studies in which status has been experimentally controlled, the status indicators most often used are age, sex, and task ability. J. C. Moore (1968) manipulated status experimentally leading some subjects to believe that their subject-partner was from a higher (or lower) status school than they themselves came from. It was assumed that status outside the group would generalize to status inside the group, and that this would affect the power roles (in this case, influence) inside the group. Findings supported these predictions even in the case of groups in which status "differentiation has no obvious or direct bearing on the task confronting the group" [J. C. Moore, 1968, p. 47]. Just as in the Hopkins' study, the high status member is likely to become the most influential role player.

1. Status Differences as the Independent Variable

There are relatively few experimental studies of families in which status differences are treated as the independent variable, yet family theories place great importance on the differentiation of statuses. Especially for researchers concerned with socialization, child development, and clinical problems of children, the status structure of the family is a central theoretical dimension. Families that deviate from the expected pattern (in which parents have high status and children low status) are predicted to socialize their children in deviant ways and to provide inadequate models for identification (Lidz, Fleck, & Cornelison, 1965; Wynne & Singer, 1963). These theories are a rich source of hypotheses about the effects of family statuses on role-taking and child development, yet these hypotheses remain largely unexamined by experimental methods.

One reason for the dearth of experimental studies investigating family status structure as the independent variable is the difficulty in selecting an indicator of status that is both theoretically meaningful and measurable. The variable selected must not be confounded with "role"; that is, it must stand for a family member's prestige relative to other members, rather than for a member's typical or expected behavior. Furthermore, in order to select families on the basis of differences in status structure, the indicator must be a quality that is relatively easily observable or measurable from outside the family. Because of these problems of selection and measurement, the indicators of status most often used to investigate effects of status structure on other dependent variables such as role-taking are the age and/or generational structure of the family.[3] An alternative solution to the selection and measurement problem is the creation of artificial families, experimentally composed on the basis of status structure.

Scott's experiment investigating the effects of different status structures on interaction in natural family groups exemplifies the first alternative (Scott, 1962). She selected families on the basis of generational patterns; three-person families (husband, wife, and one elderly parent), four-person families (parents, one grandparent, and one adolescent child) and five-person families (parents, grandparent, and two children) and

---

[3]However a two-stage study may be used in which the first stage is designed to obtain the measure of status; see Hetherington (1965), for an example of this approach.

examined power roles within each type of family. While family theories such as Parsons' (1955) would predict that status is positively related to power role, Scott shows that this relationship is a much more complicated one that is dependent on the specific status structure present in the family.

For example, only in the four-person family does the husband have the most powerful role (measured by rates of initiation); in the three-person family it is the wife who is most powerful and in the five-person family it is the oldest child. Furthermore, Scott's findings are not consistent with predictions and findings from studies of *ad hoc* groups that show age (as an indicator of status) to be a predictor of power.[4] In fact, here the grandparents are consistently powerless, whether measurement is based upon rate of interaction, rate of support, or both.

Not only did Scott examine the relationship between the status structure and individual power roles but she also looked at the patterning of power roles in each type of family group following the model provided by Mills (1953). Mills showed that the "solidary" coalition pattern (in which a pair of numbers gives each other high rates of support and there is mutual rejection between the isolate and the pair) is most stable and perhaps is more basic in *ad hoc* groups. Scott found in the three-person family that the "solidary" two-against-one pattern is common, with one important difference. The isolate is given support at a greater rate than is the isolate in Mills' *ad hoc* groups. This may be a major difference between families and *ad hoc* groups that have no history and little future. While a family member may be of low status, have little power, and take no part in decision making, he may still receive the emotional support necessary to maintain him as a member.

Mills suggests that over time, particularly in families, the power structure will tend toward the solidary or conflicting patterns; Scott shows that in approximately one-fourth of the families examined this prediction is not upheld. Furthermore, types of status structure were correlated with the power role patterns of the family. The contending or dominating power patterns are more likely to be found in large families having more adolescent children. One might hypothesize that with changes in status structures (here, the addition of a third generation of nearly adult members) not only does the power distribution shift but the power structure itself changes.

Although family theory has been relatively unspecific about the relationship between status and role patterns, usually focusing on an ideal-

---

[4]Ziller and Exline (1958) showed that age is directly related to influence, but only in groups of men.

type four-person family, it is clear from this examination of natural family groups that theory must be specified in order to account for other types of family organization. Further, questions are raised about how relationships in the family are developed and maintained in these non-typical families. If in three-person families, solidary coalitions between two members against the third exist yet the third member stays in the family, what does the powerful coalition do to maintain its position without alienating the third member? If adolescent children in some families are members of the powerful coalition what influence strategies do they use, especially in the face of normative support for powerful parental roles?

2. Status in Synthetic Families

The use of natural family groups, exemplified by Scott's study, has inherent limitations. The investigator cannot be assured of the time order of the variables and thus cannot draw conclusions about causal relationships. In these experiments we know nothing of the way in which each of these patterns developed, only that some seem to occur together. The second type of experiment is designed to solve this problem; here the status structure of a family is varied experimentally by creating types of artificial families and assuming or assuring that all other differences between families are random.

The experiment carried out by Bandura et al. (1963) is an excellent example of this approach. While the problem was set within the framework of theories of identity development, the independent variable may be restated in group structure terms as "family status." Two types of artificial family were constructed, one in which the child was clearly of low status (not only was he the youngest, but he received few rewards from the "parents") and the other in which the child was of mixed status (he was youngest, but he received more rewards from the high status "parent" than did the other "parent"). The adult male and female who played the "parent" roles were rotated across the high and low status positions. Bandura's question was this: With whom does the child identify, the parent who gives rewards or the one who receives, and does identification depend on the sex of the parent? The dependent variable measures centered on the degree to which the child imitated certain standard behaviors of each of the "parents." In general, findings show that the child imitates the high status "parent," regardless of who receives the rewards. Furthermore, cross-sex imitation occurs, particularly for girls, who are likely to imitate the male adult if he is the high status "parent." Bandura points to the possible effect of the distribution of power between parents, particularly to the effect of reversed parental roles, on the child's identity development.

It is clear that the use of "artificial families" in experimental situations has definite advantages for the study of family structure. Bandura's experiment allows us to examine certain family organizations that are important to family theory yet difficult to sample. We know from clinical evidence that some families have status structures quite deviant from the expected patterns but since these differences are not indicated by variables that can easily be seen from outside the family it is difficult to sample them. Second, the use of artificial families not only allows us to be clear about the direction of the effects of these variables, but also about the process through which the effect occurs. In natural families the effect occurred in the past; in artificial families the effect occurs in the experimental room and can be observed and measured. Bandura provides an interesting example of this latter advantage. While most of the children imitated the high status adult, a few did not; examination of these deviant cases showed that several children superimposed a cultural expectation about adult males on the situation and their imitative behavior was a response to this cultural expectation in combination with the experimental variable. These children reported that the low status male was really powerful anyway, reasoning that "He's the man and it's all his because he's a daddy. Mommy never really has things belong to her" [Bandura, *et al.*, 1963, p. 533].

3. Real and Synthetic Families

The fact that similar issues of status have been examined in similar ways (using age and sometimes sex as status indicators) in both *ad hoc* groups and families has led to questions about the differences between groups and families and has guided some investigators to select this question as the central one for empirical study. Underlying specific questions about the relationship between status structure and other variables in families and other groups is the more general question: Are family groups and *ad hoc* groups different from each other only along the dimension of "length of group history" or are there basic structural differences between them that lead, perhaps, to contradictory hypotheses and to theories based on distinctly different parameters?

Leik's (1963) experiment compares real families having known status structure with synthetic families having the same status structure; age and sex of the participants are the indicators of status. (A third set of groups is homogeneous with regard to both age and sex). Findings suggest that status is predictive of instrumental and expressive role playing in both types of groups but that predictions vary depending on whether the group is a real family or an artificially constructed family having the same status structure (i.e., a group composed of a father, mother, and daughter unrelated to each other). For example, the high status female member plays a different role in the two groups. With an artificial family she takes the emotional role and avoids the instrumental role; with strangers she behaves in the way that the common culture expects a "mother" to behave. But with her own family her role changes significantly; here she takes on much more of the task activity, equal to that of her husband, but at the same time has a high rate of emotional activity. Her daughter is low on task activity within her own family. Thus, in the artificial families it is sex that is most important in predicting role playing; in the real families it is age.

Bodin's (1965) experiment is similar in design and supports Leik's findings. Real and synthetic families do not differ in modes of accommodation to solve a game problem; each type of "family" group used more coalitions and forms of compromise than did *ad hoc* groups composed of equal status members. However, mothers in the real family situations compromised significantly less than did "mothers" in artificially constructed families. Again, there is a suggestion that adult women are likely to take a more powerful or instrumental role in their own families than in *ad hoc* groups. Ryder's (1968) findings on husband-wife dyads suggest, in general, that there are few differences in role playing between real and artificially constructed dyads. "The differences between married and split dyads seem much better described by noting that Ss treat strangers more gently and generally more nicely than they do their spouses" [Ryder, 1968, p. 237]. Since no child was present in these latter groups, it is not possible to test the relative effects of sex and age on role playing.

An interpretative thread seems to run through the empirical studies comparing real with synthetic family groups. Relationships between status and role playing may be understood in the two groups by introducing ideas of "cultural expectations," "expected behavior with strangers," and other explanations that rest on the fact that real and artificial families differ in length of acquaintance. It is assumed that individuals in a new group bring with them a set of norms for group behavior that are initially used to organize interaction. One such norm might be the following: Women are supposed to be "followers" and to play expressive roles. Over time these norms are modified and perhaps altered on the basis of experience within the group. For example, the modified norm might be: High status women have an equal part, with men, in the group decision. If one were to compare artificial families having a long history (for example, using work groups or therapy groups with the status structure of families) with real families, then, following the above argument, the artificial, long-term families and the real families should not differ in role or norm patterns. The conclusion that seems to underlie explanations of the findings in these comparative studies is that the structure of family groups and *ad hoc* groups may both be explained by the same theory. Only the dimension of "time" need be included to predict the family *vs.* group differences.

B. FAMILY ROLES

A central concern, for both small group and family theorists, is with the development and differentiation of roles. A member role is the set of behaviors and relationships that the group and the individual taking the role expect the role player to carry out. A person taking the role of "mother," for example, is expected to have both a controlling and a warm relationship with small children in the family, and she has internalized this expectation. A committee chairman is expected to take charge and a committee member to follow his lead. Within any prestige or status level there may be a number of roles, each with specific expected relationships.

While theorists have predicted that role structure of the family is related to other dimensions such as the child's identity development, experimental work on family roles seems not to have kept pace with theoretical formulations. Instead the major empirical question has been, what is the nature of the role relationships within the family, and particularly between the parental pair? Therefore, a major portion of this section will be devoted to experimental studies in which family roles are the dependent variable. Only recently have the parameters of family role structure become clear enough that role could be introduced experimentally as an independent variable and its effects examined. At the end of the section, two experiments using role as the control variable will be presented as models for further work.

1. Parsons and Bales Theories of Role Development

The major theories of role development in *ad hoc* groups and in families, those of Bales (1951) and Parsons and Bales (1955) are obviously closely related. However, there seems to be at least one major theoretical difference that would lead to differing predictions for role development in the two types of groups. Bales links the group's differentiation into instrumental and expressive roles to the two systems problems imposed on all groups: (1) the requirement that groups adapt to externally imposed tasks and at the same time (2) maintain the group as a group or deal with the integrative needs of the members. The inherent need to handle both problems leads to differentiation of roles into two specialties, task and emotional, that are assumed to be "complementary and supporting in the long run but in the short run tend in some degree to conflict with each other in a way that makes it difficult for the same man to be top specialist on both" [Parsons and Bales, 1955, p. 298]. In its most general sense, then, the theory predicts that roles will differentiate into expressive or instrumental specialties and that these two special roles will not be played by the same individual.

Parsons' theory of family role structure begins with the same assumptions as Bales' theory of group development (Parsons and Bales, 1955). Because the family as a system must deal with both adaptive (economic, political, and other extrafamily) relationships and integrative (internal) family relationships, presumably it must develop a division of labor in which instrumental and expressive roles become specialized. For the same reasons as in all groups, these special roles cannot be taken by the same individual. However, the Parsonian theory of the family adds a further qualification. Family roles are allocated according to sex, with the father taking the instrumental role and the mother the expressive role. This addition is necessary because Parsons assumes that the major function of the family is to socialize the child, and, following psychoanalytic theory, the socialization process rests on the identification of the child with the same-sex parent. Thus, if parental roles are not clearly differentiated, and further, if they are not clearly related to the sex of the parent, the child's identity will be impaired. Therefore, the theory predicts that family roles differentiate into instrumental and expressive specialties, that one individual cannot take both roles, and that the delegation of roles is based on sex.

2. Research Findings Bearing on the Theories

These theories of role development in families and groups have provided guide lines for research over a relatively long period of time. Yet there are few instances in which the theories, as they stand, are supported empirically. Until relatively recently, Slater's test of Bales' theory of role differentiation in *ad hoc* groups (Slater, 1955) stood as its major empirical support. In his groups there was clear evidence that members perceived and expected someone to be a task leader and another member to be an emotional specialist. However, as Slater points out in a later publication (Slater, 1961), the correlation between instrumental and expressive role activity was positive rather than negative as the theory would predict. This finding suggests that the major hypothesis must be qualified. It is even more difficult to find support from experimental studies of families for the Parsonian theory of family role differentiation. There are no experimental studies in which the theory is confirmed in an unqualified way.

Instead, empirical evidence from *ad hoc* group experiments indicates that the proposition that instrumental and expressive roles are taken by separate members must be modified. Burke (1968) points to group size, task orientation, legitimacy of the leader, and acceptability of the task behavior as necessary qualifications. Turk (1961) has shown that one member could be both the task leader as well as the best liked person in instances in which the task of the group is highly valued by the whole group. In groups of student nurses doing work together that they felt to be interesting and worthwhile the differentiation between instrumental and expressive leadership did not occur, perhaps because the tension raised by the task leader's instrumental activity was not so great that it could not be allayed by her own expressive acts. Burke's (1968) experiment, methodologically much more rigorous, supports the Turk findings. Role differentiation is mediated by the group members' acceptance of the "task ethic." *In groups where the task is not accepted, the members' instrumental activities are negatively correlated with the liking they receive. There is no task-liking correlation in the groups accepting the task; in these latter groups, instrumental and expressive role playing may be carried out by the same person.*

Several family experiments have indicated also that the separation between instrumental and expressive role taking, while theoretically expected, occurs only under special conditions. Kenkel (1957) in one of his series of experiments on husband-wife decision making measured "influence" (by whose idea was finally adopted by the couple), "task acts" (by Interaction Process Analysis categories), and "expressiveness" (in the same way). Husbands did give the higher proportion of task acts and wives performed most of the social emotional activity; however, the wives who were most expressive were also most influential and the least expressive wives were least influential. This latter finding suggests that one must qualify the prediction that "expressiveness" and "power over decisions" do not occur in the wife role. Leik's findings, already reported (Leik, 1963), also suggest qualifications of the theory. In real families both parents are high on task activity, and the mother and daughter are both specialists in emotional acts. Mothers therefore specialize in both task and emotional areas.

If we could stop with the Kenkel and Leik findings we might assume that the family theory should be qualified so as to account for the extension of the mother role into both instrumental and expressive areas, perhaps to be explained by the decreasing importance of the father in socialization. However, experimental findings reported by M. A. Straus (1967) add further complications. Family triads were asked to work on the task previously used in Straus' cross-cultural study. Parental power was measured by counting directive acts to other members (e.g., "Shoot the ball now") and expressiveness was measured by supportive encouraging statements. As expected, fathers were more directive of their children than were mothers; however, they were also much more supportive of children as well. But even more significant for the parental role-taking theory is the finding that "the husbands [were] predominant in both the expressive and instrumental role in the conjugal interaction sphere as well as in the parent-child sphere" [M. A. Straus, 1967, p. 18]. In the Straus study, therefore, it is the father whose role combines both major areas of role playing.

3. Role Differentiation Contingent on Situational Conditions

These experiments on *ad hoc* groups and families suggest that instrumental and expressive roles may not be inherently incompatible, and that, if there are temporary incompatibilities, groups and families develop methods for avoiding or resolving them. Freilich's discussion of the universal triad (Freilich, 1964) describes a number of situational factors which, if they are used to modify the role differentiation theories, may result in much more accurate descriptions of role playing in real families and other groups.

The prototype for the "situational" effect on role differentiation (described by Freilich, 1964) is the primitive family in which the parents' instrumental and expressive roles are played in different subsystems, thus serving to avoid the basic incompatibilities. For example, in a patrilineal family system, a father is an instrumental leader in relation to his own child but an expressive leader in relation to his nephew, allowing him to switch roles only when switching objects.

Such clearly defined situational rules do not seem to be available to families in modern society. However, the sex of the child may be one variable worth investigating in this regard. Nonexperimental literature suggests that the instrumental or expressive role chosen by the parent is dependent on the sex of the child with whom the parent interacts. Bronfenbrenner (1961) reports that in lower class families parents punish children of the same sex and indulge children of the opposite sex. O'-Rourke's (1963) experimental study, shows that fathers with female children are more expressive than mothers with female children. Since in the latter study, the parents of female children are not the same as those with male children, and since no parents were observed interacting with a male and a female child at the same time, we have little information of the process of role switching or what the child might do to elicit the switch.

The O'Rourke (1963) experiment points to a second "situational" variable that predicts the quality of role differentiation in the family. He compared a set of three-person families observed in an experimental laboratory with the same families observed in their own homes. When the family is placed in the more "artificial" laboratory surroundings, role differentiation follows the Parsonian predictions, with fathers taking the instrumental role and mothers the expressive role. At home, fathers drop in instrumentality and mothers increase; the opposite change occurs in expressive role taking.

Another situational variable that seems to be related to the selection of instrumental and expressive roles by parents is the cultural definition of the task. March (1954), in his study of political discussions between husbands and wives, used Interaction Process Analysis categories to show that instrumental and expressive role playing depended upon the cultural definition of the discussion topic as a "male" or "female" concern. When issues of foreign policy were discussed, the husband was the dominant instrumental leader; when the topic was of local politics, the wife's instrumental role became more prominent. Unfortunately, measures of expressiveness were not reported so that it is not at all certain that role differentiation actually changed form when the cultural definition of the task changed.

These experimental findings show that the clear division between instrumental and expressive roles of parents predicted by Parsonian theory must be modified, and, further, that several aspects of the parents' situation seems to be predictive of the role division. Role sharing may be associated with the sex of the child present, the cultural definition of the task, and the degree of "artificiality" of the situation.

This last dimension, "artificiality," seems to be an important one in both *ad hoc* group and family experiments. In real families there is little evidence of the instrumental and expressive role differentiation associated with the sex of the members, as is predicted by Parsonian theory. However, when families are examined under somewhat more artificial conditions (either in the laboratory rather than the home, as in O'-Rourke's (1963) study, or as artificially composed families, as in Leik's (1963) experiment, there is greater support for the Parsonian theory that relates role-playing to the sex of the parent. In *ad hoc* groups having none of the structural qualities of a family and having an even greater degree of artificiality, role differentiation is most clearly related to the sex of the group member. Strodtbeck and Mann (1956), for example, report in one of their jury studies that women exceed men in positive expressive reactions and men exceed women in attempted answers. Thus as the group becomes more artificial and *less* like a family in its structure, role differentiation becomes *more* clearly related to sex, and the Parsonian theory of family role structure more clearly supported. *Ad hoc* groups may use sex of the members as the determinant of role differentiation because of its visible and agreed-upon quality. When the group has developed specific norms of its own and when the integrative problem becomes crucial — usually this occurs when the group has its own history and has plans for a future — roles are shared, and instrumental and expressive roles become detached from the sex of the member.

4. Role as Independent Variable

Most of the preceding studies of role structure have centered on instrumental and expressive role playing and have been concerned largely with finding the parameters of family role structure and with describing the empirical patterns of roles. Few have gone on to examine role as an independent variable in order to understand the effects of different role structure on other variables such as the development of the child or the content of family norms. However, studies by Hetherington and Frankie (1967) and Mussen and Parker (1965) may serve as useful models for further experimental work in this area. Each uses some quality of the role of parent as the independent variable and asks how this role is related to the child's identification with the parent, measured in terms of imitation of parental behavior. One question that may be asked, following this model, is, if parental role differentiation is not based on sex, what are the implications for the identification of the child?

Hetherington and Frankie sampled parents on the basis of "dominant," "warm," and "conflictful" interaction with each other in an experimental situation (these measures indicate some dimension of instrumental and expressive role-taking) and then used the parents as "experimenters" in the sense that they were asked (individually) to play a role with their child that had been defined by the researchers. The dependent variable was the extent to which the child imitated the experimentally specified behavior of his mother or father. This approach falls between the artificial family method and the use of the real families since, here, there is experimental control of the independent variable (the parent role) and yet implicitly present in the parent-child interaction is the previous history of the family with all its unique qualities.

Findings show the importance of varying "parental role" experimentally and examining the effects of role on the child's behavior. The greater the warmth of the parent the greater the imitation by the child. However, when sex of the parent and sex of the child are taken into account "maternal warmth affects the girl's imitation more than it does the boy's." Also, the major prediction that—in homes in which there is high conflict between parents and little warmth from either of them, the child will identify with the dominant (aggressor) parent—is upheld; under these conditions both boys and girls imitate the dominant parent regardless of sex. However, if conflict is lower or if one parent is warm there is less tendency to imitate the dominant parent. This is true except for boys with dominant fathers; these fathers are imitated by the boys no matter whether mothers are nurturant or not. For girls, the warmth of the mother is salient for imitation.

One finding from the Hetherington study reflects on the question raised by the descriptive studies of role differentiation: If parental role differentiation is not based on sex, what are the implications for the child's identity development? In instances where the mother was the dominant parent, rather than the father, both girls and boys imitated the mother. The boys in these families may, therefore, be the ones whose identity development is impaired.

A second experiment in the same tradition adds an interesting finding that relates parental role-taking to the child's behavior. Mussen and Parker (1965) selected "nurturant" and "non-nurturant" mothers (measured by an open-ended interview) and asked each of them to take an "experimenter" role with the child. Findings showed that the degree to which the mother is nurturant (or expressive) is unrelated to the child's improvement on the task (a Porteus Maze). However, nurturance of the mother is significantly related to the child's imitation of his mother's incidental (non-task) behavior. For example, the greater the nurturance the more are irrelevant comments of the mother repeated by the child.

Once family researchers produce findings showing that parental roles are not simply divided between instrumental and expressive types, and not simply allocated at all times in all situations according to the sex of the parent, then an abstract theoretical model may be constructed to take into account these complex relationships between variables. These patterns can then be examined concurrently, either by sampling real families or by constructing artificial families, in order to test their effects on other qualities of the family or the child. The Hetherington and Mussen experiments are excellent models for investigating complex role structures in a clear and interesting way.

C. Family Norms and Sanctions

Knowledge of a family's status structure and patterns of role differentiation provides the framework for understanding interrelationships between family members. However, this structural framework shows only the skeleton of the family and tells us little about the quality of family interaction, the climate of feeling in the family, what kinds of behavior are expected and what kinds avoided or punished. The concepts that underly this latter concern are "norms" and their accompanying "sanctions." Norms are rules for behavior that regulate the family members' conduct and that function to make interaction predictable for its members; they apply, naturally, to role-taking but affect other facets of interaction as well. When family norms are broken, or when they are particularly well followed, the norms call forth positive or negative sanctions to reinforce the family's normative system.

The clinical and descriptive literature on families best exemplifies the importance of family norms in predicting other aspects of the family. Wynne, Ryckoff, Day, and Hirsch (1958), for example, describe the normative structure of families who have produced a schizophrenic child and tie these peculiar norms to the problems the child has in forming an identity. A norm found commonly in these families is that "there is to be no hostility, disagreement, or other expression of negative feelings." Sanctions that support this norm are extreme, consisting of mass denials of behavior or feelings, and family myths that threaten catastrophe if the norm is not followed. The child growing up in this family finds it difficult to perceive accurately both his own feelings of hostility and hostile feelings of others, thus isolating him from relationships outside the family and limiting the possibility of identifying with qualities of non-family members.

Wynne's use of the family norm concept implies that family norms are rules for behavior known by the members (at some level) and carrying with them a sense of moral obligation. Family members feel that it is right and good to behave in a certain way and to punish those members who do not. However, there is an alternative definition of the concept, important because it leads to quite different measurement techniques. This second conception of "norm" does not imply knowledge by members nor a sense of obligation, but instead centers on the predictability or patterning of family behavior; if a family's behavior over time is consistent then it is assumed that its interaction is governed by norms.

1. Norm as Moral Obligation or Behavioral Consistency

The particular conception of "norm" used by an investigator sets limits on the ways in which norms are investigated. When the concept implies a moral obligation known by the family member, then two experimental techniques have generally been selected to examine the relationship between family norms and other variables. Members have been asked to predict others' behavior and the accuracy of the family's predictions used as a measure of their agreement on the family norms. The more common technique, taken from earlier experimental studies of *ad hoc* groups, consists of examining sanctioning acts and inferring the nature of the norm being violated.

In contrast, when norm is understood as a consistently patterned behavior, then measurement usually involves comparing interactions across a period of time. Experiments on *ad hoc* groups provide a model for this approach. For example, Heinicke and Bales (1953) examined consistency in interaction over four meetings for a set of student groups, using Interaction Process Analysis categories as the dependent variables. Interaction in the task, or instrumental, area did not change significantly over this period, but there was a significant increase in the level of solidary and tension release acts. While we could assume a norm for instrumental acts, the norms about level of expressiveness either changed over time, or were in the process of development. It is apparent here that conclusions about norms drawn from time samples are tenuous.

Do families have consistently patterned behavior? Two experimenters have followed the Heinicke-Bales model in examining families experimentally. M. Moore (1967) asked normal and "clinic" families to interact twice across an eight to ten week period, and rated their interaction on variables such as anger, tension, warmth, cohesiveness. Correlations ranged from .12 to .74, averaging .41. Moore concludes ". . . there exists a core of interaction consistency over time for both experimental groups" [pp. 4564-4565]. The fact that the average correlation is low and the range large suggests that much more interesting normative variation hides behind the correlation values.

Ferreira and Winter (1966) also examined normal and clinic families across a six month time period, asking members to discuss "revealed differences" items. They found no significant differences across time on levels of spontaneous agreement, decision time and choice fulfillment. Here, the mode of family decision making seems to be governed by normative rules.

These measures of behavioral consistency, while clearly important in evaluating the validity of findings from the usual family experiment in which each family is measured at only one point in time, are only introductory to the more interesting and specific questions about family norms. For example, are there some behavioral consistencies that, if interrupted call forth sanctions, and others that do not? Do families differ in the scope of their norms? Is there variation within a family in the degree to which norms are internalized by its members? If there is variation in understanding (or following) of family norms, does this variation affect other dependent variables for example, the development of the child? Each of these questions calls for more specific questions about knowledge of the norms, agreement on the norms, and strength of the norms. Thus, using the "consistency of behavior" studies as background, a few experimenters have moved on to examine the rules for behavior that are obligatory and understood by the group members.

2. Relation between Normative Structure and Other Group Aspects

It is surprising that there are so few experimental studies of *ad hoc* groups that deal directly with the problem of internally developed, obligatory group norms and the relationship between normative structure and other aspects of the group. The experiments that can be said to investigate norms usually examine the conditions under which group members "conform" to an experimentally-imposed norm, such as the group norm represented by the set of trained subjects who report inaccurate line lengths in the Asch-type experiment. In none of these experiments can one assume that the norm has naturally arisen as a result of group interaction over time.

However, a recent field study of small groups is an exception to the usual focus on experimentally-imposed norms. Feldman (1968) asked children in 61 groups at a number of camps to answer questionnaire items about the norms in their groups; content of the questions was developed by pre-testing a larger number of items and, presumably, selecting those that some groups reported to be naturally developed norms of the group. Major concerns of the study are the degree to which a group is normatively integrated and the relationship between normative integration and other qualities of the group. Normative integration is the extent to which each member's report of the norms coincides with the average report of the whole group. Thus, in our terms, it is a measure of the extent to which all members can accurately report the unique norms of their group.

Feldman reports that normative integration varies significantly depending on the sex of the group members, with girls' groups having greater agreement on the norms. But, further, he shows how agreement on norms is related to other qualities of the group as a whole. Norm agreement is related to the interpersonal integration of the group, that is to the extent to which members give and receive liking choices. Thus, with greater agreement on the rules there is greater reciprocal liking. However, normative agreement is not significantly related to functional integration, that is, to agreement among group members that three separate roles are being taken by members. While we might have predicted that agreement about group norms should be directly related to agreement about role differentiation, this does not seem to be the case, at least in Feldman's groups of children.

The Feldman study is not an experimental one, yet it provides a set of methods for measuring group norms as well as a complex set of findings suggesting that the quality of the group norms is related to other aspects of group structure.

One family experiment provides for a similar conclusion. Ferreira (1963) asked: Do family members accurately perceive the amount of rejection that other members direct toward them? In other words, in the family group is there agreement about the amount of rejection that is being expressed and thus is there agreement about the norms for rejection? Further, he asked, is this agreement related to other qualities of the family group?

Normal and "clinic" family members worked alone, coloring drawings and, later, choosing or rejecting the drawings of other members of the family. Not only was each member asked to reject others' drawings but he was asked to predict who would reject his. "Rejection" and "expected rejection" measures were obtained for each whole family and for each family member. One relevant finding is that there is less agreement on norms for rejection in the clinic than in the normal families. Ferreira also reports that children are more accurate perceivers of rejection than are adults in the family. This interesting finding, that in families where norms are misperceived, and where there is a clinic patient present (usually, but not always, the child) it is the child who is the most accurate perceiver. Thus, it is not simply that children who misperceive the norm are abnormal, or that children who are abnormal misperceive the norm. Instead, the parents are more likely to be the misperceivers, and this raises the question of the effect of lack of understanding or lack of consensus about family norms on the developing child. Further, what is the consequence of lack of normative consensus on family role structure? These are questions that might be investigated more systematically with the use of "artificial" families and experimentally introduced normative consensus.

3. Responses to Deviations from Norms

Several experimental studies of *ad hoc* groups provide another model for investigating the family norms. Here, group sanctions are observed and their effects measured; in most instances the experimenter insures that sanctions will occur by introducing a role player who plays a deviant role. Schachter's (1951) early experiment on deviation and communication indicated that a member whose opinion deviated considerably from the modal opinion was sanctioned first by group pressure to change, and later by being ignored. Other experimenters have built on this model and have asked more specific questions about sanctioning. Sampson and Brandon (1964) placed deviant members in experimentally constructed groups but varied the type of deviance; in some groups the role player differed from members only in her opinion about a group issue; in other groups, the role player deviated in terms of her social background and previous experience. Thus "opinion" and "role" deviation represent different degrees of deviation and the findings can be understood in these terms. A mild deviator (the opinion deviant) receives more communications, and especially more hostility and questions than do conformists and extreme deviants. The group attempts to understand his position and pressures him to change his opinion. The role deviant, whose differences are seen as more pervasive and long-term, is not the target of communication or hostility, but instead is simply ignored. Sampson and Brandon suggest, "Perhaps, like the stigmatized individuals to whom Goffman refers, the role deviate is permitted to exist as long as we, the normals, the role conformants, do not have to look at him" [p. 281].

Wiggins, Dill, and Schwartz (1965) investigated another dimension of sanctioning by groups, again by introducing deviance experimentally and examining the resulting behavior. They found that the extent to which a deviant is rejected is related to the extent to which deviance threatens the attainment of the group goal as well as to the status or importance of the deviant member. A high status member is punished less for a minor deviation but more for a major deviation, perhaps because he is not felt to be subject to the norms of the group.

Each of these experiments on sanctioning in *ad hoc* groups raises questions and problems relevant to families. In a family group, is extreme deviation sanctioned by ignoring the family member? If this member happens to be the child, what effect do these sanctions have on the child's role or development? Is it true, as Wiggins *et al*, might predict, that a high status family member is not sanctioned for minor deviations? Does this imply that the hostility aroused in the family will be displaced from the parent to a lower status member? Or, does this pattern of sanctioning change when the group has the age and sex structure of a family?

4. Effects of Parental Sanctions

These questions on sanctioning in families have only recently been put to experimental test. While the small group experiments that we have reviewed here deal with several aspects of deviance and sanc-

tioning—degree of deviation, type of deviation, and status of deviants—
the family experiments have looked only at the effect of sanctions by
parents on the child. In some instances the parental sanctioning is
known to reflect the real family interaction patterns; in others, sanctioning methods are imposed by the experimenter.

Stevenson, Keen, and Knights (1963) investigated the effects of
positive sanctions administered by parents as compared with strangers,
on the behavior of their own young children. While children transferred
marbles from one box to another the adult present gave verbal rewards
("That's very good," etc.) Stevenson found that the strangers had a
greater effect on the child's production (number of marbles transferred)
than did his own parents; only girls rewarded by their own mothers increased production. With strangers, all children increased production
except for boys with male strangers. Stevenson does not relate these
experimentally defined positive sanctions to the specific mode of sanctioning used by each of the parents at home.

Patterson's (1965) experiment took this latter factor into account by
selecting parents according to their use of punishing sanctions in the
home and then asking them to serve as "experimenters" and to follow
standard "punishing" behavior in the experimental situation. From interviews with mothers only, the child's home was judged, using rating
scales, to be warm or hostile, restrictive or permissive. A parent was
then cued by earphones to make a critical or negative comment on his
child's performance on a task similar to that used by Stevenson. The
effect of these comments on the child's performance was found to be
contingent on the usual modes of sanctioning used by the parent at
home. The greater the restrictiveness in the home the more the child
changed his behavior as a result of negative sanctions by his parent.
However, this applies only to boys; there are few significant differences
for girls. Futhermore, the degree of warmth reported in the home is unrelated either to boys' or girls' responses to punishment. The effects of
combinations of warmth and permissiveness on the punishment of the
child are not examined. These more complex relationships could be of
great interest in understanding the effects of preferred mode of sanctioning of children by parents.

As we have pointed out the major experimental concern has been
largely with sanctions directed from parent to child. Recent theoretical
interest in the effect of the child's behavior on the parents (Bell, 1968)
and the use of sanctions by all family members to maintain a particular
normative system (Wynne et al., 1958) has not reached experimental
test. However, it is clearly a simple step to extend the above experi-

mental techniques to other family relationships. Patterson has shown that it is possible to take into account the unique family norms and sanctioning patterns and to reproduce them experimentally in order to examine their effects. The small group literature provides considerable technical help in operationalizing sanctions and measuring effects; however, with the exception of some descriptions from group therapy, it is the clinical analyses of families that describe normative content and the sanctioning strategies in detail and these may be most fruitfully examined with experimental methods.

D. SUMMARY

In the course of examining small group and family experiments concerned with structural dimensions there are several points at which findings reflect back upon a general issue: Are families and *ad hoc* groups simply two sub-types of a general type, "group?" Findings from several of the experiments comparing status structure in families and groups suggest that status is related to other variables in similar ways in both types of groups, with the empirical differences that occur explained by "length of acquaintance" or "situational" variables. A similar issue is raised, but not resolved, in the experiments on role-playing. There we noted that the theoretical formulation of family role-taking seemed to be a better explanation of role-taking in *ad hoc* groups than in the family; while expressive and instrumental roles are allocated on the basis of sex in the *ad hoc* groups this differentiation seems to break down in families, perhaps replaced by a role structure negotiated on the basis of the unique family situation and composition. An important empirical question, as yet unanswered, is whether or not a similar change in role structure occurs in long-term *ad hoc* groups.

Just as was the case in the experiments on social background and personality, when family structure is examined by looking at behavior in a controlled situation rather than simply by using interview or test data to infer structure, the resulting theory must be much more complex and specific. This is clearest in the experiments on family role-playing where it was found that a number of additional concepts must be included in the theory in order to explain the fact that parental roles are not neatly divided into instrumental and expressive types. Once the theory becomes more specified and includes behavioral concepts, then it is possible to put these specific hypotheses to the test in artificial families or through experimental intervention into the family system.

IV. Family Process

While experimental studies of families have provided a complex picture of family structure at one point in time there has been little concern with change. Yet as one looks at families interacting, either in the laboratory or under natural conditions, it is clear that change constantly occurs. Roles are switched, the balance of power shifts, sanctions are imposed, and family members learn new strategies and techniques for interacting with each other. Furthermore, these changes are not random ones but, instead, seem to be related to other variables such as the imposition of a new task on the family, the maturation of the child, or the loss of a member.

Experiments on family structure attempt to control for "change" variables by examining families at comparable points in the family history and under well controlled conditions. Suppose, however, that change or process were the major theoretical variable; how could it be approached experimentally? The few family experimenters concerned with this issue examine it in two different ways. The first reflects an interest in macro-process. Families are seen from an historical perspective and questions have to do with structural changes throughout the life cycle of the family. The second approach is through examination of micro-process, or change within an experimental session. Here the concern is with how one member's interaction affects a second and how the second, in turn, affects the first. Not only is there interest with whether change occurs, but also with exactly how this change comes about at the level of discrete behaviors.

A. FAMILY MACRO-PROCESS

Longitudinal and clinical studies of families suggest that major structural changes come about as a result of normal development (aging of parents, maturation of the children, changes in the extended family) as well as from less predictable circumstances (illness). At first glance it might be assumed that these long-term changes are not amenable to cross-sectional experimental methods and would be more appropriately tested in a longitudinal study. A sample of families might be examined over the course of the family history and changes in structure related to other variables. R. Q. Bell (1959-1960) discusses the difficult problems of longitudinal, or prospective, studies of families and also suggests an alternative technique in which the time required is telescoped by use of matched samples of families each of which is seen at only two points in its history. If theoretical concern were with the changes in family structure occurring when children reach adolescence and then begin to leave home, and if change were assumed to occur throughout this entire time period, a valid picture of change during this ten-year period in the family's history could be obtained in the following way: One sample of families might be examined when the child is 12 and again at age 14; another matched sample is examined when the child is 14 and again at 16, and so on. If samples are appropriately matched then conclusions can be drawn about the nature of the full-sequence of structural changes and the variables that predict these changes; further specific within-family changes may be measured and described.

Another technique that avoids some of the problems inherent in longitudinal studies consists of selecting families at particular "crisis" or change periods and examining them at several points in time during this relatively short period in the family history. Rapoport's study of families during the "engagement through first child" period (Rapoport & Rapoport, 1964) may allow conclusions about structural change and methods of handling it that can be generalized to other points in the family history as well.

Neither Bell's accelerated longitudinal method nor the "crisis" sampling technique has been used in combination with an experimental method to examine family process. However in a few instances appropriate sampling has been used as a substitute for the time dimension. Scott's (1962) experiment exemplifies a design in which cross-sectional materials are used to draw conclusions about structural changes in the family over a relatively long period of time. While her experiment was not specifically designed for this purpose the selection of family groups might be ordered so that cells of the experimental design stand for periods in the life history of a "family." The four-person family (parents, one child, and one grandparent) may be assumed to stand for an early point in the family history; a later stage of development is represented by the five-person family, the family having an additional child. In discussing this study earlier we suggested that differences in power structure between types of families may be related to the normal changes expected when an adolescent child matures. Scott reports that while parents are high power members in the smaller family it is the adolescent child who is high power in the larger, five-person family. If this experimental study had been specifically designed to examine the developmental process, a number of methodological requirements would have had to be met: one would want accurate matching of families on all variables assumed to be related to structure as well as careful selection of experimental groups on the basis of their "stage of family development."

A second experimental approach, taking an even longer historical perspective, asks about the relationships between family structure across three generations of related families. Borke (1966–1967) selected a small sample of families and collected interactional data from all three generations. Interaction profiles were compared and conclusions drawn about the ways in which motivations for acting might be transmitted from one generation to another. While an attempt was made to observe families under relatively standard conditions and while objective measures were applied to their interaction, the fact that each of the families was seen at a different point in its own history means that one must assume the stability of family structure over a long period of time in order to draw conclusions about the effects of socialization from one generation to the next.

The cross-sectional experimental methods represented by Scott's and Borke's work is designed to answer questions about gross changes in family structure over time. It does not tell us how these changes might have come about. For example, if the adolescent child does become a powerful member of the family, perhaps replacing the parent in some situations, we know only that this occurs, not how the adolescent takes over the role nor how the parents respond. It is the experimental studies of micro-process that add to our understanding of these mechanisms of change.

B. Family Micro-process

The studies of micro-process in families, in most cases, examine the relationship between one act and the response that immediately follows it. Thus, findings about "change" or "process" refer to very short periods of time relative to the family's history, yet the conclusions drawn from process findings may contribute to our understanding of how families go about maintaining or altering structural patterns throughout their entire history.

In this area there are few precedents in small group theory or measurement from which family experimenters may draw. The vast majority of small group experimenters has collapsed the process of interaction into an aggregate measure for the whole experimental session. Then, from the total proportions of certain kinds of actions investigators sometimes draw inferences about the sequence of actions between group members; the sequence itself usually remains unmeasured.

In his early work on equilibrium in groups, Bales (1953) discussed techniques for measuring micro-process and provided a theoretical rationale for interest in the minute sequences of actions. Using his proaction-reaction method each act is examined and the question asked, what is the quality of the next action? These two-act sequential patterns are then summed for the whole group session and the proportions compared. For example, Bales reports that for experimental groups of students, when an act of antagonism occurs the act that will most probably follow is a tension-releasing act (laughter, joking). This sequence is seen as one that functions to maintain the equilibrium of the group.

While Bales does not spell out in detail the relationship between specific sequences of interactions and the maintenance of equilibrium in the group, the theory seems to assume that sequential patternings of acts will follow rules similar to other equilibrium-maintenance mechanisms on other theoretical levels. Just as role differentiation serves to maintain the equilibrium of a group so do particular sequences of acts, and the specific content at these two levels should be the same. Instrumental acts tend to elicit more instrumental acts, just as an instrumental role player tends to carry on this role over time. If the instrumental act does not elicit another instrumental act it will most likely elicit a positive expressive act, in the same way that an instrumental role player is complemented by an expressive role player who functions to restore positive interpersonal relationships among members.

In the history of small group experimentation this concern with micro-process seems to have dropped out, probably as a result of the difficulties of measurement and analysis. Not until computers were able to take over the job of counting sequences or even of coding interaction into categories (the General Inquirer is an example of the latter technique; see Stone, Dunphy, Smith, & Ogilvie, 1966) was there a return to the issue. And as yet the return is largely in terms of problems of measurement and analysis, not more abstract theoretical development.

Raush's (1965) field study of groups of normal and hyperaggressive boys living on an inpatient hospital ward is one important model for studies of family micro-process. He reports findings that show clearly the theoretical relevance of examining interaction sequences and demonstrates statistical techniques for handling measurement of behavior contingency patterns. The boys' interactions were sampled by selecting time periods and their actions coded as either "friendly" or "unfriendly." Measures derived from information theory were used to ask several questions about the sequential patterning of interaction. In one analysis, longer chains of interactions were measured, and theoretically expected sequential patterns were compared with actual sequences.

The relevance of this method to theories of equilibrium is apparent in some of the findings. In the group of hyperaggressive boys, hostile responses increase in successive steps; "more interesting is a suggested trend for hostility to increase beyond its expected theoretical course" [Raush, 1965, p. 495]. These children appear to have no corrective ability, no techniques that will turn off the hostility once it has begun. In contrast, among normal children "the chain of interactive events . . . would be expected to wind up in a friendly fashion more than half the time. . . . At each successive step, the normal boys showed greater friendliness than would be expected from the hypothetical curve based on the initial transition matrix" [p. 496]. Raush suggests that the group of normal children had a fund of strategies for switching the tone of its interaction and, furthermore, knew at what point in time to use them. For example, shifts of topics, recognitions of others' feelings, conventional encouragements are used by normal children to alter the direction of group processes. Raush's findings show that the examination of micro-process is useful for understanding behavioral contingencies, for discovering the implicit rules of response, and for examining the conditions under which these rules are modified. The application of methods like these to family groups should provide a way for determining the unstated norms for behavior, that is, the behavioral mechanisms used to maintain equilibrium, as well as a method for determining how structural changes occur.

Two groups of family experimenters have examined interaction at the micro-process level. Each asks the question, is the behavior of one member contingent (in specified ways) on the behavior of another member? Further questions that may be asked are: Under what conditions are contingency patterns changed? How? Who does it? Perhaps these questions may lead back to an issue raised by Scott's study, namely, how does the adolescent child gain a powerful position in his family? How do parents respond to these attempts? How does a new power equilibrium get worked out in these families?

Haley (1964, 1967a) asked about family micro-process by examining only one quality of the sequence of interaction, a non-content one. From samples of three and four-person families in laboratory discussions, he obtained the sequence of speakers, i.e., who follows whom. He then asked to what extent the actual speaker-responder sequence departed from a random sequence and assumed that departure from randomness stands for the use of a family rule about who should respond to whom. In three-person families (some normal and some having a psychiatric patient member), all families deviated from the random pattern; that is, for all there was a relatively predictable "process" of interaction. Furthermore, deviation from randomness was greater for the abnormal than the normal families. When four-person families were examined the departure from randomness was greater for both types of families yet there were no differences between the normal and patient family types. It was the normal families' sequences that become more predictable when a second child was added. The addition of a fourth member (or perhaps a second child) thus seems to be one of the conditions under which family equilibrium and structure is changed (see Waxler & Mishler, 1970, for a discussion of problems of measuring sequences of speakers).

Haley's experiment opens up the possibility for micro-process measures of family interaction that extend far beyond the simple sequence of speakers. Is the content of interaction patterned? Do families have patterned processes for handling power strategies? Does a family respond with a predictable set of actions to disagreements? One experimental investigation of family micro-process that concerns itself not only with the sequence of speakers but also with the predictability of certain types of interactions was carried out by Mishler and Waxler (1969). Here families with schizophrenic children and families with normal children interacted in a controlled experimental situation; each speech was classified with a number of content analysis systems (methods are reported in Mishler & Waxler, 1968b). In the micro-process study adjacent speeches were coded according to which family member made the speech, which family member responded to it, and the degree to which the response "acknowledged" the immediately preceding speaker. As in the Raush study, techniques of multivariate informational analysis were used to examine the predictability of sequential acts.

The investigators found significant differences between the normal and schizophrenic families in the extent to which it is possible to predict the second speaker's acknowledgment from the knowledge of who that person is and to whom he is responding. The level of patterning or predictability of acknowledgment, or responsiveness, is greater in the families with schizophrenic children than in the normal families, yet neither of these types of families has a predictability level that departs from chance. (Both types of families, however, have patterns deviating from a purely random one and thus these findings confirm those reported by Haley.) The schizophrenic-normal differences in predictability reflect on certain theories about schizophrenic family structure; for example, Wynne and Singer (1963) suggest that the rigidity of the family system is one factor pushing the child into inadequate identity development, and perhaps schizophrenia.

Mishler and Waxler go beyond the question: "Are there specific contingency patterns in family interaction?" to ask "Does the level of predictability change across time?" Time here is limited to the time the family is seen in the experimental session. Therefore the question is, does the patterning in family process shift during the experimental hour? For example, with regard to the sequence of speakers, while the schizophrenic families were shown to have a more predictable process of acknowledgment, they were also shown to shift more quickly and more often into and out of high levels of patterning of speaker sequences than did the normal families. In contrast the normal families, whose overall levels of predictability were relatively low, maintained longer sequences of high level predictability of speaker sequences once it had been reached. The schizophrenic family's erratic and perhaps fragmented interaction, within the context of a rigid patterning, recalls a number of clinical observations of these families. While Raush showed that the hyperaggressive boys seem not to have the skills or capacities for stopping actions once they had begun, the schizophrenic families seem not to have the mechanisms for maintaining a predictable sequence of events. Normal families are able to keep a predictable process going for a longer period of time even though their general norm is for flexibility and change.

Once this question regarding the shifts of predictability within the experimental hour has been asked, then one may ask which variables cause the change in process. As yet this question has not been asked empirically. However, the method represented by the Raush, Haley, Mishler and Waxler studies, provides the appropriate technique for obtaining the answer.

C. SUMMARY

The methods for examining family macro-process, that is, changes that occur throughout the long history of a family, center on appropriate sampling techniques that may be combined with observation of family interaction in an experimental situation. Some experimenters, however, have chosen to examine change in microcosm in order to investigate questions originally raised by theorists concerned with long-term family development. If Bales is right (Bales, 1953) in assuming that each of the theoretical levels of group analysis involves the same basic conflicts and the same ways of resolving them, then we might hypothesize, for example, that families that alter the patterning of sequential acts as a result of an increase in power acts by a low-power member will alter family structure in comparable ways when an adolescent child matures. Thus, perhaps long-term changes in family structure may be fruitfully examined in microcosm, at the level of discrete act sequences.

V. Family Structure and Outcome

A large proportion of family experiments have as their explicit concern possible causal linkages between family structure and some outcome variable. The outcome of most interest is the development of the child. Further, with the exception of a few experiments relating family structure to the child's attitudes or values, all are concerned with linkages between family structure and the child's illness, his identity development, or his cognitive style.

A. The Problem of Direction of Causation

A brief summary of one of these experiments will serve to raise the major methodological issue that appears when questions about the outcome of family interaction are raised. Stabenau *et al.* (1965) asked families of three different types to interact in an experimental situation. The families were selected on the basis of characteristics of one child member; the child was either schizophrenic, a juvenile delinquent, or normal. In selecting the dependent variables Stabenau was concerned with the qualities of family structure and the characteristics of interaction that may or may not have led to the development of schizophrenia. Thus, the quality of the child was the major independent variable and the normal and delinquent families were included as control groups with the assumption that if families with a schizophrenic child differed in modes of interaction from the other two types these patterns of interaction might be causal agents in the schizophrenia. Stabenau *et al.* (1965) conclude, after presenting findings, "These data support, but do not prove, the hypothesis, that differing patterns of interaction between parents and child are causally related to the development of psychopathology and the establishment and maintenance of mental health" [p. 59].

But isn't it equally likely that the pattern of family interaction found, measured after the fact of the child's illness, is a response to the illness and not in any sense a causal agent, or even a stable pattern that existed prior to the illness? Inherent in the experimental study of natural groups is this problem of the time-order of the variable effects and the problem is magnified when the major concern is with a cause-effect relationship that may have taken place in the past. In contrast, experimental studies of *ad hoc* groups avoid this difficulty by constructing new groups of known structure at Time$_1$ and examining the outcome of interaction at Time$_2$. Only recently have family experimenters followed this lead by constructing artificial families and using other experimental interventions to assure that the time-order is clear (this issue is discussed further in Waxler, 1970).

Of the experiments relating family structure to the child's illness the largest number examine families having schizophrenic children (see Caputo, 1963; Cheek, 1965; Farina, 1960; Lerner, 1965; Mishler & Waxler, 1968b; Reiss, 1967; Sharan, 1966). Similar questions have been asked experimentally about families with a child diagnosed as having childhood schizophrenia (Lennard *et al.*, 1965), families with a disturbed child (Hutchinson, 1967), or with a non-adjusted child (O'Connor, 1967). Within this set of experiments the way in which the time-order problem is handled ranges from simply assuming that the family's structure is stable and was causal in the child's illness to attempting a direct test of this question.

Within the limits set by the choice of real families as experimental subjects, two design strategies have been used to shed light on the time-order of family effects. One consists of the selection of appropriate control groups with which to compare the experimental family type. Specifically, the strategy involves using a type of family having a known outcome and having a known time order for the "family structure" and "child outcome" variables. For example, O'Connor (1967) chose families with mentally retarded children as the control group with which to compare families having "adjusted" or "non-adjusted" children, the second control and experimental group, respectively. In the mentally retarded group (assuming clear diagnosis) there is reasonable assurance that the child's illness existed prior to deviations from "normal" family structure. Farber's work on families having mentally retarded children suggests that these families are pressed to alter the family system in *response* to the handicapped child (Farber, 1960). Therefore, in the mentally retarded control group the time-order of the "family structure" and "child outcome" variables is known. If the experimental group of families of "non-adjusted" children is shown to have structural and interactional patterns no different from the control group of retarded families and both are different from the "adjusted," or normal, group then the most parsimonious explanation of the findings must be that the "non-adjustment" of the child has existed prior to, and may have called forth, the changes in family structure.

Unfortunately the findings are not so clear. O'Connor found that on most family structure variables the mentally retarded families fell between the other two. Thus, if we assume that an essentially "normal" family has been forced to respond differently to a mentally retarded child, we might also conclude that a "non-adjusted child" family has been forced to respond in an even more abnormal way to the non-adjusted child. However, since the non-adjusted families also differ significantly from the mentally retarded families it is equally likely that other effects are operating, and perhaps that the family structure has caused the adjustment problems of the child.

The selection of appropriate control groups for comparison purposes is, as the O'Connor experiment shows, only a partial solution to the tangle of cause-effect relationships that may have occurred in the family's past. A second, partial, solution consists of the addition of a different type of control, the observation of parents in interaction with a well sibling of the sick child. Mishler and Waxler (1968b), Reiss (1967), and Stabenau *et al.* (1965) use this technique, each in somewhat different ways, and in combination with other control groups, to ask whether the family system is the same with all children. This question reflects on the assumptions about stability of the family structure across long periods of time; if the structure is dependent on which child is present then it is precarious to assume that the structure seen at present, when the sick child interacts with his parents, reflects a much earlier pattern. However, if the family structure is "different from normal" only when the sick child is present then one may limit further questions about cause and effect to one, rather than all, children in the family.

In the Mishler and Waxler experiment (1968b) in which parents of two types of schizophrenic children (those having good and poor premorbid adjustments) and parents of normal children were asked to interact, each set of parents was seen once with the sick child and once with a well sibling of that child. The general trend of the findings support the notion of the specificity of patterns. On many of the measures of interaction in the experimental situation, parents interacting with their sick child differ from normal parents with their normal child. Further, the parents of schizophrenic children, when interacting with a normal child in their family, behave like normal parents. The "abnormal" structure of the schizophrenic family is, in general, limited to the relationship between parents and the one schizophrenic child.

This led the investigators to ask questions about possible time relationships between the family structure and the schizophrenia variables. One explanation, the etiological one, suggests that, since the parents' interaction patterns are different only when the sick child is present, they may have preceded and possibly caused the child's illness; this explanation follows some clinicians' speculations that parents "select" one child in the family to become the schizophrenic and all further pathological relationships are limited to that child. A second explanation, the responsive one, concludes that, since the parental differences occur only with the sick child present, that child may have elicited these differences by his pathological behavior.

It is clear that neither the addition of control groups nor the addition of control subjects to an experimental design is likely to provide positive evidence that the family structure did or did not precede the child's developmental problem or illness. Inferences may be drawn from the findings and some explanations reasonably ruled out, yet no clear conclusions can be drawn.

B. EXPERIMENTAL TESTS OF CAUSATION DIRECTION

Experimental construction of artificial families and the introduction of experimentally controlled variables into a family may be better solutions to these questions of family outcome. Groups having the abstract qualities of a "family" may be formed, asked to interact, perhaps in a specified way, and the effects on the child measured. Turning the question around it is also possible to investigate the effects of the child on family by controlling the child variable and examining the resulting behavior of the parents. If the theoretical framework is clear and concretely framed, it is thus possible to test the effects of family on child, or child on family, in such a way that generalizations may be made about real families.

We have previously referred to the work of Bandura *et al.* (1963), who constructed "families" with varied status distributions and asked which "parent" the child imitated, or identified with. In this instance family structure is the causal variable. Bandura found that the child generally imitates the high status adult, regardless of the reward structure in the "family." The experimental model could easily be expanded to include "parental affective relationship patterns, differences in participation of "parents," and varied distributions of instrumental and expressive roles, all of which have been singled out as important in the development of the child's personality.

It is also possible to test the effects of child on parents in an experimental situation. An experiment by Siegel and Harkins (1963) asked if it is reasonable to conclude that family structure may change as a response to a sick or deviant child. Here the child's behavior is the independent variable and "parental" behavior the dependent variable. An institutionalized retarded child was placed in an experimental room with a housewife (not his mother) and the housewife was asked to teach the child to do a puzzle; there was an unstructured time period as well. Whether the child was known by the housewife to have "high" or "low" verbal ability had no effect on her level of verbal interaction with the child; thus her response to the child was not simply a result of a label having been attached to the child. Instead, when data were combined with those from a previous study there was considerable evidence "suggesting that the actual verbal level of the children . . . may have been the crucial determinant of adult verbal behavior." Adults interacting with these children consistently use lower type-token ratios with the "low verbal ability" children than with the "high ability" children. Thus it is reasonable to conclude that one aspect of mental retardation in the child—restricted vocabulary usage—elicits specific responses from an adult, and thus, presumably from the child's own parents.

The relationships between family structure and outcome for the child may be estimated, as we have shown, by comparing the real families in question with control groups for which the time-order is known, or by constructing artificial families and experimentally introducing the "family structure" or "child outcome" variables. One family experiment, now in process, is designed to combine the advantages of these two approaches (see Haley, 1968, for a related experiment). In this study (Waxler, 1967) natural family groups, rather than experimentally trained role players, are used, yet experimental manipulations are introduced so as to insure that the time order of the family structure and child outcome variables is known. The problem follows directly from findings of the earlier experiment on families having schizophrenic children (Mishler & Waxler, 1968b).

In the present experiment, families having schizophrenic, normal and chronically ill children (the latter with non-psychiatric illness) are sampled. For the same reason that O'Connor (1967) included the families of mentally retarded children, the medically ill families are included here; it is assumed that the illness preceded any "abnormalities" in family structure. Following the sampling of family types, artificial families are created by pairing parents of one type with a child of another type. Schizophrenic parents interact with a normal child and at another time with a schizophrenic child (not their own) or with a medically ill child. Normal parents and medically ill parents also interact with different types of children. (The control situation in which parents interact with children of the same type as their own, but not their own, is designed to standardize degree of acquaintance between parents and child for all interaction situations.)

The logic of this experimental design allows one to ask which is most likely to stand as a causal variable, the child's schizophrenia or the parents' modes of interaction. If, for example, the schizophrenic child's behavior remains the same across different parents and the normal parents' behavior moves toward the "abnormal" when interacting with that schizophrenic child, then it is reasonable to conclude that the schizophrenic child may have elicited "abnormal" responses from parents. If, on the other hand, the schizophrenic parents' behavior remains stable across children and the normal child with whom they interact changes his behavior in the direction of the schizophrenic, we could conclude that the schizophrenic parent can elicit "abnormal" responses from the normal child. Within the design itself there is no way in which findings will reflect on the true cause-effect relationship in the past family history; findings may, however, suggest which relationship is more likely and thus which theory and experimental strategy to focus on in the future.

C. SUMMARY

One of the major questions raised when the experimental method is used to investigate the effects of family interaction on certain "outcome" variables is which came first, the structure or the outcome? This problem becomes most obvious at this point, where the sequence of events is a crucial part of the theory; however, time-order is also an issue in all of the earlier experimental studies in which real families are used. Our discussion here has centered on solutions involving experimental designs. When real families are used, certain control groups or control subjects are added in order to provide a comparison group having a known time-order. When artificial families are used, the effect of "parent" on "child" or "child" on "parent" may be measured by experimentally controlling the causal variable, either the "parent" or the "child" role, and allowing the effect variable to vary.

There are however strategies other than the creation of special experimental designs to investigate the cause-effect relationships within a family. One consists of a particular mode of data analysis applied to interactions in real families, where, by their very nature, the cause and effect in which one is interested has occurred in the past. The microprocess analysis described in the section on change is one example. The investigator may examine the relationship between one act and the next and may assume that each preceding act is the "cause" of the following act. He may then look for cause-effect patternings in the family's interaction during the course of the experimental hour and may draw conclusions about the effect of parent on child, or child on parent, that may also be relevant to the earlier cause-effect relationship that occurred in the family's past.

VI. Conclusion

In the course of reviewing selected experimental studies of families we have pointed a number of times to two recent methodological trends, the tendency toward greater experimental intervention into the family system and the use of synthetic family groups as sources of data on the family. The movement in this direction carries with it certain implications for further study of the family which we will discuss here.

Experimental intervention into the family system was most apparent in family outcome studies. There intervention offers one solution to the problem of the time-order of structure and outcome variables. However, increasing experimental control over the family's situation and interaction was also apparent in other conceptual areas. As we have reported, some experimenters have asked real family members to interact with strangers (Bodin, 1965; Leik, 1963; Ryder, 1968; Waxler, 1967); others have trained parents to play an experimentally defined role with their own child (Hetherington & Frankie, 1967; Mussen & Parker, 1965; Patterson, 1965; Stevenson *et al.*, 1963); and some families have been asked to interact in artificial and unusual situations (Reiss, 1967; Haley, 1968). This variety of experimental interventions insures that the independent variable is clearly controlled and standard in all experimental treatments. Further, intervention may serve to control some aspects of the family situation so as to allow for more exact and clearer measures of other aspects. Findings from the experiments that have been reviewed suggest that, when hypotheses warrant them, that experimental manipulations of families generate findings permitting more precise statements of relationships than can be derived from the correlational findings of observational studies. Particularly when theoretical concerns move from the description of family structure to examining how it is formed or changed (i.e., to the understanding of cause-effect relationships), experimental intervention into the family system is a valuable approach.

The second major trend, the use of artificial or synthetic families, is apparent in almost all content areas. "Families" with defined status structures have been composed and the effect on the child measured (Bandura *et al.*, 1963); "families" having a deviant child have been examined and the parental response to deviance observed (Siegel & Harkins, 1963). The obvious advantages of this approach—experimental control over the important variables and the assurance that other variables are randomly distributed, permitting more direct tests of cause-effect hypotheses—have been discussed earlier. A question often raised, however, is whether findings from artificially composed families have any relevance to real families. In particular, investigators who have had clinical experience with families, wonder whether it is possible to generalize to processes in real families when the people observed are strangers without the long, intimate experience with each other that is unique to the family.

No experimenter assumes an identity between real and experimentally constructed families; in fact a number of experiments have been designed to investigate differences between them (Bishop, 1951; Bodin, 1965; Leik, 1963; Ryder, 1968). Instead, findings from artificial families are assumed to be relevant to an abstract theoretical conception or model of the family. Findings from experimentally composed families serve to test and to modify the abstract theory just as do findings from real families. Neither real nor artificial families meet all of the theoretical requirements of a model and thus, for both, variations from what the model would predict must be accounted for by introducing other dimensions thus further complicating the model. For artificial families one might, for example, introduce the dimension of "acquaintance," while for real families one might introduce such dimensions as "developmental level" or "position in the economy" to explain variations from the abstract model. Thus, those experimenters who have worked with experimentally composed families do so, in most instances, with the awareness that they are contributing to the specifications of a theory of interpersonal relations in families and other groups rather than to empirical generalizations about "real" families.

These trends toward the use of synthetic families and intervention into the real family system are representative of the more general methodological changes in experimental family study. These changes were briefly outlined in the introduction of this paper. Now that the family experiments have been reviewed it is clear that such methodological changes have more general implications for family theory and research. When a researcher chooses to test an hypothesis in an experimental situation the evidence presented here suggests that the addition of behavioral data forces family theorists to add concepts and qualifications to the original family theory. The most common effect on theory consists of clarification of the behavioral processes that intervene between a quality of a member, or of the whole family, and some structural or outcome variable. Thus, for example, the predictions about the relationship between generally valued and expected role differentiation between parents and a certain outcome for the child may vary depending on the demands of particular situations; whether the situation is public or private is one source of variation.

In addition to this impact on theory, the experimental technique has led to greater concern with all aspects of measurement. In this paper we have stressed issues of experimental design and particularly the logic of control group comparisons and have not dealt directly with problems of measurement; however the decision to test hypotheses experimentally forces greater attention to the theoretical relevance of and the scientific adequacy of measuring instruments. One general trend has been a decreased reliance on measures taken directly from *ad hoc* group studies (for example, Interaction Process Analysis) and a movement toward the construction of interaction codes and other outcome measures that indicate concepts found in theories of family structure (see studies by Farina, 1960; Mishler & Waxler, 1968b; Riskin, 1964; for theoretically derived interaction coding systems). Also, many of the hypotheses being tested are centered on the processes or mechanisms that occur in families and seem to explain the relationship between a quality of the family and some outcome measure. Because of this increasing focus on family processes, measurement has moved from the use of global ratings or single outcome scores to codes of ongoing interaction. Thus, instead of judging who "won" a family discussion, code categories are applied to every statement in order to understand "how" the discussion proceded and "how" a member won.

Closely intertwined with this trend toward theoretically related measures is an increasing concern with the scientific adequacy of measuring instruments. There has been greater attention paid to objectivity of measurement; for example, investigation has moved from observer's ratings of "mother's warmth" to computer-counted measures of members' speeches (Haley, 1967a). Problems in the reliability of measurement have also received more attention (Waxler & Mishler, 1966). A number of these aspects of measurement have been discussed by M. Straus (1964).

The complexity of theoretical questions has been paralleled by an increased complexity of analytic methods. In a number of places in our discussion we have pointed to the utility of analysis of variance methods, or to other techniques that allow certain variables to be controlled and effects of others to be examined; movement beyond correlational analysis is required by the nature of current hypotheses. Further, there has been increasing interest in separating the effects of individual family members from the effect of the family as a whole. This has led to the development of multi-level data analysis techniques; both Reiss (1967) and Mishler and Waxler (1968c) have shown this to be a useful approach to the complexity of family structure. Finally, with increasing concern with family processes rather than simply with "before" and "after" measures of families, the development of statistical techniques for analyzing act sequences (Raush, 1965) may prove to be particularly fruitful.

While experimental studies of families have contributed to theoretical clarification and methodological rigor in family research, there has been continuing concern with the ethical issues. This is understandable in the light of increasing interest in ethical problems in all areas of human experimentation. Family experimenters are acutely aware that the same care must be exercised here as occurs in other small group experimentation; an additional problem, however, is the fact that family members continue their relationship far beyond their experience in an experiment. Is it harmful for them to participate in an experiment, particularly one in which the experimenter intervenes or controls some aspect of the family interaction?

Information about effects of the experimental experience on the family — positive or negative, present or absent — is notably absent from most research reports. Neither do many family researchers describe the measures taken to explain the research or to answer questions after it is over, a standard procedure in most small group experiments. Thus there is little objective information about the possible harmful effects of this experience. However, if one places a family's experimental experience in the context of that family's history, then the experimental hour becomes what it probably is for the family, a trivial though hopefully interesting experience. Our own experience indicates that where the usual care expected in responsible scientific work is exercised, a family's participation in an experiment is neither threatening nor harmful. There is no evidence that would suggest that these problems are different from those in other areas of research on human behavior.

REFERENCES

Bales, R. F. *Interaction process analysis.* Cambridge, Mass.: Addison-Wesley, 1951.

Bales, R. F. The equilibrium problem in small groups. In T. Parsons, R. F. Bales, & E. Shils (Eds.), *Working papers in the theory of action.* Glencoe, Ill.: Free Press, 1953. Pp. 111-161.

Bandura, A., Ross, D., & Ross, S. A comparative test of the status envy, social power and secondary reinforcement theories of identificatory learning. *Journal of Abnormal and Social Psychology,* 1963, **67**, 527-534.

Beaver, A. P. The initiation of social contacts by pre-school children. *Child Development Monographs,* 1932, No. 7.

Bell, N. W., & Vogel, E. F. *The family.* Glencoe, Ill.: Free Press, 1960.

Bell, R. A reinterpretation of the direction of effects of socialization. *Psychological Review,* 1968, **75**, 81-95.

Bell, R. Q. Retrospective and prospective views of early personality development. *Merrill Palmer Quarterly of Behavior and Development,* 1959-1960, **6**, 131-144.

Berkowitz, L. Personality and group position. *Sociometry,* 1956, **19**, 210-222.

Bishop, B. M. Mother-child interaction and the social behavior of children. *Psychological Monographs,* 1951, **65**, 11.

Bodin, A. Family interaction, coalition, disagreement, and compromise in problem, normal and synthetic family triads. Mimeo., Mental Research Institute, Palo Alto, Calif., 1965.

Borke, H. The communication of intent: A systematic approach to the observation of family interaction. *Human Relations,* 1967, **20**, 1.

Bronfenbrenner, U. Toward a theoretical model for the analysis of parent-child relationships in a social context. In J. C. Glidewell (Ed.), *Parental attitudes and child behavior.* Springfield, Ill.: Thomas, 1961. Pp. 90-109.

Burke, P. J. Role differentiation and the legitimation of task activity. *Sociometry*, 1968, **31**, 404-411.

Caputo, D. V. The parents of the schizophrenic. *Family Process*, 1963, **2**, 339-356.

Cheek, F. E. Family interaction patterns and convalescent adjustment of the schizophrenic. *Archives of General Psychiatry*, 1965, **13**, 138-147.

Farber, B. Perceptions of crisis and related variables in the impact of a retarded child on the mother. *Journal of Health and Human Behavior*, 1960, **1**, 108-118.

Farina, A. Patterns of role dominance and conflict in parents of schizophrenic patients. *Journal of Abnormal and Social Psychology*, 1960, **61**, 31-38.

Feldman, R. A. Interrelationships among three bases of group integration. *Sociometry*, 1968, **31**, 30-46.

Ferreira, A. Rejection and expectancy of rejection in families. *Family Process*, 1963, **2**, 235-244.

Ferreira, A., & Winter, W. Stability of interactional variables in family decision-making. *Archives of General Psychiatry*, 1966, **14**, 352-355.

Framo, J. L. Systematic research on family dynamics. In I. Boszormenyi-Nagy & J. L. Framo (Eds.), *Intensive family therapy: Theoretical and practical aspects.* New York: Harper & Row, 1965. Pp. 407-462.

Freilich, M. The natural triad in kinship and complex systems. *American Sociological Review*, 1964, **29**, 529-540.

Gates, G. S. The effect of an audience upon performance. *Journal of Abnormal and Social Psychology*, 1924, **18**, 334-342.

Greenberg, P. J. Competition in children: an experimental study. *American Journal of Psychology*, 1932, **44**, 221-248.

Haley, J. Research on family patterns: an instrument measurement. *Family Process*, 1964, **3**, 41-65.

Haley, J. Speech sequences of normal and abnormal families with two children present. *Family Process*, 1967, **6**, 81-97. (a)

Haley, J. Cross-cultural experimentation: an initial attempt. *Human Organization*, 1967, **26**, 110-117. (b)

Haley, J. Testing parental instructions to schizophrenic and normal children: a pilot study. *Journal of Abnormal Psychology*, 1968, **73**, 6, 559-565.

Haythorn, W., Couch, A., Haefner, D., Langham, P., & Carter, L. The behavior of authoritarian and equalitarian personalities in groups. *Human Relations*, 1956, **9**, 57-74.

Heinicke, C., & Bales, R. F. Developmental trends in the structure of small groups. *Sociometry*, 1953, **16**, 7-38.

Hetherington, E. M. A developmental study of the effects of the sex of the dominant parent on sex-role preference, identification, and imitation in children. *Journal of Personality and Social Psychology*, 1965, **2**, 188-194.

Hetherington, E. M., & Frankie, G. Effects of parental dominance, warmth, and conflict on imitation in children. *Journal of Personality and Social Psychology*, 1967, **6**, 119-125.

Hoffman, M. L. Power assertion by the parent and its impact on the child. *Child Development*, 1960, **31**, 129-143.

Hoffman, M. L. and L. W. Hoffman. *A review of child development research.* Vol. 1. New York: Russell-Sage Foundation, 1964.

Hopkins, T. K. *The exercise of influence in small groups.* Totowa, N. J.: Bedminster Press, 1964.

Hutchinson, J. G. Family interaction patterns and the emotionally disturbed child. Mimeo. Presented at the meeting of the Society for Research in Child Development, New York, March 1967.

Katz, I., Goldston, J., & Benjamin, L. Behavior and productivity in biracial work groups. *Human Relations*, 1958, 11, 123-141.

Kenkel, W. F. Influence differentiation in family decision making. *Sociology and Social Research*, 1957, 42, 18-25.

Kenkel, W. F. Dominance, persistence, self-confidence, and spousal roles in decision-making. *Journal of Social Psychology*, 1961, 54, 349-358.

Leik, R. K. Instrumentality and emotionality in family interaction. *Sociometry*, 1963, 26, 131-145.

Lennard, H. L., Beaulieu, M. R., & Embrey, N. G. Interaction in families with a schizophrenic child. *Archives of General Psychiatry*, 1965, 12, 166-183.

Lerner, P. M. Resolution of intrafamilial role conflict in families of schizophrenic patients. I: Thought disturbance. *Journal of Nervous and Mental Disease*, 1965, 141, 342-351.

Lidz, T., & Fleck, S. Family studies and a theory of schizophrenia. In American Psychiatric Association, *The American family in crisis*. Des Plaines, Ill.: Forest Hospital Publ., 1965.

Lidz, T., Fleck, S., & Cornelison, A. R. Family studies and a theory of schizophrenia. In T. Lidz *et al.* (Eds.), *Schizophrenia and the family*. New York: International Universities Press, 1965. Pp. 362-376.

Liu, W. T. Family interactions among local and refugee Chinese families in Hong Kong. *Journal of Marriage and Family*, 1966, August, 314-323.

March, J. Husband-wife interaction over political issues. *Public Opinion Quarterly*, 1954, 17, 461-470.

Milgram, S. Group pressure and action against a person. *Journal of Abnormal and Social Psychology*, 1964, 69, 137-143.

Mills, T. M. Power relations in three-person groups. *American Sociological Review*, 1953, 18, 351-357.

Mishler, E. G., & Waxler, N. E. (Eds.) *Family processes and schizophrenia: Theory and selected experimental studies*. New York: Science House, 1968. (a)

Mishler, E. G., & Waxler, N. E. *Interaction in families: An experimental study of family processes and schizophrenia*. New York: Wiley, 1968. (b)

Mishler, E. G., & Waxler, N. E. Family interaction patterns and schizophrenia: a multi-level analysis. In J. Romano (Ed.), *The origins of schizophrenia*. Amsterdam: Excerpta Medica Found., 1968. (c)

Mishler, E. G., & Waxler, N. E. Interaction sequences in normal and schizophrenic family triads: applications of multivariate informational analysis. Unpublished manuscript, Boston: Massachusetts Mental Health Center, 1969.

Moore, J. C. Status influence in small group interactions. *Sociometry*, 1968, 31, 47-63.

Moore, M. Consistency of interaction in normal and clinic families. *Dissertation Abstracts, Section A*, 1967, 12, 1, 4564-4565.

Mussen, P., & Parker, A. Mother nuturance and girls incidental imitative learning. *Journal of Personality and Social Psychology*, 1965, 2, 94-97.

O'Connor, W. A. Patterns of interaction in families with high adjusted, low adjusted, and mentally retarded members. Unpublished doctoral dissertation, University of Kansas, 1967.

O'Rourke, J. Field and laboratory: The decision making behaviors of family groups in two experimental conditions. *Sociometry*, 1963, 26, 422-435.

Parsons, T., & Bales, R. F. *Family socialization and interaction process*. Glencoe, Ill.: Free Press, 1955.

Patterson, G. R. Parents as dispensers of aversive stimuli. *Journal of Personality and Social Psychology*, 1965, 2, 844-851.

Rapoport, R., & Rapoport, R. New light on the honeymoon. *Human Relations,* 1964, **17,** 33-56.

Raush, H. Interaction sequences. *Journal of Personality and Social Psychology,* 1965, **2,** 487-499.

Reiss, D. Individual thinking and family interaction. *Archives of General Psychiatry,* 1967, **16,** 80-93.

Riley, M. W. *Sociological research.* Vol. 1. *A case approach.* New York: Harcourt, Brace & World, 1963. Pp. 700-739.

Riskin, J. Family interaction scales. *Archives of General Psychiatry,* 1964, **11,** 484-494.

Ryder, R. Husband-wife dyads versus married strangers. *Family Process,* 1968, **7,** 233-238.

Sampson, E., & Brandon, A. The effects of role and opinion deviation on small group behavior. *Sociometry,* 1964, **27,** 261-281.

Schachter, S. Deviation, rejection, and communication. *Journal of Abnormal and Social Psychology,* 1951, **46,** 190-207.

Schutz, W. C. *FIRO: A three-dimensional theory of interpersonal behavior.* New York: Rinehart, 1958.

Scott, F. G. Family group structure and patterns of social interaction. *American Journal of Sociology,* 1962, **68,** 214-228.

Sharan (Singer), S. N. Family interaction with schizophrenics and their siblings. *Journal of Abnormal Psychology,* 1966, **71,** 345-353.

Siegel, G. M., & Harkins, J. P. Verbal behavior of adults in two conditions with institutionalized retarded children, *J. of Speech and Hearing Disorders Monographs,* 1963, **10,** 39-47.

Slater, P. Role differentiation in small groups. *American Sociological Review,* 1955, **20,** 300-310.

Slater, P. Parental role differentiation. *American Journal of Sociology,* 1961, **67,** 296-311.

Stabenau, J. R., Tupin, J., Werner, M., & Pollin, W. A comparative study of families of schizophrenics, delinquents, and normals. *Psychiatry,* 1965, **28,** 45-59.

Stevenson, H. W., Keen, R., & Knights, R. M. Parents and strangers as reinforcing agents for children's performance. *Journal of Abnormal and Social Psychology,* 1963, **67,** 183-186.

Stone, P. J., Dunphy, D. C., Smith, M. S., & Ogilvie, D. M. *The General Inquirer: A computer approach to content analysis.* Cambridge, Mass.: M.I.T. Press, 1966.

Straus, J., & Straus, M. A. Family roles and sex differences in creativity of children in Bombay and Minneapolis, *Journal of Marriage and Family,* 1968, February, 46-53.

Straus, M. A. Measuring families. In H. T. Christensen (Ed.) *Handbook of marriage and the family.* Chicago: Rand McNally, 1964. Pp. 335-400.

Straus, M. A. The influence of sex of child and social class on instrumental and expressive family roles in a laboratory setting. *Sociology and Social Research,* 1967, **52,** 7-21.

Straus, M. A. Communication, creativity, and problem-solving ability of middle- and working-class families in three societies. *American Journal of Sociology,* 1968, **73,** 417-430.

Strodtbeck, F. L. Husband-wife interaction over revealed differences. *American Sociological Review,* 1951, **16,** 468-473.

Strodtbeck, F. L. Family interaction, values and achievement. *In* D. McClelland (Ed.), *Talent and society.* Princeton, N.J.: Van Nostrand, 1958.

Strodtbeck, F. L., & Mann, R. D. Sex role differentiation in jury deliberations. *Sociometry,* 1956, **19,** 3-11.

Turk, H. Instrumental values and the popularity of instrumental leaders. *Social Forces*, 1961, **39**, 252-260.

Waxler, N. E. Families and schizophrenia: studies in deviance. Mimeo., Boston: Massachusetts Mental Health Center, 1967.

Waxler, N. E., Families and schizophrenia: alternatives to longitudinal studies. *In* M. Levitt (Ed.), *The mental health field: A critical appraisal*. Forthcoming publication, 1970.

Waxler, N. E., & Mishler, E. G. Scoring and reliability problems in interaction process analysis: a methodological note. *Sociometry*, 1966, **29**, 28-40.

Waxler, N. E., & Mishler, E. G. Sequential patterning in family interaction. *Family Process*, 1970, **9**, 211-220.

Wiggins, J., Dill, F., & Schwartz, R. On status liability. *Sociometry*, 1965, **28**, 197-209.

Winch, R. *Mate selection*. New York: Harper, 1958.

Wynne, L. C., Ryckoff, I., Day, J., & Hirsch, S. Pseudomutuality in the family relations of schizophrenics. *Psychiatry*, 1958, **21**, 205-220.

Wynne, L. C., & Singer, M. T. Thought disorder and family relations of schizophrenics. I. A research strategy. *Archives of General Psychiatry*, 1963, **9**, 191-198.

Ziller, R. C., & Exline, R. V. Some consequences of age heterogeneity in decision-making groups. *Sociometry*, 1958, **21**, 198-211.

Subject Index

A 8
B 9
C 0
D 1
E 2
F 3
G 4
H 5
I 6
J 7